THE NATIONAL GEOGRAPHIC SOCIETY

100 YEARS OF ADVENTURE AND DISCOVERY

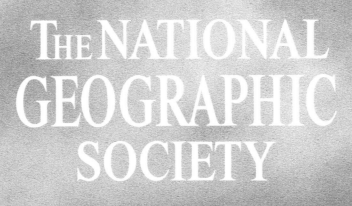

THE NATIONAL GEOGRAPHIC SOCIETY

100 YEARS OF ADVENTURE AND DISCOVERY

C. D. B. BRYAN

HARRY N. ABRAMS, INC., PUBLISHERS NEW YORK

Page 1: Lightning dances across the mountains behind Tucson, Arizona.
Pages 2–3: African elephants find drink and sanctuary in Namibia's Etosha National Park.
Pages 4–5: The gaping crater of Washington State's Mount St. Helens a year after the May 18, 1980, eruption.
Pages 6–7: The face behind the mask of an Asaro Valley, Papua New Guinea "Mudman."
Pages 8–9: A frog-eating bat swoops down at a poisonous toad.
Pages 10–11: Afghan horsemen play the national sport, *buz kashi,* a wild version of polo.
Pages 12–13: The roof of the world: the Himalayas in Nepal.
This page: Four large mullet in Japan's Izu Oceanic Park.
Overleaf: The Great Pyramids at Giza, Egypt.

Editor: Edith M. Pavese

Art Director: Samuel N. Antupit

Library of Congress Cataloging-in-Publication Data
Bryan, C.D.B. (Courtlandt Dixon Barnes)
    The National Geographic Society.

    Bibliography: p.
    Includes index.
    1. National Geographic Society (U.S.) I. Title.
G3.N37B79 1987   910'.6   87–1397
ISBN 0–8109–1376–3

Times Mirror Books

Printed and bound in the United States of America

Contents

18   Introduction

22   **1888–1900**
       "A Society for the Increase and Diffusion of Geographical Knowledge"

50   North and South Poles Conquered
       "The Most Cherished of Geographical Prizes"

80   **1900–1905**
       "The End of the Beginning"

96   Natural Disasters
       "Lord, that's enough now. Please stop it."

116   **1905–1920**
       The Language of the Photograph

136   Archeology
       "Ruins, Lost Cities, and Bones"

160   **1920–1930**
       A Love Affair with Automobiles, Airplanes, and Autochromes

176   Glorious Expeditions
       "Something lost behind the Ranges. Lost and waiting for you. Go!"

200   **1930–1940**
       A Revolution in Color Photography

220   The Depths
       "These Marvelous Nether Regions"

242   **1940–1950**
       A Household Institution in the Theater of War

262 **War and Conflict**
"Young Men Put Into the Earth"

286 **1950–1957**
The Red Shirt School of Photography

306 **The Heights**
"Because it's there."

328 **1957–1967**
"Like a Cluster of Rockets on a Quiet Night"

352 **Space**
"The Choice Is the Universe—or Nothing"

376 **1967–1977**
"Holding Up the Torch, Not Applying It"

400 **Conservation**
"Riders on the Earth Together"

426 **1977–1988**
"Broadening the Outreach"

454 **Inner Space**
"The Stuff of Stars Has Come Alive"

474 Acknowledgments

476 Bibliography

477 Index

483 Credits

Introduction

Writing this book could not have been anything but fun. To have the opportunity to explore nearly one hundred years of the National Geographic Society's history is like becoming a child again—a child who, confined indoors on a languorous, rainy, endless summer afternoon, discovers in the attic of a rented beach cottage the elaborate illustrated family albums of a somewhat eccentric, always fascinating, and truly remarkable clan.

Since its founding in 1888, the Society has been offering "a window on the world" to millions of its members. Before color photography, before movies and television, the familiar *National Geographic Magazine* provided—through riveting eyewitness accounts and dazzling photographs—the primary means by which generations of armchair explorers first discovered the distant wonders, exotic customs, and strange people throughout our world. A century of exploding scientific knowledge and expanding ecological awareness is reflected in the pages of the Magazine, as are a hundred years of American intellectual curiosity and the Society's sometimes rose-colored political attitudes and reporting. Books, articles, and films disseminate the results of diverse scientific researches. And its Explorers Hall displays its heroes' memorabilia.

What are five-score years of the Society's journal but nearly twenty-dozen memory-crowded scrapbooks in which stirring adventures, awesome disasters, breakthrough discoveries, and daring explorations have been carefully recorded? How can one not feel affection for a Society that has played so significant a role in the dreams of so many millions?

Where else but in the pages of the *Geographic* might one read the entirely serious late-nineteenth-century report of a man who proposed attaching hydrogen-inflated balloons to himself so that he might bound across the frozen Arctic wastes to reach the North Pole?

Where else but at the Society could one browse unrestrained through the correspondence of one of this nation's legendary editors—Gilbert Hovey Grosvenor—and discover that two of his Salem forebears had been hanged as witches? Or learn that a French Ambassador had awarded him the Legion of Honor for having done so much to "vulgarize geography"? Or find such a gem of dismissal as this terse Grosvenor note written in 1930 to an editor about a man seeking Society support: "Mr. Graves, this man is a windbag! He can

Opposite: *Bradford and Barbara Washburn map the heart of Grand Canyon, a monumental seven-year task completed in 1970.*

use more words & say less than a Tibetan prayer wheel. I wouldn't answer him further."

During the year and a half I lived and worked among the men and women of the National Geographic Society, I was able to interview them freely. I read their articles and books, pored over their photographs, paintings, and films. I attended their lectures and press conferences, sat in on their division meetings, and first suffered and then appreciated their researchers' meticulous dedication to facts. I lunched with them, laughed with them, shared their visions and their stories. There really is nothing like the National Geographic Society in the world.

Where else could one find a nineteenth-century English explorer's account of his adventures in Africa as charming as this excerpt taken from a January 1899 *National Geographic Magazine* report, "Lloyd's Journey Across the Great Pygmy Forest":

> ...I reached the Belgian frontier post of M'beni on October 1, and then entered the great dark forest. Altogether I was twenty days walking through its gloomy shades...At one little place in the middle of the forest...I came upon a great number of pygmies who...told me that, unknown to myself, they had been watching me for five days, peering through the growth of the primeval forest at our caravan. They appeared to be very frightened, and even when speaking covered their faces. I slept at this village, and in the morning I asked the chief to allow me to photograph the dwarfs. He brought ten or fifteen of them together, and I was enabled to secure a snapshot. I could not give a time exposure, as the pygmies would not stand still. Then with great difficulty I tried to measure them, and I found not one of them over four feet in height... Their arms and chests were splendidly developed, as much so as in a good specimen of an Englishman.

Albert B. Lloyd, one of those Englishmen who hitherto seemed to have existed only in our comic imaginations, continued his journey down the Belgian Congo's (now Zaire) Aruwimi River, where, he wrote,

> Personally I was received most kindly by these cannibals. They are, it is true, warlike and fierce, but open and straightforward. I did not find them to be of the usual cringing type, but manly fellows who treated one as an equal. I had no difficulty with them whatever. At one place I put together the bicycle I had with me, and, at the suggestion of these people, rode round their village in the middle of a forest. The scene was remarkable, as thousands of men, women, and children turned out, dancing and yelling, to see what they described as a European riding a snake.

How could there not be a special satisfaction in having the chance to read, firsthand, explorer Robert E. Peary's 1909 account of becoming the first to reach the North Pole; or Hiram Bingham's discoveries in 1912 at Machu Picchu; or Richard E. Byrd's 1926 and 1930 articles on being the first to fly over the North and South Poles; or the Army balloonists' 1935 daring, record-setting ascent into the stratosphere? Or Barry C. Bishop's sense of accomplishment upon his successful ascent of Everest in 1963 despite the loss of his toes?

How could there not be a thrill in sharing, even vicariously, Matthew Stirling's excitement in 1939 at unearthing in the Mexican jungle a colossal and mysterious twenty-ton stone head? Or Jacques Cousteau's exhilaration diving with Aqua-lungs to sunken ships in the 1950s? Or Jane Goodall's pleasure in a baby chimp's affection in the 1960s? Or Louis and Mary Leakey's excitement with their discoveries during the 1960s and 1970s of our early ancestor's fossils and footsteps in East Africa? Or Robert D. Ballard's awe in the 1980s upon seeing the huge, rusting, sunken remains of the once-grand luxury liner *Titanic*, lost in a collision with an iceberg nearly three-quarters of a century before. In spirit and support, the Society was with them all.

Members visiting the Geographic's marble and glass headquarters in Washington, D.C., rarely get beyond Explorers Hall on the first floor. Visitors on their way back from viewing one of Peary's heavy sledges or the patched, taffeta American flag he wore wrapped around his body during his dash to the Pole, might see Society staff passing through the lobby or waiting at the elevators—the men, conservatively dressed in jackets and ties, politely stepping aside to let equally properly-dressed women staff members enter the elevators first. But the staff's deceptive appearance would give no hint of the adventures they have experienced and the stories they have to tell. Tales of bandit raids and angry mobs, of walking away from airplane and helicopter crashes, of being flung from boats into icy rapids, of having cameras bitten by crocodiles or retrieved only through tugs-of-war with venomous sea snakes, of surviving shark bites, capricious imprisonments, and mysterious fevers.

Where else but at the Society would one learn of freelance photographer Alan Root, who, while on assignment in Mzima Springs, Kenya, in 1974, was attacked by a bull hippopotamus. "The next thing I knew," Root said, "he had my right leg in his mouth. The hippo then shook me like a rat." After skin grafts, treatment for gangrene, and a month's convalescence in a Nairobi hospital, Root recovered.

Where else might a guest be privy to such amusing anecdotes as the one told me by former Geographic staffer Edwards Park, who, upon learning he was being sent on assignment to Machu Picchu, had sought Illustrations Editor Kip Ross' advice and was told, "Watch out for snakes." Ted Park, hating snakes, pressed Ross for details and learned that Ross, while riding mule back up to the high ridge upon which the Lost City of the Incas had been built, had heard little rustling noises in the grass along the trail and asked his guide, "Pedro, what is that?"

"Señor," the guide said, "that is a fer-de-lance."

"Oh my God," Ross said. The fer-de-lance, he knew, was a species of large, extremely dangerous pit viper that infests those parts. Ross and the guide rode together in silence for a few minutes then Ross asked, "Pedro, what happens if I get bitten by a fer-de-lance? What do I do?"

Pedro thought for a moment and answered, "Señor, you *compose* yourself."

With annual receipts in excess of $350 million, with nearly eleven million members in 170 out of the 174 nations that now exist, the National Geographic Society is the largest non-profit scientific and educational membership institution in the world. Why has the Geographic been so successful?

National Geographic Editor Wilbur E. Garrett asked himself that same question in 1983. "I believe it is primarily because we still fill the same need felt by that small group of thoughtful men who gave us our start almost a century ago," he wrote, "—a need to address the insatiable human curiosity to know what makes this world tick."

Within the pages of this centennial volume that celebrates the National Geographic's insatiable curiosity with the world around it, it is my hope that readers may gain some insight into what makes the Society "tick" as well.

"A Society for the Increase and Diffusion of Geographical Knowledge"

Washington, D.C., January 10, 1888.

Dear Sir:

You are invited to be present at a meeting to be held in the Assembly Hall of the Cosmos Club, on Friday evening, January 13, at 8 o'clock P.M., for the purpose of considering the advisability of organizing a society for the increase and diffusion of geographical knowledge.

Very respectfully, yours,

Gardiner G. Hubbard.
A.W. Greely,
    Chief Signal Officer, U.S.A.
J.R. Bartlett,
    Commander, U.S N.

Henry Mitchell,
    U.S. Coast and Geod. Survey.
Henry Gannett,
    U.S. Geol. Survey.
A.H. Thompson,
    U.S. Geol. Survey
And Others.

On January 13, 1888, a damp, chilly Friday evening in the nation's capital, thirty-three gentlemen made their way by foot, on horseback, or by elegant private carriages through the fog to the Cosmos Club, a club notable over the preceding decade for its ability to draw upon Washington's scientific élite for its members. The thirty-three gentlemen—bearded, mustached, in heavy dark suits with high stiff collars and somber four-in-hand ties tucked into waistcoats crossed by thick gold chains—had been invited to gather at the clubhouse, situated on Lafayette Square diagonally across from the White House, that evening at eight to consider "the advisability of organizing a society for the increase and diffusion of geographical knowledge."

They were geographers, explorers, military officers, lawyers, meteorologists, cartographers, naturalists, bankers, educators, biologists, engineers, geodesists, topographers, inventors. As one of their own put it, they were the "first explorers of the Grand Canyon and the Yellowstone, those who had carried the American flag farthest north, who had measured the altitude of our famous mountains, traced the windings of our coasts and rivers, determined the distribution of flora and fauna, enlightened us in the customs of the aborigines, and marked out the path of storm and flood."

There was the prematurely aged Brigadier General Adolphus Washington Greely, Chief Signal Officer of the United States Army, who seven years earlier as a thirty-seven-year-old lieutenant with no Arctic experience had led an expedition of men farther north than any had gone before and become stranded. Not rescued until 1884, only Greely and six of the original twenty-five men were found still alive.

There was the heroic, pain-wracked geologist and explorer John Wesley Powell, who lost his right arm in the bloody Battle of Shiloh and seven years

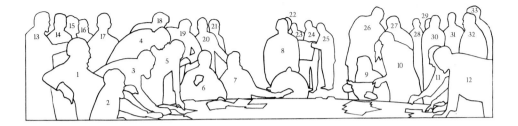

Key to painting identifies:
1. Charles J. Bell, banker
2. Israel C. Russell, geologist
3. Commodore George W. Melville
4. Frank Baker, anatomist
5. W. B. Powell, educator
6. Brig. Gen. A. W. Greely, Polar explorer
7. Grove Karl Gilbert, geologist and a future Society President
8. John Wesley Powell, geologist, explorer of the Colorado River
9. Gardiner Greene Hubbard, lawyer and first President of the Society, who helped finance the telephone experiments of Alexander Graham Bell
10. Henry Gannett, geographer and a future Society President
11. William H. Dall, naturalist
12. Edward E. Hayden, meteorologist
13. Herbert G. Ogden, topographer
14. Arthur P. Davis, engineer
15. Gilbert Thompson, topographer
16. Marcus Baker, cartographer
17. George Kennan, explorer of Arctic Siberia
18. James Howard Gore, educator
19. O. H. Tittmann, geodesist and a future Society President
20. Henry W. Henshaw, naturalist
21. George Brown Goode, naturalist
22. Cleveland Abbe, meteorologist
23. Comdr. John R. Bartlett
24. Henry Mitchell, engineer
25. Robert Muldrow II, geologist
26. Comdr. Winfield S. Schley
27. Capt. C. E. Dutton
28. W. D. Johnson, topographer
29. James C. Welling, educator
30. C. Hart Merriam, Chief, U.S. Biological Survey
31. Capt. Rogers Birnie, Jr.
32. A. H. Thompson, geographer
33. Samuel S. Gannett, geographer

later, in 1869, led the first expedition down the Colorado River through the Grand Canyon in boats—a hazardous, remarkable journey of some 900 miles.

There was brisk, assertive lawyer-capitalist Gardiner Greene Hubbard, friend and adviser to presidents, statesmen, and scientists, financer and promoter of his son-in-law Alexander Graham Bell's experiments and subsequently trustee of the Bell Telephone Company. (Though not a founder himself, Bell was one of the Society's 165 original members.)

There was the dashing student of Russian affairs George Kennan, who, during his eleven-week crossing of Siberia, had, he reported in *Tent Life in Siberia*, "changed dogs, reindeer, or horses more than two hundred and sixty times and had made a distance of five thousand seven hundred and fourteen miles, nearly all of it in one sleigh." There was Henry Gannett, the distinguished chief geographer of the U.S. Geological Survey and a pioneer mapmaker of Colorado and Wyoming's Rocky Mountains in addition to other parts of those states. And there was tenacious O. H. Tittmann, leader of numerous surveying expeditions to the opening frontiers, who in 1900 would be appointed Superintendent of the U.S. Coast and Geodetic Survey.

There was naturalist William H. Dall, pioneer explorer of Alaska; geologist Robert Muldrow II, who would measure Mount McKinley; and meteorologist Edward E. Hayden, who had lost his leg in a landslide.

There was the noted physician and naturalist C. Hart Merriam, who had shortly before been made Chief of the U.S. Biological Survey; and stolid, persevering Navy Commodore George W. Melville, chief engineer on a relief expedition ship that four years earlier had finally broken through the ice to retrieve the by-then near-starved and frozen, half-crazed Greely.

As they gathered on that foggy, wintry evening in the nation's capital, it is, perhaps, an interesting comment on those times that so momentous an assembly of notable men had been achieved in response to an invitation issued but three days before. Washington was smaller then, its population only about a third of what it is now. That January 1888 marked the last year of Grover Cleveland's first presidential term; the year William II would become Kaiser of Germany; the year Van Gogh would paint *The Night Café* and Toulouse-Lautrec, *Cirque Fernando: The Equestrienne*; the year Tchaikovsky's Fifth Symphony and Rimsky-Korsakov's *Sheherazade* would premier in St. Petersburg; the year Oscar Wilde would publish *The Happy Prince and Other Tales* and "Jack the Ripper" would murder at least six women in London; the year George Eastman in America would introduce the Kodak box camera, Thomas Edison refine the phonograph, and Hiram Maxim produce the first satisfactory fully automatic machine gun.

Now as Melville and Greely, Hubbard and Powell, Kennan and Gannett, Tittmann and Merriam, Dall and Muldrow and Hayden and the others collected about the huge, round mahogany table in the Cosmos Club's Assembly

Hall with its dim portraits, its gas-lit chandeliers, and its large standing globe with so much of its landmasses and oceans as yet unexplored, they agreed unhesitatingly about the necessity of a geographic society in Washington, D.C., the center then not only of the nation's politics but also its science.

Warmed by after-dinner coffee, brandy, and the blazing hardwood fire, they readily approved passage of the resolution offered by the U.S. Geological Survey's A. H. Thompson that the Society be organized "on as broad and liberal a basis in regard to qualifications for membership as is consistent with its own well-being and the dignity of the science it represents." A nine-man committee was appointed to prepare a draft constitution and plan of organization, both to be presented at the second meeting, to be held the following week, and the National Geographic Society was born.

Two weeks later the founders of the Society would elect Gardiner Greene Hubbard to lead them. In his introductory speech, Hubbard would tell them that he was "not a scientific man," nor could he "claim to any special knowledge that would entitle [him] to be called a 'Geographer.' "

"I owe the honor of my election as President of the National Geographic Society," Hubbard continued, "simply to the fact that I am one of those who desire to further the prosecution of geographic research. I possess only the same general interest in the subject of geography that should be felt by every educated man."

"By my election," Hubbard declared, "you notify the public that the membership of our Society will not be confined to professional geographers, but will include that large number who, like myself, desire to promote special researches by others, and to diffuse the knowledge so gained, among men, so that we may all know more of the world upon which we live."

Hubbard's opening oratorical humility was not lost on his audience. Although he was not a scientist himself, no one in that room was unaware that Hubbard had long had a keen interest in science, eagerly promoted science, and had many friends in the scientific community. Nor would any present question Hubbard's assumption that his interests were those "that should be felt by every educated man," for the educated man of that time was committed to science.

Eventually, Hubbard's basic egalitarian ideal—*to increase and diffuse* what would become *geographical* knowledge (defined only in its broadest, least-binding sense) with any interested citizen—became the fundamental tenet which would both guide the National Geographic Society's policies and provide the essential explanation for the Magazine's astounding popularity throughout the next hundred years. At that time, however, Hubbard was only reflecting late-nineteenth-century America's love affair with science.

If, in the 1880s, this was a nation staggering under the onerous burden of strangling monopolies, tariff favoritism, political patronage, corrupt city bosses, ruthless robber barons, economic depressions, federal troops and hired thugs called out to combat strikes by fledgling national labor unions, child-labor abuses, despairing farmers, and such vast waves of immigration that in some major cities fewer than half the population was American-born, it was also a nation of exhilarating optimism, unbridled energy, boundless ambition, insatiable curiosity, and a passionate belief in science's ability to provide the cure for society's ills.

One need look no further than the Society's remarkable, self-educated

Opposite: *First President of the National Geographic Society, Gardiner Greene Hubbard, and his wife Gertrude McCurdy Hubbard, on the veranda at Twin Oaks, their Washington, D.C., summer home.*

"By my election," declared Hubbard in his February 17, 1888, introductory speech to the new Society's assembled members, "you notify the public that the membership of our Society will not be confined to professional geographers, but will include that large number who, like myself, desire to promote special researches by others, and to diffuse the knowledge so gained, among men, so that we may all know more of the world upon which we live."

Geologist and explorer John Wesley Powell, one of the Society's founders, speaks with a Paiute Indian in this 1873 photograph taken on the Kaibab Plateau on the north rim of the Grand Canyon in northern Arizona. Powell, who lost his right arm in the Civil War Battle of Shiloh in 1869, led the first expedition in boats down the Colorado River through the Grand Canyon.

geologist and anthropologist W J McGee to find a spokesman for that era's awesome confidence in science. For it was McGee, one of the *National Geographic Magazine*'s earliest editors, who would rhapsodize in 1898:

> In truth, America has become a nation of science. There is no industry, from agriculture to architecture, that is not shaped by research and its results; there is not one of our fifteen millions of families that does not enjoy the benefits of scientific advancement; there is no law in our statutes, no motive in our conduct, that has not been made juster by the straightforward and unselfish habit of thought fostered by scientific methods.
>
> —From "Fifty Years of American Science," *Atlantic Monthly*, September 1898

The Society's first magazine, published in October 1888, was a slim, tall, octavo-sized scientific brochure with a somewhat forbidding-looking terra-cotta-colored cover. An "Announcement" published in its opening pages articulated the Society's aims:

> The "National Geographic Society" has been organized "to increase and diffuse geographic knowledge," and the publication of a Magazine has been determined upon as one means of accomplishing these purposes.
>
> It will contain memoirs, essays, notes, correspondence, reviews, etc., relating to Geographic matters. As it is not intended to be simply the organ of the Society, its pages will be open to all persons interested in Geography, in the hope that it may become a channel of intercommunication, stimulate geographic investigation and prove an acceptable medium for the publication of results....
>
> As it is hoped to diffuse as well as to increase knowledge, due prominence will be given to the educational aspect of geographic matters, and efforts will be made to stimulate an interest in original sources of information.
>
> In addition to organizing, holding regular fortnightly meetings for presenting scientific and popular communications, and entering upon the publication of a Magazine, considerable progress has been made in the preparation of a Physical Atlas of the United States.
>
> The Society...has at present an active membership of about two hundred persons. But there is no limitation to the number of members, and it will welcome both leaders and followers in geographic science, in order to better accomplish the objects of its organization.

Over the years, the earliest volumes of the Magazine have been unfairly criticized for being "dreadfully scientific, suitable for diffusing geographic knowledge among those who already had it and scaring off the rest." Consistently, the example given to buttress such censure is W J McGee's scholarly Vol. I, No. 1, article "The Classification of Geographic Forms by Genesis." ("...The second great category of geologic processes," McGee wrote, "comprehends the erosion and deposition inaugurated by the initial deformation of the terrestrial surface.") But buried within that first issue's pages was Everett Hayden's report of The Great Storm of March 11–14, 1888.

Certainly this article on the now notorious "Great Blizzard" was scientific; certainly it contained meteorological charts with isobars, isotherms, and arrows whose feathers indicated the force of the winds; and certainly one had to wade through much of the text before learning that the storm dumped forty inches of snow over the northeastern United States, brought freezing temperatures and seventy mph winds. But the text also included a compelling account of the gallant efforts by which the New York pilot boat No. 3, *Charles H. Marshall*, survived the violence and long duration of the storm (see p. 30).

When, seven months after the first issue, the second *National Geographic Magazine* appeared, in April 1889, its lead piece, "Africa, its Past and Future," by Society President Gardiner G. Hubbard, was notable for its unromantic depiction of the African continent's geography:

> …a vast, ill-formed triangle, with few good harbors, without navigable rivers for ocean-vessels, lying mainly in the torrid zone. A fringe of low scorched land, reeking with malaria, extends in unbroken monotony all along the coast, threatening death to the adventurous explorer.…

And its brutal assessment of the dangers an "adventurous explorer" risks:

> Some [travelers] who have entered [Africa] from the Atlantic or Pacific coasts have been lost in its wilds, and two or three years after have emerged on the opposite coast; others have passed from the coast, and have never been heard from. Zanzibar has been a favorite starting-point for the lake region of Central Africa. Stanley started from Zanzibar on his search for Livingstone with two white men, but returned alone. Cameron set out by the same path with two companions, but, upon reaching the lake region, he was alone. Keith Johnson, two or three years ago, started with two Europeans: within a couple of months he was gone. Probably every second man, stricken down by fever or accident, has left his bones to bleach along the road.

Hubbard's subsequent argument against the "sinister effects" of the slave trade and slavery—"the great curse of Africa," he called it—may have been somewhat weakened by his assertion that the Negro's "temper and disposition…make him a most useful slave. He can endure continuous hard labor, live on little, has a cheerful disposition, and rarely rises against his master." Still, Hubbard's observation on the slave trade was remarkable for its forthright journalism.

Of the four to five million slaves estimated by Hubbard to have been imported to America (even the larger of the two figures was believed by Hubbard to be an underestimate) he wrote, "12½ per cent were lost on the passage, one-third more in the 'process of seasoning;' so that, out of 100 shipped from Africa, not more than 50 lived to be effective laborers." And Hubbard unflinchingly provided eyewitness accounts by those familiar with the problem:

> [The explorer] Cameron says that Alrez, a Portuguese trader, owned 500 slaves, and that to obtain them, ten villages, having each from 100 to 200 souls, were destroyed; and of those not taken, some perished in the flames, others of want, or were killed by wild beasts. Cameron says, "I do not hesitate to affirm that the worst Arabs are angels of mercy in comparison to the Portuguese and their agents. If I had not seen it, I could not believe that there could exist men so brutal and cruel, and with such gayety of heart." Livingstone says, "I can consign most disagreeable recollections to oblivion, but the slavery scenes come back unbidden, and make me start up at night horrified by their vividness."

That sort of judgmental reporting would not last long in the Magazine; even over the next ten years one has to search hard to find it.

By the turn of the century, straightforward, unfettered journalism had disappeared from the Magazine almost entirely and would reappear only inter-

Founder Adolphus Washington Greely aboard the Thetis *about two weeks after founders George W. Melville and Winfield Scott Schley, of Schley's relief expedition, reached Greely's Greenland camp on June 22, 1884, and found the explorer more dead than alive.*

Greely, later to become a member of the Society's Board of Managers and one of its first vice presidents, had led his twenty-four-man Lady Franklin Bay Expedition into the Arctic in 1881 for purposes of scientific study and exploration—especially an attempt to push north. In 1882 four of Greely's men, before rejoining Greely's party, did reach latitude 83° 24' N—four miles farther north than any rival expedition had gone before but still 440 miles short of the Pole. In late summer 1883, after promised supply and relief ships failed to arrive, Greely led his expedition 500 miles south to Cape Sabine, which they reached after fifty-one days of difficult travel. By June 1884, when Greely and his party were finally rescued, only seven of the original twenty-five were still alive.

"FIGHTING FOR LIFE AGAINST THE STORM"

When [the *Charles H. Marshall* was] about 18 miles S.E. from the lightship, a dense fog shut in, and it was decided to remain outside and ride out the storm. The wind hauled to the eastward toward midnight, and at 3 A.M. it looked so threatening in the N.W. that a fourth reef was taken in the mainsail and the foresail was treble-reefed. In half an hour the wind died out completely, and the vessel lay in the trough of a heavy S.E. sea, that was threatening every moment to engulf her. She was then about 12 miles E.S.E. from Sandy Hook lightship, and in twenty minutes the gale struck her with such force from N.W. that she was thrown on her beam ends; she instantly righted again, however, but in two hours was so covered with ice that she looked like a small iceberg. By 8 A.M. the wind had increased to a hurricane, the little vessel pitching and tossing in a terrific cross-sea, and only by the united efforts of the entire crew was it possible to partially lower and lash down the foresail and fore-staysail. No one but those on board can realize the danger she was in from the huge breaking seas that rolled down upon her; the snow and rain came with such force that it was impossible to look to windward, and the vessel was lying broadside to wind and sea. A drag was rigged with a heavy log, anchor, and hawser, to keep her head to sea and break the force of the waves, but it had little effect, and it was evident that something must be done to save the vessel. Three oil bags were made of duck, half filled with oakum saturated with oil, and hung over the side forward, amidships, and on the weather quarter. It is admitted that this is all that saved the boat and the lives of all on board, for the oil prevented the seas from breaking, and they swept past as heavy rolling swells. Another drag was rigged and launched, although not without great exertion and danger, and this helped a little. Heavy iron bolts had to be put in the oil bags to keep them in the water, and there the little vessel lay, fighting for life against the storm, refilling the oil bags every half hour, and fearing every instant that some passing vessel would run her down, as it was impossible to see a hundred feet in any direction. The boat looked like a wreck; she was covered with ice and it seemed impossible for her to remain afloat until daylight. The oil bags were replenished every half hour during the night, all hands taking turn about to go on deck and fill them, crawling along the deck on hands and knees and secured with a rope in case of being washed overboard. Just before midnight a heavy sea struck the boat and sent her over on her side; everything movable was thrown to leeward, and the water rushed down the forward hatch. But again she righted, and the fight went on. The morning of the 13th, it was still blowing with hurricane force, the wind shrieking past in terrific squalls. It cleared up a little towards evening, and she wore around to head to the northward and eastward, but not without having her deck swept by a heavy sea. It moderated and cleared up the next day, and after five hours of hard work the vessel was cleared of ice, and sail set for home. She had been driven 100 miles before the storm, fighting every inch of the way, her crew without a chance to sleep, frostbitten, clothes drenched and no dry ones to put on, food and fuel giving out, but they brought her into port without the loss of a spar or a sail, and she took her station on the bar as usual.

—From Everett Hayden's report of The Great Storm of March 11–14, 1888, which appeared in the first issue of the Magazine, October 1888

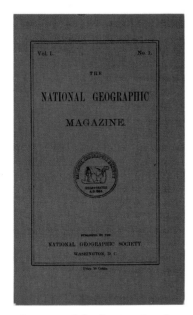

The cover of the first National Geographic Magazine. *Membership numbered 165.*

mittently until its eighty-ninth year, by which time Gilbert H. Grosvenor's grandson, Gilbert M. Grosvenor, was occupying the editor's desk.

Vol. I, No. 3, published in July 1889, contained "The Rivers and Valleys of Pennsylvania," inaugurating what would be one hundred years of the Magazine's coverage of the states; and Vol. I, No. 4, appeared three months later with "Across Nicaragua with Transit and Machéte" by R. E. Peary, a young naval civil engineer who would later gain fame in less temperate climes.

Peary's description of the brutal hardships involved in clearing the Nicaraguan jungles in order that engineers might take transit sightings and determine elevations for the then-proposed trans-Nicaraguan canal, closed with this surprisingly eloquent passage:

> After the day's work comes the dinner, the table graced with wild hog, or turkey, or venison, or all. After dinner the smoke, then the day's notes are worked up and duplicated and all hands get into their nets. For a moment the countless nocturnal noises of the great forest, enlivened perhaps by the scream of a tiger, or the deep, muffled roar of a puma, fall upon drowsy ears, then follows the sleep that always accompanies hard work and good health, till the bull-voiced howling monkeys set the forest echoing with their announcement of the breaking dawn.

Two years later the Society published in its April 1891 issue "Summary of Reports on the Mt. St. Elias Expedition," which, although jointly sponsored with the U.S. Geological Survey, was the Society's first "venture in exploration."

Under Society sponsorship, an expeditionary force of ten men, led by the geologist Israel C. Russell, had explored Mount St. Elias, the highest point on the boundary between Alaska and Canada, discovered Mount Logan (subsequently determined to be, at 19,524 feet, North America's second-highest peak), and a huge frozen river of ice they named Hubbard Glacier after the Society's first president.

"In several ways, this first expedition set a pattern for the Society's 200 explorations and researches that have followed over the years," Melvin M. Payne, the National Geographic's now-retired Chairman of the Board of Trustees, would write in 1963. "It triumphed over the forces of nature. It added to man's knowledge of his world. And it established a tradition of close cooperation between the National Geographic Society and agencies of the United States Government."

In addition, Russell's first-person adventure-narrative in the following month's *Geographic* established the pattern for reporting explorations:

> Darkness settled and rain fell in torrents, beating through our little tent. We rolled ourselves in blankets, determined to rest in spite of the storm. Avalanches, already numerous, became more frequent. A crash told of tons of ice and rock sliding down on the glacier. Another roar was followed by another, another, and still another. It seemed as if pandemonium reigned on the mountains.
>
> Looking out, I saw rocks as large as one's head bounding within a few feet of our tent. One struck the alpenstock to which the ridgerope was fastened. Our tent "went by the board," and the rain poured in. Before we could gather our soaked blankets, mud and stones flowed in upon them. We retreated to the edge of the glacier and pitched our tent again. Wet and cold, we sought to wear the night away. Sleep was impossible.

—From *Great Adventures with National Geographic*

ANNUAL REPORT OF THE TREASURER.

FOR THE YEAR ENDING DEC. 27, 1888.

THE TREASURER, in account with the NATIONAL GEOGRAPHIC SOCIETY.

1888.

Dec. 27.	To cash received from life members		$100 00
	" " for annual dues year 1888		1025 00
			$1125 00

1888.			
Apr. 16.	By Cash—M. F. Peake & Co. (20 chairs)		$ 60 00
	" Paid Columbian University, rent of hall		20 00
Oct. 31.	" Paid Tuttle, Morehouse & Taylor, for printing and binding vol. I of Magazine	$ 190 56	
	" Norris Peters, for lithographing storm plates for Magazine	58 00	
	" Sundry expenses of Magazine	6 35	254 91
Dec. 27.	" Paid Cosmos Club, rent of hall		18 00
	" " for miscellaneous expenses:		
	" " " Printing	74 50	
	" " " Stationery	28 35	
	" " " Postage	29 15	
	" " " Sundries	13 39	145 39
Balance on hand (Bank of Bell & Co.)			626 70
			$1125 00

C. J. BELL,

Treasurer.

December 28, 1888.

To the National Geographic Society :

The undersigned, having been appointed an Auditing Committee to examine the accounts of the Treasurer for 1888, have the honor to make the following report :

We have compared the receipts with the official list of members and find complete agreement. We have compared the disbursements with the vouchers for the same and find them to have been duly authorized and correctly recorded. We have examined the bank account and compared the checks accompanying the same. We have compared the balance in the hands of the Treasurer as shown by the ledger ($626.70) with the balance as shown by the bank book ($644.70) and found them consistent, the difference being explained by the fact that a check for $18 drawn in favor of the Secretary of the Cosmos Club has not yet been presented for payment. We find the condition of the accounts entirely satisfactory.

Very respectfully,

S. H. KAUFMANN.

G. K. GILBERT.

13

As this December 27, 1888, Annual Report of the Treasurer shows, the National Geographic Society's total assets at the end of its first year were $626.70.

By January 1896 the erratically issued *National Geographic Magazine* became a monthly. In the hopes of increasing its disappointing circulation the Board of Managers offered copies at 25 cents each through newsstands and began to accept advertising. The Magazine's terra-cotta cover was discarded in favor of a buff cover, with the contents, editors, authors, and the legend "An Illustrated Monthly" superimposed upon a latitudinal and longitudinal lined globe.

The January 1896 issue contained Hubbard's article "Russia in Europe" in which he did for the Russian peasant what he had done for the African Negro:

> The peasants wear the same clothes night and day...and are required by their priests to bathe every Saturday evening....They lead idle, listless lives in winter, and when winter ends are little inclined to work....
>
> We can scarcely comprehend such a people or such a life and are not surprised to learn that they resort to cards and drink as the only relief from the dullness of the interminable winter. They never hurry, for time is not money.

One relief from the dullness of Washington's interminable winter was the Society's lecture series; although to Hubbard and his contemporaries, the increase and diffusion of geographical knowledge was serious business. For, as WJ McGee had dourly reported to the Board of Managers: "In choosing popular speakers on current topics, preference is given either to actual explorers or original investigators who are known to treat geography as a branch of science, and such speakers arrange and present their matter freely, save that *the excessive use of picture and anecdote is discouraged*—the object is to instruct as well as entertain." [Author's italics.]

McGee's admonition to the lecturers was assumed to relate to the *Geographic* as well. And yet an article in the next issue "The Recent Earthquake Wave on the Coast of Japan" by geographer Eliza Ruhamah Scidmore, an Associate Editor of the Magazine and the only woman then to serve on the Board—with its photographs of a floating corpse, splintered houses and beached, demasted ships—ignored McGee's caveat entirely:

> The barometer gave no warning, no indication of any unusual conditions on June 15, and the occurrence of thirteen light earthquake shocks during the day excited no comment. Rain had fallen in the morning and afternoon, and with a temperature of 80° to 90° the damp atmosphere was very oppressive. The villagers on that remote coast adhered to the old calendar in observing their local fêtes and holidays, and on that fifth day of the fifth moon had been celebrating the Girls' Festival. Rain had driven them indoors with the darkness, and nearly all were in their houses at eight o'clock, when, with ...a roar, and the crash and crackling of timbers, they were suddenly engulfed in the swirling waters. Only a few survivors on all that length of coast saw the advancing wave, one of them telling that the water first receded some 600 yards from ghastly white sands and then the Wave stood like a black wall 80 feet in height, with phosphorescent lights gleaming along its crest. Others, hearing a distant roar, saw a dark shadow seaward and ran to high ground, crying "*Tsunami! tsunami!*" Some who ran to the upper stories of their houses for safety were drowned, crushed, or imprisoned there, only a few breaking through the roofs or escaping after the water subsided....Ships and junks were carried one and two miles inland, left on hilltops, treetops, and in the midst of fields uninjured or mixed up with the ruins of houses, the rest engulfed or swept seaward....Many survivors, swept away by the waters, were cast ashore on out-lying islands, or

Opposite: "*As the rain became heavier, the avalanches, already alarmingly numerous, became more and more frequent....It seemed as if pandemonium reigned on the mountains.... Looking out, I saw rocks as large as one's head bounding past within a few feet of our tent.*"

Thus reported Society founder Israel C. Russell on the thundering avalanche that clipped his tent on Marvine Glacier in Alaska during the first expedition organized by the Society. Russell's joint Society–U.S. Geological Survey team then pressed on across ice fields and treacherous passes toward Mount St. Elias, mapped several hundred square miles of the region and discovered and named Mount Logan, Canada's highest peak (a fact then unknown to them).

Russell's first-person account, "*An Expedition to Mount St. Elias, Alaska,*" appeared in the May 1891 issue of the National Geographic Magazine *and set the pattern for reporting future explorations.*

seized bits of wreckage and kept afloat. On the open coast the wave came and withdrew within five minutes, while in long inlets the waters boiled and surged for nearly a half hour before subsiding. The best swimmers were helpless in the first swirl of water, and nearly all the bodies recovered were frightfully battered and mutilated, rolled over and driven against rocks, struck by and crushed between timbers. The force of the wave cut down groves of large pine trees to short stumps, snapped thick granite posts of temple gates and carried the stone cross-beams 300 yards away. Many people were lost through running back to save others or to save their valuables.

One loyal schoolmaster carried the emperor's portrait to a place of safety before seeking out his own family. A half-demented soldier, retired since the late war and continually brooding on a possible attack by the enemy, became convinced that the first cannonading sound was from a hostile fleet, and, seizing his sword, ran down to the beach to meet the foe.

And immediately following a turgid piece on "The Economic Aspects of Soil Erosion" in the October issue, there appeared a report on the Nansen Polar expedition which began:

On the 17th day of June, 1896, as some of the men of the English Jackson and Harmsworth expedition, in Franz Josef land, were looking out over the ice they discovered a weird figure advancing towards them, with long straggling hair and beard and garments covered with grease and blood stains, who proved to be none other than Dr Fridhjof Nansen, who fifteen months previous had left his ship, the *Fram*, at 83° 59' north latitude and 102° 27' east longitude in order to push on with sleds, boats, and dogs towards the Pole. In a shelter some distance off was Dr Nansen's companion, Lieutenant Johansen.

But perhaps more surprising still was the appearance in the November 1896 *Geographic* of a photograph of a half-nude Zulu bride and bridegroom in their wedding finery, both staring full face into the camera—the groom with obvious courage and pride, the bride looking somewhat less stoic. He is holding her hand as if they were sealing their troth with a handshake. It is a nice photograph, even a tender photograph, and the reader must have wondered what to make of its accompanying text: "These people are of a dark bronze hue, and have good athletic figures. They possess some excellent traits, but are horribly cruel when once they have smelled blood."

Despite its facelift, despite running its first photographic bare-breasted woman, despite becoming "An Illustrated Monthly," and despite its occasional more popular than technical journalistic reporting, the Magazine was still being edited by committee and its determinedly academic approach to geography remained essentially unchanged. Circulation rose only slightly.

Gardiner Greene Hubbard died in 1897, and his son-in-law Alexander Graham Bell was, as Bell would later recount, "forced to become president of the [National Geographic Society] in order to save it." Bell, after considerable persuasion, accepted the presidency in January 1898.

Alexander Graham Bell was a stout, fifty-one-year-old, bearded bear of a man, a daring, innovative, somewhat disorganized, eccentric genius with a childlike enthusiasm for and fascination with the world around him. It was

This 1896 Japanese print depicts the earthquake-generated wave that struck the coast of Japan on June 15 of that year. Eliza Ruhama Scidmore's report on the devastating eighty-foot-high, phosphorescent-crested Sanriku tsunami appeared in the September 1896 National Geographic.

through his work with the deaf that Bell, then twenty-five years old, met Gardiner Greene Hubbard in Northampton, Massachusetts, in 1872. Hubbard's then-fifteen-year-old daughter, Mabel, had been left totally deaf from a severe attack of scarlet fever ten years before. Hubbard had sought Bell's advice on improving her speech. Bell became Mabel Hubbard's teacher, they fell in love, and five years later—only after Bell felt he had proved himself through his invention of the telephone—they married.

For their first two years of marriage, the Bells' worst quarrels were over the inventor's preference for working late at night when the quiet gave him peace. Bell would not retire until four o'clock in the morning, after making sure his wife and children were shielded from moonlight. (Bell's one superstitious fear, his biographer Robert V. Bruce disclosed, was having moonlight fall on himself or others while asleep.)

In early 1892, the insatiably curious Bell reported to Mabel that after an evening spent reading Jules Verne he had turned to "my usual night reading, Johnson's Encyclopedia. Find this makes splendid reading matter for night. Articles not too long—constant change in the subjects of thought—always learning something I have not known before—provocative of thought—constant variety."

During the first year of Bell's presidency of the National Geographic Society, he seems to have been too absorbed in his experimental work on flying machines to give the Magazine much time or thought. And by Bell's first anniversary as President, membership had increased only from 1,140 to 1,400, a dismayed Board was faced with the Society's $2,000 debt, and the Magazine was on the brink of bankruptcy.

Bell was not discouraged. He approached the Board and outlined his plan. In effect he told them, "Geography is a fascinating subject, and it can be made interesting. Let's hire a promising young man to put some life into the magazine and promote the membership. I will pay his salary. Secondly, let's abandon our unsuccessful campaign to increase circulation by newsstand sales. Our journal should go to *members*, people who believe in our work and want to help." Bell recognized that the lure was not the Magazine, but *membership* in a society that made it possible, as Ishbel Ross would later write in a 1938 *Scribner's* article, for "the janitor, plumber, and loneliest lighthouse keeper [to] share with kings and scientists the fun of sending an expedition to Peru or an explorer to the South Pole."

Bell's first step was to search out a full-time editor, someone who would devote his entire time to the editorial work required by the Magazine and the promotion of Society membership. Up to now these duties had been carried out by a committee whose distinguished members performed their work without pay. As Bell was to write later, "But in starting out to make a magazine that would support the Society, instead of the Society being burdened with the magazine, a man was of the first necessity; if we did not get the right man the whole plan would be a failure...."

He already had a young man in mind.

Among the noted scholars and scientists invited to lecture before the Society in 1897 was the distinguished historian Professor Edwin A. Grosvenor, who, before accepting his appointment at Amherst College, had been a professor of history at Robert College in Constantinople for twenty-three years. While a guest in the Bells' Washington home, Professor Grosvenor had spoken glowingly of his identical twin sons, Gilbert and Edwin. The Bells' attractive teen-age daughter Elsie had listened with interest and when later that spring Mrs. Grosvenor invited Elsie and her sister, Daisy, to attend the twins' graduation from Amherst, the girls accepted.

Throughout their years as students and until they graduated and separated, their father, as if to accentuate their twin-ness, insisted that Edwin and Gilbert dress alike; if one of the boys were to come down to dinner wearing a tie different from that of the other, he would be sent immediately upstairs to change to a tie that was the same (see p. 39).

Gilbert Grosvenor so enjoyed being an identical twin that he felt it was "next to a wife...the greatest favor the Lord can give a man."

That summer following the Grosvenor boys' graduation from Amherst, Elsie, who had no difficulty in telling the twins apart, found herself attracted to Gilbert. Aware that her father was searching for a "promising young man," she reminded him about the twin sons of his friend Professor Grosvenor. Perhaps one of them might be interested in working in Washington?

"Dr. Bell, who greatly admired my father, embraced the idea as his own," Gilbert H. Grosvenor later wrote. "Soon he was at his desk, writing to his friend. Would either of the twins, Edwin and Gilbert, be interested in a job that might be 'a stepping stone to something better?'"

Bell then sent a brief personal note to each of the twins along with a copy of the letter he had written their father. Bell's offer appealed strongly to young Gilbert Grosvenor who, from early childhood, had been interested in editing and writing, and had helped with the proofs and layout of his father's erudite two-volume 1895 *Constantinople,* according to Grosvenor, the first scholarly work to be profusely illustrated with photographs. Authors and editors were common visitors in the Grosvenor household, and young Bert eagerly took part in their literary discussions. But he wanted to make sure that Edwin was genuine in his denial of having any interest in running a magazine. Edwin was sincere. He was interested in the law and recommended that Gilbert follow his heart, take the job, and be near the young lady in Washington.

At that time, Gilbert Grosvenor was studying for his master of arts degree and was in his second year of teaching at the Englewood Academy for Boys in New Jersey. Hired as an instructor in French, German, Latin, college algebra, chemistry, public speaking, and debating, Grosvenor would later recall, "Compared with this program, a job as editor seemed very easy."

He wrote Dr. Bell that the editing job did interest him and was invited to a meeting at Bell's home. Present also were the inventor's wife and his daughter Elsie.

Bell showed Gilbert copies of some of the leading magazines of that period—*Harper's Weekly, McClure's, Munsey's, The Century,* among others—and asked Grosvenor, "Can you create a geographic magazine as popular as these, one that will support the Society instead of the Society being burdened with the magazine?"

"Yes, I believe I can," Grosvenor replied, "but I must proceed slowly and feel my way."

Aware that Dr. Bell expected quick results, Grosvenor found he had to emphasize repeatedly that the Magazine would have to go through a period of evolution; and because of the magnitude of the task, Dr. Bell should not expect it to become a success overnight.

The article that accompanied this November 1896 National Geographic *photograph "ZULU BRIDE AND BRIDE- GROOM" (the first bare-breasted natives to appear in the* National Geographic*) reported, "These people are of a dark bronze hue, and have good athletic figures. They possess some excellent traits, but are horribly cruel when once they have smelled blood."*

The twins Edwin Prescott Grosvenor and Gilbert Hovey Grosvenor (about ten years old) with their elder brother, Asa Waters Grosvenor (right to left), photographed in their home near today's Istanbul, Turkey. The twins, born in what was then Constantinople in 1875, lived there until 1890 while their father, Edwin A. Grosvenor, taught history at Robert College.

"Before the future president of your Society went to America," later wrote Maynard Owen Williams, for many years chief of the Geographic's *foreign staff, "his eyes were focused on scenes of many lands and his ears tuned to the babel of tongues then spoken on Galata Bridge....*

"His nurse was an Armenian, Kurdish porters toiled up the cobbled paths carrying provisions to his home. Albanians, Bulgarians, and Greeks were his classmates....

"Little wonder that geography seemed to Gilbert Grosvenor a dramatic series of living pictures, rather than mere dots on a chart."

Bell accepted Grosvenor's cautions then proposed something that immediately made Grosvenor wary: Some years earlier, Dr. Bell and Gardiner Hubbard had spent $87,000 in a vain attempt to establish a magazine called *Science*, which had failed after two unprofitable years. Bell told the young man he was willing to back the *National Geographic Magazine* with funds equal to those poured into *Science*.

Much to the relief of Mrs. Bell, who was not in the least happy at the thought of her husband losing a vast sum in a magazine again, Grosvenor firmly told Dr. Bell he would accept the job only if the inventor limited his gift to a $100 monthly salary—a figure even less than Grosvenor had been earning as a teacher.

"I knew that sheer weight of money would not accomplish what he wanted," Grosvenor would later write of that fateful meeting with Bell. "I also realized that, despite Dr. Bell's good will, a youth of 23 was not prepared to administer so large a sum. Older men, men unwilling to experiment, inevitably would push me aside, and I would have little opportunity to create and to try new ideas. Yet, without imagination and a new approach, there could be no hope for the magazine."

Dr. Bell reluctantly accepted Grosvenor's conditions and the meeting was concluded. In the brief moment that Elsie had alone with Gilbert by the door before he departed, she whispered, "I told Papa you had the talent he sought and would like to come to Washington!"

On April Fool's Day, 1899, Bell brought Gilbert H. Grosvenor to the Society's headquarters opposite the United States Treasury building. The "headquarters" Grosvenor later wrote, was but "half of one small room (the other half occupied by the American Forestry Association), two rickety chairs, a small table, a litter of papers and ledgers, and six enormous boxes crammed with *Geographics* returned by the newsstands."

Bell looked about the room. "No desk!" Bell exclaimed. "I'll send you mine."

That afternoon deliverymen brought Grosvenor a handsome rolltop desk made of Circassian walnut. When he sat down before it, Grosvenor was looking at the only visible property of the National Geographic Society. The treasury was not just empty; the Society was nearly $2,000 in debt.

Poor in funds though the Society may have been, it was extraordinarily rich for its most recently acquired asset: Gilbert Hovey Grosvenor, its first full-time employee.* However, it was not, as Dr. Bell's letter had proposed, with the title of Managing Editor, but as Assistant Editor that Grosvenor was hired—and only for three months with the understanding that a more permanent engagement would be made at the end of that time if the appointment was satisfactory to the Society's Editorial Committee.

John Hyde was Editor of the Magazine. And although Hyde was employed as statistician to the U.S. Department of Agriculture, he clearly had considerably more influence over the selection of the materials and editing of the Magazine than Grosvenor edited *Geographic* histories would have one believe.

During his first few days at the *Geographic*, Dr. Bell took pains to introduce Grosvenor to his distinguished colleagues of the Society, all men consid-

---

*There was, in addition to Grosvenor, one part-time Assistant Secretary, a clerk who, Grosvenor later recalled, "had just had a baby—his first—and about every half hour he'd go out and drink to its health."

THE GROSVENOR TWINS AS COLLEGE SPORTS: GAME, SET, AND MATCHED

Neckties were at the root of some of the social games the twins played at college. In those days of dance cards young men would not be permitted to sign up for more than one dance with a young lady or else their attentions might appear unseemly. Gilbert would arrive at the dance in a black dress tie, Edwin in white. If one of the twins had been granted a dance with a young lady whom the other desired, they would swap ties and the girl would not only be unaware of the difference, she would be rather pleased with herself for noticing that the color of the necktie was the key to the twin's identity. But as the October 7, 1900, *Boston Sunday Post* article on the twins reported:

"The only hitch came when both preferred to dance with the same partner, and then it was a contest of wits, and each was unscrupulous in stealing marches on the other. This necessitated carrying an extra necktie to use when an exchange was refused."

There were those times, however, as in college athletics, when dressing alike was demanded by fashion. And, as the Boston newspaper article concluded:

"Even in the college sports the twins had their opportunity. Both were exceptional tennis players, and so even was their relative ability that in contests against each other victories were evenly distributed. But it was as partners in doubles that the boys made their reputation, and so skilful were they that they won the college championship and represented Amherst in doubles

Edwin P. Grosvenor (left) and Gilbert H. Grosvenor (right) shown with fellow members of the Amherst College Class of 1897 elected to Phi Beta Kappa in April 1896.

in the intercollegiate tournament at New Haven. Dressed alike in tennis shirts, duck trousers and white canvas shoes, they were more indistinguishable than ever. One excelled in service and the other in net play, and their college mates used to assert that they took advantage of this to have one serve for both, since neither opponents nor spectators could tell the difference. This charge they always repudiated, but it was a standing joke among their friends..."

erably older than the twenty-three-year-old former boys' school teacher, and then Bell left for his summer home and laboratory at Baddeck, Nova Scotia. Bell's family, Elsie included, soon followed.

Grosvenor's first assignment was to prepare an index of the previous *National Geographic Magazines*, but the major task confronting him as he settled into his new job was to find the means to revive the Society's depressed membership (and with it the Society's bank account). From the beginning, Grosvenor had enthusiastically endorsed Bell's doctrine of membership-subscription: the lure of joining was not just in receiving a subscription to a magazine, but in becoming a *member* of a geographic society whose functions were the "increase and diffusion of geographical knowledge" through its flagship publication the *National Geographic Magazine*, its monographs, bulletins, afternoon and evening lectures, courses, and so on.

Grosvenor began by asking prominent men to nominate their friends for membership. "Initially that meant seeking nominations from my father and Dr. Bell," Grosvenor later wrote, "and my letters reveal that I badgered them unceasingly, unmercifully."

He was no less persistent with the officers and Board members of the Society; and they, too—once their initial misgivings about direct-mail soliciting were appeased—responded generously. As soon as a nomination was received, the Admissions Committee would approve it and Grosvenor would write a personal letter to each new nominee inviting him or her to join. He made certain, however, that the letter was on engraved, embossed bond paper. "People don't pay any attention to cheaply written letters," he later said. "I had already decided we should have the finest stationery when we wrote to people inviting them to join the Society."

Grosvenor's personal letter to each new nominee would begin, "I have the honor to inform you that you have been recommended for membership...."

Such an opening may have had a somewhat pompous ring to it, but it was effective. Membership created a bond with other "educated men" who, like the late Gardiner Hubbard, desired only "to promote special researches by others, and to diffuse the knowledge so gained, among men, so that we may all know more of the world upon which we live."

Scores of such letters went out each week and Grosvenor's campaign began quickly to gain new members for the Society. "You see," Grosvenor wrote his father at about this time, "we want to make [much of] the fact of an election to the National Geographic Society so that people will think that they are complimented by nomination."

So appreciative was the Society of Grosvenor's efforts that even before his first month on the job had ended, Alexander Graham Bell had written:

> I am happy to inform you... that the Editorial Committee have been so much pleased with you that they consider it unnecessary to wait for the expiration of the three months before making a more permanent engagement with you...and I now have great pleasure in confirming your appointment as Assistant Editor of the Magazine, for one year instead of three months, dating from the first of April, 1899, your salary to be one hundred dollars per month....
>
> I have written to Mr. Hyde suggesting that your name should appear in the June issue of the Magazine as Assistant Editor under his.*

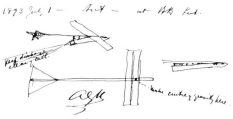

Top: *Bell's 1892 notebook sketch for a "vacuum jacket," a precursor of the iron lung.* Above: *An 1893 sketch of a rocket-powered airplane. Bell's "little model had tin wings and tail and a hollow brass fuselage packed with fuel." The inventor recorded that his experimental model rose thirty feet and flew seventy-five. His note, "Keep discharge clear of tail," was a reference to a lesson learned from an earlier model whose tail was buckled by the rocket blast.*

---

* *When Grosvenor's name did appear on the June issue's cover as Assistant Editor it appeared beneath Hyde's—and under those of the twelve Associate Editors as well.*

Shortly after Gardiner Greene Hubbard's death, in 1897, his son-in-law Alexander Graham Bell agreed to become the second president of the National Geographic Society. Inventor of the telephone, Alexander Graham Bell is shown here with Helen Keller at Baddeck, his summer estate on Nova Scotia's Cape Breton Island, where he bred twin-bearing sheep, designed solar stills to provide drinking water from seawater, and carried on his experiments with hydrofoil boats, man-carrying tetrahedral kites, jet- and rocket-powered blades resembling helicopter rotors, airplanes with hinged-surface wing-tip ailerons. And yet, when asked his profession, Bell would invariably reply that he was "a teacher of the deaf." Behind Bell and Miss Keller sits Alexander Melville Bell, the inventor's father.

Helen Keller was taken to Bell as a six-year-old, blind, deaf, and unable to speak. Bell helped to direct her education, and years later she dedicated her autobiography to him and wrote of their first meeting, "I did not dream that that interview would be the door through which I would pass from darkness into light, from isolation to friendship, companionship, knowledge, and love."

Alexander Graham Bell, photographed in 1892 at the opening of the New York–Chicago long-distance telephone service just sixteen years after he had invented the telephone.

The following day Grosvenor received a note from Bell's personal secretary saying that Bell had wanted Grosvenor to know that the "Editorial Committee passed a resolution giving you great scope in the matter of securing Corresponding Members and pushing The Magazine—therefore go ahead and do what you think best in that direction, and if any question arises regarding expense in getting out circulars, etc., remember you have Mr. Bell behind you. The point is to materially increase the corresponding membership."

To "materially increase the corresponding membership" was synonymous with increasing the Society's revenues.

As a sales device, as a means of achieving subscription renewal, and the establishment of a permanent pool of subscribers, the concept of membership in a society whose magazine was but a tangible bonus was an inspired marketing device. It would remain so, however, only so long as the Magazine itself remained interesting. To attract readers, Grosvenor would have to change the public's attitude toward geography, which he knew was regarded "one of the dullest of all subjects, something to inflict upon schoolboys and avoid in later life." The Society's key to success, a popular approach to geography, was missing.

He began by studying other geographic journals then being published by geographic societies throughout the world. He next turned to those books in which geography played an important part, books that endured like Charles Darwin's *Voyage of the Beagle*, Richard Henry Dana Jr.'s *Two Years Before the Mast*, Joshua Slocum's *Sailing Alone Around the World*, and Herodotus's travels, written 2,000 years before. What was there in Herodotus's *History*, Grosvenor wrote, "that gave the book such life that it had survived 20 centuries and was still going strong?" What did those geographic books to which readers returned again and again have in common?

The answer, Grosvenor became convinced, was that "each was an accurate, eyewitness, firsthand account. Each contained simple, straightforward writing—writing that sought to make pictures in the reader's mind."

As his first summer as Assistant Editor drew to a close, Grosvenor managed to increase both the number of pages in the Magazine and its print order; but he continued to have to subordinate his tastes to that of Editor John Hyde and to the tastes of those twelve, unpaid Associate Editors—each of whom was a member of the Society's Board of Managers and identified on the *Geographic*'s cover by name and title: the Chief Signal Officer, U.S. Army; the Chief Geographer of the U.S. Geological Survey; the chiefs of the Bureau of Statistics and the U.S. Weather Bureau; the Assistant Secretaries of State and the Navy, and so on.

Many of them continued to want to insert technical, unintelligible-to-the-layman articles into the Magazine; and it was often difficult for Grosvenor to edit or simplify such material without giving offense. Grosvenor's ally, Dr. Bell, was away and not always aware, the young editor felt, of the stresses Grosvenor was under, nor appreciative of his problems. Nevertheless, all that summer Bell was sending Grosvenor letters of advice (the Magazine should have "a multitude of good illustrations and maps"), ideas on how to increase circulation, and the note that since scientific agencies of the federal government were so close to the Society's headquarters, Grosvenor should take advantage of being able to pick up considerable material from them at no cost.

According to Robert V. Bruce, Bell wrote the best writers would need to be paid well, and that meant advertising revenues would need to increase. The way to do that was through increased circulation, and the way to increase circulation was through a magazine that was not only scientifically reliable but of popular interest. Timeliness would help, too, Bell added. The *Geographic* should have material stockpiled in advance so that if the world's attention shifted to a new place, the appropriate information would be available. The Society could also, Bell suggested, publish "popular books on geographic subjects."

What did Bell consider "geographic subjects"?

Bell replied, "THE WORLD AND ALL THAT IS IN IT is our theme."

Bell would not have received any quarrel from Grosvenor who would write, "I thought of geography in terms of its Greek root: *geographia*—a description of the world. It thus becomes the most catholic of subjects, universal in appeal, and embracing nations, people, plants, animals, birds, fish. We would never lack interesting subjects."

In September 1899, Grosvenor volunteered to take charge of the Society's lecture series and invited former Secretary of State John W. Foster to give the first lecture. Foster spoke as an authority on "The Alaskan Boundary," a topic of intense interest at that time because of the international controversy involving the United States, Great Britain, and Canada over exactly where Alaska's southeastern boundary lay. At stake was access to the Klondike gold fields discovered just three years before. When Grosvenor published the Secretary's speech in the November *Geographic*, long editorials appeared about it in newspapers nationwide.

"Thus we entered 1900 with rising prospects—or so Dr. Bell and I thought," Grosvenor later wrote. "However, the Executive Committee, a five-man minority of the Board but a determined one, believed our plans hopelessly optimistic. These men wanted to stop 'undignified' membership promotion and sell the magazine on newsstands, although that approach previously had led to a debt-ridden Society."

Among the various prominent editors and publishers whose suggestions for improving the financial position of the Magazine Bell had sought, the most influential was S. S. McClure, publisher of *McClure's Magazine*,* which had begun publication four years after the *National Geographic* and in seven years had achieved an attention-commanding circulation of 370,000.

McClure advised the Executive Committee that since it was impossible to establish a popular magazine in Washington, they should: 1) move the *National Geographic* to the center of magazine publishing, New York; 2) change its name to something simpler; 3) abandon the idea of attracting subscribers by offering memberships; 4) beef up its advertisements and depend on newsstand sales to increase revenue and circulation; and, 5) *never* mention the name National Geographic Society, since people abhor geography.

Grosvenor disagreed with every one of McClure's suggestions, but to his deep dismay McClure's argument powerfully influenced the Executive Committee. Fortunately the Board of Managers was not swayed. They reaffirmed their confidence in Bell's and Grosvenor's membership-subscription concept.

It was about this same time that Grosvenor was called to Dr. Bell's home.

* *The Hubbard family had invested heavily in McClure's, but no dividends were ever received. Dr. Bell, himself, had granted an interview for its first issue (June 1893).*

The Society's 1900 Excursion to Norfolk, Virginia.

In the spring of 1900 a total eclipse of the sun was to be readily visible from Norfolk. So on May 27 the Society chartered a steamer to transport some 250 of its members to view it. In the party were Alexander Graham Bell, his wife, Mabel Hubbard Bell, astronomer Simon Newcomb, various distinguished scientists, and the young Gilbert Grosvenor.

"The ladies all got out and spread white sheets over the dock...the idea being to watch the shadow of the moon as it traversed over these sheets," Grosvenor later recalled. "And then Simon Newcomb stationed himself to count the seconds—the length of the shadow, and...other men were taking pictures of the sun, itself. Well...as [Newcomb] counted out the seconds he tapped on a piece of wood so that they would hear. Later on, when they developed the photographs, there wasn't one that was any good, because the distinguished scientist had not realized that the wood he was tapping was a tripod and every tap shook the tripod. But we had a wonderful time...."

Right: *The* Silver Dart *with John A.D. McCurdy at the controls takes to the frozen air.*

"On the morning of February 23, 1909, Dr. Bell and nearly 150 townsmen gathered on frozen Baddeck Bay to witness Canada's first flight. Presently a sleigh towed Silver Dart *into position. While four skaters kept pace, the plane ran forward on its tricycle landing gear, a surprising forerunner of the modern undercarriage. Taking off from the ice,* Silver Dart *flew across the wintry landscape for half a mile."* Silver Dart *was the fourth "aerodrome" (the word Bell preferred to "aeroplane") to be built by the Aerial Experiment Association created by Alexander Graham Bell, McCurdy, F.W. (Casey) Baldwin, Glenn Curtiss, and Army Lt. Thomas Selfridge. McCurdy's historic flight "above the ice of Baddeck Bay," the* Geographic *reported in August 1956 had been "trailed by a crowd of excited skaters and a horse-drawn sleigh bearing his snowy-bearded chief, Dr. Bell."*

Bottom left: *Dr. and Mrs. Bell demonstrate the lightness of both the inventor's tetrahedral structure and their hearts.*

Bottom center: *Gilbert H. Grosvenor made this photograph himself as the five-ton torpedo-shaped* HD-4 *designed by Bell rose up on its thin steel hydrofoil blades to roar across Baddeck Bay, Nova Scotia. This hydrofoil set a world speedboat record of 70.86 mph on September 9, 1919—a record that remained unbroken for ten years.*

Bottom right: *"DEDICATORY BUNTING FLIES FROM A TETRAHEDRAL TOWER ATOP BEINN BREAGH: AUGUST 31, 1907," the* Geographic *noted in an article honoring Bell. The eighty-foot-tall viewing stand had been built by the inventor as a demonstration of the tetrahedron's strength. Assembled on the ground then jacked up into place, the seemingly fragile tower weighed less than 10,000 pounds and was able to support considerable weight.*

The inventor, obviously upset, asked the young editor, "What trouble have you been getting into?"

"Trouble?" Grosvenor asked. He couldn't think of any.

Bell sat frowning and then he said, "The Chairman of the Executive Committee has asked for an interview to talk about some serious matters connected with the conduct of Mr. Grosvenor."

Later that afternoon Bell summoned Grosvenor again. Seeing the young editor's trepidation, Bell roared with laughter. He was a large man, weighing at least 250 pounds, and his laughter was as big as his frame. "The Executive Committee thought the Chairman should tell the President that Mr. Grosvenor was paying more attention to his daughter than to the *National Geographic Magazine*!"

The Bells did not object to Grosvenor's courtship of Elsie, though Mrs. Bell seemed somewhat puzzled by the form it took. She was disturbed by how Grosvenor could be "so perfectly self-possessed when his lady-love is near him. I am sure you weren't," she told her husband. Mabel Bell decided that that was just probably the way Bert Grosvenor was. Besides, Elsie not only did not seem upset by her suitor's *sangfroid,* she apparently copied it: "Elsie says she thinks she will probably decide to marry Gilbert," Mabel Bell wrote to her husband the month the McClure proposal discussions were taking place, "but she is so perfectly matter-of-fact about it that I am sure she is not a particle in love with him."

Despite her reservations, Mabel Bell had purchased a new trunk for her daughter and had it monogrammed E.M.G. instead of E.M.B. When Elsie had shyly asked her mother why, Mrs. Bell had replied, "I think for the next journey you make those initials will be right for you."

When Elsie told young Bert, his heart leapt.

The Bells sailed for Europe on June 30, 1900; immediately thereafter, Gilbert Grosvenor's difficulties with John Hyde and the Executive Committee escalated.

Just prior to Bell's departure Hyde had written:

> My dear Mr. Bell:
>
> At a meeting of the Executive Committee this afternoon, I reported your exceedingly generous offer to continue your contribution toward Mr. Grosvenor's salary for another year and also your expression of opinion that the Society would be in danger of losing Mr. Grosvenor's services unless it should make some addition to the salary he is now receiving. That brought on a consideration of the entire question of the assistant editorship with a special reference to the arrangement with the McClure Company, nearly all the details of which have been agreed upon, subject, of course, to ratification by the Board. In view of the large possibilities opening up in connection with this agreement and of other circumstances, *the Committee are of the opinion that it would be unwise for the Board to take any action that might seem to commit it to Mr. Grosvenor* [author's italics], either when your generous contribution should cease to be available or when the increasing importance of the magazine should make larger demands upon its editorial management....
>
> With great respect, I am, my dear Mr. Bell,
>
> Very faithfully yours,
> John Hyde

"My position for this year, with regard to the Magazine, naturally came up to be settled, and I found that one or two men...wanted to turn me out, or at least make things very unpleasant for me," Grosvenor wrote his father three weeks before Bell left. "I can't help feeling that Mr. Hyde wants to retain the editorship after all, and thought that if he could get someone else in my place, he would be sure to stay in. I do not like to suspect him, but when he tells Mr. Bell he loves me like a son, and with the same breath that I am lacking in business ability and weak in proof-reading and hence do not deserve any increase in salary (not in these words, but to that effect), it is funny to say the least.... My position was referred to the Executive Committee, by the Board, with the unanimous recommendation that I receive a good promotion. The Executive Committee is comprised of [Henry] Gannett, [John] Hyde and [Alfred J.] Henry, and I shall hear from them today what they propose...." Six weeks later, on July 27, 1900, the Executive Committee offered Grosvenor his same job; he was given neither a promotion nor a raise.

Grosvenor's father, aware of the young editor's continuing problems with Gannett and Hyde, was counselling his son that troublesome summer in Washington, "Just be patient, be patient." But Grosvenor wrote his father that he would resign in the fall if the ideas of the Executive Committee and McClure prevailed.

Grosvenor had not wanted to bother Dr. Bell with his difficulties, but on August 6, 1900, knowing the inventor was preparing to return to Washington, he wrote that the battle with the Executive Committee was Bell's, too, and that soon he would have to face it again. "Naturally I am very much distressed with the committee," Grosvenor wrote, "but as I firmly believe that they are working not against me personally...but against your plans for the Society and for their individual interests, I do not intend to get out of their way, as they plainly hint they want me to."

An unexpected dividend to the Executive Committee's opposition to the young editor was that it angered Elsie Bell enough in late August for her to write Grosvenor that she would be glad to marry him, if he hadn't changed his mind. By return mail Grosvenor assured Elsie that he hadn't.

Dr. Bell telegraphed his congratulations, and shortly thereafter Mrs. Bell wrote Gilbert Grosvenor's mother:

> Of course, Elsie would not have written to Gilbert as she did at this time without Mr. Bell's and my full approval. We feel that Gilbert has proved his mettle in this summer's trials and deserves the reward Elsie wants to give him. He has certainly had a hard summer—meeting treachery where he expected loyal help and friendship—but I doubt whether Elsie would have been as sure of her own mind if all her love and sympathy had not been aroused by her indignation at the attacks upon him. So perhaps after all, Mr. Hyde has unwittingly done Gilbert a great service, and his late troubles are blessings in disguise.

Upon his return to Washington in September, Dr. Bell called together a meeting of the scientists, professionals, explorers, and businessmen who comprised the Society's Board of Managers. The Executive Committee did not have the support of this distinguished group; and in a resolution passed by the Board of Managers on September 14, 1900, they made a point of praising Grosvenor's work, unanimously reaffirmed his permanent status with the Society, named him Managing Editor, and increased his salary to $2,000 a year—the additional $800 to be paid by the Society beginning January 1, 1901.

Hyde was given the title of Editor-in-Chief but remained without pay.

The following day Grosvenor wrote his father:

...the resolutions...are, I think, all one would wish....Gannett and Hyde were thunderstruck and neither went to the meeting....But as Henry said when recording [the resolution], "they did it because of my work and because they believed in me and not because I was to be Mr. Bell's son-in-law." And I am going to think that true or mostly true.

"I had interpreted the title Managing Editor as the controlling man in the organization," Grosvenor later said. "And from that time on, I was the chief executive of the office, being responsible for the Magazine, the makeup of the Magazine, and the responsibility of increasing the membership of the Society."

Convinced that the rift between himself and the Executive Committee had been mended, Gilbert joined Elsie in London. They were married on October 23, 1900—the anniversary of his parents' wedding—and immediately set out on their honeymoon tour of Europe. On their way to Constantinople, the young couple had reached Vienna, when Grosvenor found himself worrying about the Magazine. Instinct warned him that something was wrong. He and Elsie returned to Washington, arriving early in December 1900.

Something indeed was wrong: in Grosvenor's absence the Executive Committee had arranged to have the Magazine printed not as it had been at Judd & Detweiler in Washington, but at McClure, Phillips & Company in New York. Evidently, Dr. Bell, who had gone abroad at the same time as the young Grosvenors but had returned before them, had participated in the lengthy discussions preceding the move and had opposed neither the negotiations, nor the contract when it was signed. Grosvenor confronted Bell with the New York printing bill for the January 1901 issue, pointing out that not only was the charge twice as expensive as the previous Washington bill, but that McClure, Phillips had failed to acquire for the *Geographic* either new members, new advertising, or revenues. Bell responded, "Well, Bert, the Board made you the Managing Editor. You are responsible now."

As Managing Editor Grosvenor was responsible; furthermore, he had the authority to act. He immediately went to New York. However, both the January and February 1901 issues were printed there before he could reverse the Executive Committee's actions and return the printing to Washington. But return it he did and "knew," he later wrote, "I had saved something more important than dollars: [I had saved] Dr. Bell's original plan to enlist members who would help us create a great educational institution."

In what with hindsight can be seen as the pivotal point in the young Magazine's history, Grosvenor had overturned the Executive Committee's decision, overruled his father-in-law's concurrence, and convinced the Board of Managers not only of the need to retain the Magazine's link with the Society through membership-subscription, but also to eventually disregard every one of McClure's suggestions. From that point on the National Geographic Society was run by Gilbert Hovey Grosvenor and not the other way around.

Grosvenor, offered the job less than two years earlier as "a stepping stone to something better," was to become the driving force behind the National Geographic Society for the next sixty-six years. Under his leadership the *Geographic* was transformed from an irregular, often dowdy technical journal with a circulation of a few hundred, into a glossy, color-packed popular publication with a circulation at the time of Grosvenor's death of over 5,000,000—large enough for one month's edition to make a stack twenty-five miles tall.

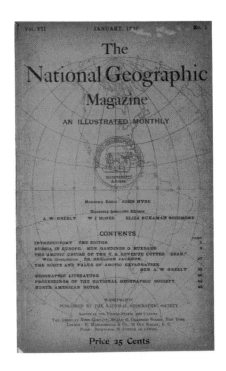

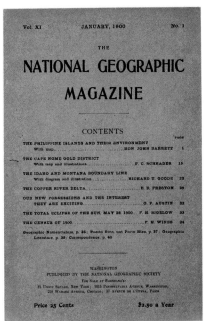

Top: *In January 1896, the erratically issued* National Geographic Magazine *became a monthly, and its forbidding terra-cotta cover was discarded in favor of the one shown here.*

Above: *By the end of 1900 the* National Geographic Magazine's *cover had changed for the third time, and circulation had risen to more than 2,200.*

"The Most Cherished of Geographical Prizes"

"The Pole at last!!! The prize of 3 centuries, my dream & ambition for 23 years. Mine *at last. I cannot bring myself to realize it. It all seems so simple..."*
—From Robert E. Peary's diary, April 6, 1909

"So we arrived and were able to plant our flag at the geographical South Pole. God be thanked!"
—From Roald Amundsen's diary, December 15, 1911

I n the annals of Polar exploration probably no man was more obsessed with attaining the Polar prize than America's Robert E. Peary. In 1880, when Peary was twenty-four years old, he had written to his mother, "I shall not be satisfied that I have done my best until my name is known from one end of the world to the other." The following year Peary entered the Navy as a civil engineer and began a tour in Nicaragua surveying sites for the proposed Atlantic-Pacific canal route across that country. While in Nicaragua, Peary chanced to come across an account by Adolf Erik, Baron Nordenskjöld, of his 1879 voyage through the Northeast Passage across the top of Europe and Asia to the Pacific Ocean. Peary immediately recognized that Arctic exploration had potential, that, as L.P. Kirwin wrote in *A History of Polar Exploration,* it could be Peary's "avenue to fame, an avenue leading to the most cherished of all geographical prizes, the North Geographical Pole."

In April 1891, before departing on his 1891–92 Greenland expedition, Peary addressed a gathering of Society members. "At the end of [his] talk," Benton McKaye later recalled in a letter written to Gilbert Hovey Grosvenor,

> President Hubbard, striding across the stage, picked up a large American flag, unfurled it, and stood facing Peary, to whom he addressed a brief speech of encouragement, then spoke these final words: *"Now take this flag and place it as far north on this planet as you possibly can!"*
>
> Peary gripped the flag and made his bow, while we of the audience roared forth a bon voyage for this start of Peary's career as an explorer.

During this 1891–92 Greenland expedition Peary walked some 600 miles across the mile-high ice cap and back again from what would be subsequently named Peary Land. On the way, he made many valuable scientific observations, established that Greenland was an island and not continental land reaching the Pole, and so enhanced his reputation that he became a popular lecturer, earning most of the money to finance future expeditions.

During the 1890s, Peary returned four more times to Greenland, perfecting the techniques that would enable him to reach the Pole. In 1898, three years after the Norwegian Fridtjof Nansen reached a record 86° 12′N, 229.5 miles from the Pole, Peary made a brief effort, losing eight of his toes. In 1900,

Preceding spread: *An icy bath befell one of the Steger 1986 Polar expedition's sled dogs when it slipped into a narrow lead. However, as the* Geographic *noted in its September 1986 "North to the Pole" report, "Bred for hardship and well insulated by their thick coats, the dogs easily withstood such common mishaps. But had a sled gone through the ice, the expedition would have ended immediately."*

Opposite: *"Eyes intent on a long-sought goal, Comdr. Robert E. Peary reached the North Pole over the ice on April 6, 1909—a feat never accomplished before or since," wrote the* Geographic *of this photograph in its 1888–1946 Index. But on May 1, 1986, the Steger International Polar Expedition's five men and one woman arrived at the North Pole, the first dog-sled expedition to make it to the top of the world without resupply since Peary's triumph seventy-seven years before.*

Overleaf: *Crowded with sled dogs, walrus blubber, and seventy tons of whale meat, the 186-foot ship* Roosevelt *carrying Peary north for his dash to the pole contained a "choking stench" that its captain, Bob Bartlett, never forgot.*

Italian Navy Lt. Umberto Cagni of the Duke of the Abruzzi's expedition penetrated twenty-two miles closer to the Pole than had Nansen. Peary made a brief try in 1901, and again in 1902, but fell far short at 84° 17'N. Finally, in 1906 on his fourth try, Peary reached 87° 6'N, 175 miles from the Pole and attained the record for "farthest north."

On the evening of December 15, 1906, at the annual white tie banquet of the National Geographic Society, attended by 400 members, various foreign dignitaries, U.S. government officials, and other distinguished guests, President Theodore Roosevelt presented the Society's first Hubbard Medal to guest of honor Commander Robert E. Peary. There was prolonged applause as Peary rose to his crippled feet to accept. Peary, his penetrating ice-blue eyes sparkling over his walrus mustache and hatchet nose, looked with intense personal satisfaction about the room then down at the gold medal. On one side was the seal of the Society, and on the reverse a map of the Arctic upon which was inscribed: "The Hubbard Medal, awarded by the National Geographic Society to Robert E. Peary for arctic exploration. Farthest north, 87° 6'. December 15, 1906." A blue sapphire star marked the northernmost point Peary had attained.

Peary looked back up at the President and the Society's officers and guests, and as the room quieted down he began to speak:

> …The true explorer does his work not for any hope of rewards or honor, but because the thing he has set himself to do is a part of his very being, and must be accomplished for the sake of accomplishment, and he counts lightly hardships, risks, obstacles, if only they do not bar him from his goal.
>
> To me, the final and complete solution of the Polar mystery, which has engaged the best thought and interest of the best men of the most vigorous and enlightened nations of the world for more than three centuries, and today quickens the pulse of every man or woman whose veins hold red blood, is the thing which *should* be done for the honor and credit of this country, the thing which it is intended that *I* should do, and the thing that *I must do* [author's italics].

The National Geographic Society awarded Peary $1,000, its first grant, and the Peary Arctic Club of New York built and outfitted a ship for him. In July 1908, not quite a year and a half after receiving the Hubbard Medal, Peary left New York in the middle of a heatwave and two months later his expedition's ship, *Roosevelt*, rounded the northeastern tip of Ellesmere Island in the Arctic Ocean and punched through the ice to 82° 30'N, the northernmost point reached by a ship under its own power. As the ice closed in about the *Roosevelt*, supplies were off-loaded at a quickly built igloo village and, five months later, sledged ninety miles to the Cape Columbia base camp established at the northern tip of Ellesmere Island, 413 nautical miles from the Pole.

Peary's polar team consisted of 24 men and 133 dogs. Seventeen of the men were Eskimos; the remaining seven were Peary, the leader; Robert A. Bartlett, the *Roosevelt*'s captain; Dr. John W. Goodsell, a surgeon; Ross Marvin, a civil engineer; George Borup and Donald MacMillan, two young Arctic enthu-

"ESKIMO SEXTETTE. MAIN DECK, ROOSEVELT" was the caption given this photograph, presented to the Society by Mrs. Josephine Peary, the explorer's widow.

siasts; and Peary's remarkable black associate, Matthew Henson. Peary had first encountered Henson as a clerk in a hat shop and hired Henson to accompany him on a second survey to Nicaragua in 1887. Henson, who had been with Peary ever since, was, by now, a sledge driver with expertise comparable to that of the Eskimos.

On February 28, 1909, Peary's advance party, consisting of the ship's master, Bartlett, and two Eskimos, left with their dogs and sledges and struck northward over the frozen Arctic sea. The next morning, four other parties followed with Peary's group in the rear.

Peary's method of moving men and equipment over the ice was revolutionary. His theory was to work the supporting parties to the limit in order to keep the final assault party fresh. The men whom he expected from the beginning to form the last team, therefore, had things made as easy as possible for them on the way up.

Small advance parties carved out a trail through the jumble of ice and pressure ridges for the main expedition to follow, built igloos for sleeping, and deposited supplies. It was an exhausting task, and these groups were constantly rotated to keep the men and dogs strong and healthy. At the beginning, Peary's progress was frustratingly slow. Once for a day, another time for a week, his advance was halted by black expanses of open water—channels, or "leads," up to a quarter mile wide. Despite the terrible cold and wind and jagged terrain, by March 15, the date scheduled for surgeon Goodsell's return to the *Roosevelt*, the party had advanced about 100 miles. Soon MacMillan was forced to turn back with a frostbitten heel. By March 26, all except one support team had returned to the base camp. Peary's force was down to himself, Henson, Bartlett, six Eskimos, seven sledges, and sixty dogs. They had reached 86° 38' the day before. Peary recounted:

> Next day, March 27, we met heavy, deep snow—a smothering mantle lying in the depressions of rubble ice. I came upon Bartlett and his party, fagged out and temporarily discouraged by the heart-racking work of making a road....I rallied them a bit, lightened their sledges, and sent them on.
>
> During the next march we fought a biting wind from the northeast and came upon Bartlett camped beside an open lead with water in three directions. We built our igloos 100 yards away so we would not wake him.
>
> I was dropping off to sleep on my bed of deerskins when a movement of the ice and a shout brought me to my feet. Through the peephole of the igloo I saw a rapidly widening strip of black water between our igloo and Bartlett's. One of my dog teams had barely escaped being dragged in; another had just avoided being crushed by the ice blocks that were jamming up over them. Bartlett's igloo was drifting east on an ice raft. Kicking out my door, I called to the captain's men to get ready for a quick dash.
>
> At last their ice sheet crunched against ours, and we got Bartlett's party onto the floe with us. For the rest of the night, and during the next day the ice suffered the torments of the damned, surging together, opening out, groaning and grinding while the open water belched its dark mist like a prairie fire. Then the motion ceased, the open water closed, the atmosphere to the north cleared, and we hurried on before the lead should open again.

On April 1, 1909, Bob Bartlett and the last supporting party turned back, leaving Peary and Henson alone with their four Eskimos to race forty dogs 133 miles to the Pole.

Above: *Peary's patched flag is now on display in Explorers Hall at the Society's Washington, D.C., headquarters. Corner pieces had been placed at Cape Morris Jesup on the north coast of Greenland during Peary's 1900 expedition; a triangle cut from the center had been put in the ice at latitude 87° 06' N to mark the "farthest north" achieved by Peary in April 1906. The rectangle at the left of the flag was planted at Cape Columbia, Ellesmere Island, June 1906; the top piece was left at Cape Thomas Hubbard that same month. And the diagonal band was deposited by Peary at the North Pole in April 1909.*

Opposite: *"The Pole at last!!!" Peary wrote in his diary. To commemorate his arrival at the North Pole on April 6, 1909, Peary took this photograph showing his black companion, Matthew Henson (center), and (from left to right) Ooqueah, Ootah, Egingwah, and Seegloo, the four Eskimos who had accompanied him on the final dash, standing before a pressure ridge at the North Pole.*

Now was no time for reverie. I turned to the problem for which I had conserved my energy on the upward trip, trained myself as for a race, lived the simple life—for which I had worked 23 of my 52 years. In spite of my age I felt fit for the demands of the coming days....My five remaining men were as responsive to my will as the fingers of my right hand.

Henson, with his years of experience, was almost as skillful at handling dogs and sledges as the Eskimos themselves. He and Ootah had stood with me at my farthest north, three years before. Two other Eskimos, Egingwah and Seegloo, had been with that expedition and were willing to go anywhere with me....

My 40 remaining dogs were the pick of the 133 which had left Cape Columbia. Almost all were powerful males, hard as iron, in good spirits. My last five sledges, fully repaired, were ready to go. I had supplies for 40 days. By eating the dogs themselves, we could last 50....

I decided to strain every nerve and sinew to make five marches of 25 miles each, crowding them in such a way that the fifth march would end long enough before noon to allow an observation for latitude....Underlying all calculations was the knowledge that a 24-hour gale would open leads, knock my plans into a cocked hat, and put us all in peril.

I hit the trail a little past midnight on April 2, leaving the others to break camp and follow. As I climbed the pressure ridge back of our igloos I set another hole in my belt, the third since I started. Every man of us was lean and flat-bellied as a board, and as hard.

It was a fine morning. The wind had subsided and the going was the best I had hit—large, clear floes surrounded by stupendous pressure ridges, all of which were easily negotiated....

The years seemed to drop from me, and I felt as I had when I led my party across the great Greenland icecap, leaving 20, 25 miles behind my snowshoes day after day, sometimes stretching it to 30 or 40.

A short sleep and we were off again. The going was practically horizontal....The weather remained fine. At the end of ten hours we were halfway to the 89th parallel.

Again a few hours' sleep. We hit the trail before midnight. The weather and going were even better. We marched something over ten hours, the dogs often on the trot. Once we rushed across a frozen lead 100 yards wide which buckled under our sledges and broke as the last sledge left it. We stopped near the 89th parallel, in temperature of 40° below.

Scant sleep, and on our way once more. We ran on smooth young ice, and the dogs sometimes broke into a gallop. Bitter air, keen as frozen steel, burned our faces until they cracked. Even the Eskimos complained. We needed more sleep this time, and took it. Then on again. Up to now our fears of an impossible lead had increased. At every rise I had found myself hurrying forward breathlessly, fearing it would mark a lead. At the summit I would catch my breath in relief—only to hurry on. Now this fear fell away.

Before turning in after that last march, I took an observation which gave our position as 89° 25′ [35 miles from the Pole]. A dense, lifeless pall hung overhead. The horizon was black, the ice beneath a ghastly shell white. I strained my eyes across it, trying to imagine myself already at the Pole.

Japanese adventurer Naomi Uemura who, aided by Society funds, made the first solo trek to the North Pole, arriving on April 29, 1978. The photograph, taken by Uemura at the North Pole, shows him stringing up flags of nations in which he had found support.

Above: *"ENTERING THE PLANE FOR THE FLIGHT TO THE POLE"* was the title for this photograph of Lt. Cmdr. Richard E. Byrd at Spitsbergen, with Pilot Floyd Bennett at his left and Lt. G.O. Noville at his right, when it appeared in the September 1926 Geographic. The caption of this photograph, which appeared as part of Byrd's article "The First Flight to the North Pole," went on to point out that the parkas worn by the men had been carried 500 miles by dogsled from Nome to Fairbanks, Alaska, and were made of "reindeer skin with an inner parka of squirrel skin." Their boots and gloves were of reindeer skin, too, but their trousers were "of polar-bear skin with the fur outside...."

Right: *"The dog sledge must give way to the aircraft; the old school has passed,"* Lt. Cmdr. Richard E. Byrd told a National Geographic audience in 1926 upon returning from his epic first flight over the North Pole. This romanticized 1926 painting by N.C. Wyeth of Byrd's trimotor Josephine Ford *flying toward the North Pole hangs in Hubbard Memorial Hall, the Society's first headquarters building.*

On the fifth march a rise in temperature to 15° below reduced the friction of our sledges. The dogs seemed to catch our spirit. They tossed their heads and yelped as they ran with tight-curled tails.

I had now made my five marches and on April 6 was in time for a hasty noon observation: 89° 57'. Three nautical miles from the magic 90°. With the Pole practically in sight, I was suddenly too weary to take the last steps. I turned in for a nap, then pushed on with two Eskimos and a light sledge for another ten miles and made another observation.

I was now beyond the Pole.

"The Pole at last!!! The prize of 3 centuries, my dream & ambition for 23 years. *Mine* at last," Peary wrote in his diary. "I cannot bring myself to realize it. It all seems so simple...."

Peary, Matthew Henson, and the Eskimos Ootah, Egingwah, Seegloo, and Ooqueah remained at the North Pole thirty more hours in "side trips taking [latitude] observations and photographs, planting flags, depositing records," Peary wrote. "The weather was flawless: minimum temperature 33° below, maximum 12° below. On the afternoon of April 7, we double-fed the dogs, repaired the sledges for the last time, discarded spare clothing, and started back...."

Sixteen days later, having force-marched the old trail in perfect weather, Peary arrived at the Cape Columbia base camp; after sleeping for two days and making two more marches he boarded the *Roosevelt*. There he learned that Ross Marvin, who had turned back with his support party on March 25, had drowned in a big lead en route to Cape Columbia. His death, Peary wrote, "was a bitter flavor in the cup of our success."

On Sunday night, September 5, 1909, the *Roosevelt*, delayed by the absence of its captain, who was out on the Greenland icecap, next by sea ice, and then by a stop to take the Eskimos home, reached Indian Harbor, Labrador. The following morning Peary and Bartlett rowed ashore for their first opportunity to send messages. Peary's cables went to *The New York Times*, to the Associated Press, to the Secretary of the Peary Arctic Club, and the following to his wife: HAVE MADE GOOD AT LAST. I HAVE THE D.O.P.* AM WELL. LOVE....

"This was the news the world had been waiting to hear for 300 years," Peary later wrote, "the discovery of the top of the earth."

On September 8, Peary sent the following telegram to President William Howard Taft:

HAVE HONOR TO PLACE NORTH POLE AT YOUR DISPOSAL.

The President wired Peary in reply:

THANKS FOR YOUR GENEROUS OFFER. I DO NOT KNOW EXACTLY WHAT I COULD DO WITH IT. I CONGRATULATE YOU SINCERELY ON HAVING ACHIEVED, AFTER THE GREATEST EFFORT, THE OBJECT OF YOUR TRIP, AND I SINCERELY HOPE THAT YOUR OBSERVATIONS WILL CONTRIBUTE SUBSTANTIALLY TO SCIENTIFIC KNOWLEDGE. YOU HAVE ADDED LUSTRE TO THE NAME "AMERICAN."

* *The New York Times*, upon learning of Peary's cable to his wife, asked her what "D.O.P." meant. She replied, "Damned Old Pole."

Beneath a twilit, autumnal sky,
stately emperor penguins surveyed
their bleak and unforgiving Antarctic
kingdom.

If the President's response seems somewhat muted, one explanation might be that Taft had received the following telegram from Copenhagen only four days before:

> I HAVE THE HONOR TO REPORT TO THE CHIEF MAGISTRATE OF THE UNITED STATES THAT I HAVE RETURNED HAVING REACHED THE NORTH POLE.
>
> FREDERICK A. COOK

Over the years the controversy has continued to rage as to whether or not Cook beat Peary—or whether either Cook or Peary even made it to the Pole. Both men were asked to submit their records and observations to a group of National Geographic scientists for verification. Peary did; Cook would not.

As recently as December 1983, CBS television aired "Cook and Peary: the Race to the Pole," a drama purporting to prove that Cook—by all accounts a charming man—had reached the Pole a year before Peary, but had been denied his claim by a vicious and paranoid Robert E. Peary. It was further suggested that the National Geographic Society had taken part in perpetrating a fraud upon the American public. Letters from members asked what really happened. Gilbert M. Grosvenor in the Magazine's March 1984 President's page responded that the CBS program was "blatant distortion of the historical record, vilifying an honest hero and exonerating a man whose life was characterized by grand frauds."

Grosvenor recounted some of Cook's deceptions: that he had published another man's life work as his own, that his false claim to have climbed Mount McKinley had resulted in his expulsion from the Explorers Club, of which he had been a founder and President, that he had served a prison sentence as a result of his part in an oil-stock swindle. Grosvenor also cited the lack of any evidence or witnesses that Cook had, in fact, reached the Pole, then concluded:

> Peary's claim is backed by astronomical observations made by others as well as himself, by soundings taken through the ice, and by the testimony of his companions, including Eskimos. Even his severest critics cannot deny he came close to his goal, and his supporters have no doubt that he made it. It is small wonder that he felt his life's achievement had been stolen by a con man.

Perhaps the best evaluation of the controversy can be found in an aphorism popular during that era: "Cook was a gentleman and a liar; Peary was neither."

The seduction of the Poles for explorers was—and is still—their inaccessibility, the challenge of getting there, but most of all getting there first. Norway's Roald Amundsen, the restless, driven, consummate professional, who was the first to traverse, and thereby prove the existence of the Northwest Passage, was in Norway in September 1909, preparing an expedition to the North Pole when news of Peary's accomplishment reached him. Amundsen immediately and secretly switched plans. Nine months later, when already at sea, he turned his ship southward, in order to make an attempt on the South Pole. On September 13, the *Times* of London announced that Captain Robert Falcon Scott of the Royal Navy was arranging his own expedition to the South Pole that "will, it is hoped, start in August next." Not quite two years later both Scott and Amundsen were in Antarctica about to set out on a race that was to

Above: *"Alone in a hut beneath the Antarctic ice, Byrd spent four bitter winter months in 1934. He had moved to the advance station to make weather observations, but fumes from a defective stove soon made him ill. Realizing his comrades would risk their lives to save him if they knew his peril, he sometimes crawled on hands and knees to make regular radio reports. Finally alerted by his faltering signals, three men broke through the polar night and rescued him."* Caption from the April 1962 Geographic tribute *"The Nation Honors Admiral Richard E. Byrd."*

Right: *Richard E. Byrd's triumphant May 9, 1926, flight over the North Pole was seconded three years later when, on November 28, 1929, he and three companions flew their Ford trimotor from their Little America Antarctic base to the South Pole and back. Byrd's broadcasts, originating from his unearthly Antarctic base, were as thrilling to a previous generation as were the first transmissions from astronauts on the moon.*

be both remarkable and glorious, terrible and tragic, a race which, though won by the best man, ironically saw the vanquished emerge the hero.

On October 19, 1911—a dull, foggy day with a maliciously shifting wind—Amundsen and his finely honed Norwegian team set off from their Bay of Whales base camp on the Ross Ice Shelf for the perilous 1,244-mile trip by ski and sledge to the Pole. Eleven days later and sixty miles farther from the Pole, Scott and his ill-prepared party left their Cape Evans winter camp above McMurdo Sound on foot behind two impractical motorized sledges and nine ponies unsuited for the Antarctic temperatures. The race for the South Pole— one of the last classic expeditions with men and animals and no outside support—had begun.

Not quite two months later, at a little after three in the afternoon, on December 14, 1911, Amundsen, together with his four companions, was standing at the South Pole under clear skies and in calm air and a temperature of −7.6°F.

In his diary Amundsen simply noted, "So we arrived and were able to plant our flag at the geographical South Pole. God be thanked!"

Just over a month later, on January 15, 1912, Scott's exhausted and frostbitten party arrived and first found one of Amundsen's black flags marking a miles-wide circle enclosing the Pole, which they mistook for his mark at the Pole itself. It mattered little, the flag was proof the race had been lost. As Scott despondently noted in his diary:

> The Pole. Yes, but under very different circumstances from those expected. …Great God! this is an awful place and terrible enough for us to have laboured to it without the reward of priority.

The next morning members of Scott's assault team saw, in the distance, a black speck which upon investigation turned out to be Amundsen's tent with the Norwegian flag and one from Amundsen's ship, *Fram*, flapping and trembling over their goal. But disappointment at not being first made Scott's achievement of reaching the Pole a hollow one. None of Scott's party survived the agony, the hunger, the exhaustion of the storm-ridden, freezing two-month walk back toward the base camp. The glory Scott had sought in his lifetime came to him only in his death when his diaries, found seven months later with his body, were published. Amundsen had won the Pole, but Scott's carefully edited diaries made him the legend. As his biographer Roland Huntford noted in *The Last Place on Earth*, Scott's achievement was to perpetuate the romantic myth of explorer as martyr and, in a wider sense, to glorify suffering and self-sacrifice as ends in themselves."

Two years after Scott's death, World War One began and the age of Polar exploration, as typified by Amundsen and Peary, was near its end. Within a decade, radios, airplanes, and powerful motorized tracked vehicles began to replace dogs, sledges, and Polar adventurers isolated from the rest of the world.

In 1926, not long after Lt. Cmdr. Richard E. Byrd and his co-pilot, Floyd Bennett, guided by a sun compass invented by the Society's chief cartographer

"Navy icebreaker Atka," *shown in a February 1965* Geographic *article, "smashes through frozen Antarctic seas, sculptured by raging winds into* *fantastic crags. She keeps channels open in the summer season (September to March) to support Americans quartered at the bottom of the world."*

71

Above: *This wind-tattered flag shown at the Amundsen-Scott South Pole Station on March 22, 1957, flies at half-mast in honor of Richard E. Byrd, who had died just eleven days before.*

Right: *The sun seems to skip rather than to rise or set in this six-exposure photograph taken over the course of three hours in November 1957 by Geographic staffer Thomas J. Abercrombie, the first civilian photographer to reach the South Pole Station.*

Above: *"Waddling through the snows ...a gentoo penguin wears a radio backpack that provides monitoring biologists with data on blood flow and pressure,"* explained the caption of this November 1971 *"Antarctica's Nearer Side"* Geographic *photograph. "This project—helping man understand penguin physiology and adaptation to a harsh environment— is part of the multi-nation Antarctic research program that began with the 1957–58 International Geophysical Year [IGY]."*

The crossing of Antarctica: Sir Vivian Fuchs' great motorized Polar trek of 2,158 bleak wind-lashed miles from Shackleton Base at the edge of the Weddell Sea over hidden crevasses and iron-hard sastrugi across the South Pole to Scott Base at McMurdo Sound marked the completion by the British Commonwealth Trans-Antarctic Expedition of 1957–1958 of what Antarctic pioneer and Hubbard Medal winner Ernest H. Shackleton had called "the last great Polar journey that can be made."

Albert H. Bumstead, had been the first to fly to the North Pole and back, Byrd told a National Geographic audience, "The dog sledge must give way to the aircraft; the old school has passed."

With the North Pole behind him—and an unsuccessful attempt the following year at beating Charles Lindbergh in becoming the first to cross the Atlantic non-stop for the $25,000 Orteig prize—Byrd turned his attention to the Antarctic, which remained one of the world's great mysteries.

At 3:29 A.M., on November 28, 1929, at Little America, the Byrd expedition's Antarctic headquarters near Amundsen's former base camp overlooking the Bay of Whales, Byrd and three companions, pilot Bernt Balchen, co-pilot Harold June, and cameraman Capt. Ashley McKinley, climbed aboard a stripped down, high wing Ford trimotor airplane for an attempt to fly to the South Pole. Byrd had named the aircraft the *Floyd Bennett* after his beloved companion, who had been badly injured in a trial flight for their transatlantic endeavor and then, the following year, while on a rescue mission in Canada, had contracted pneumonia and died. Byrd and Bennett had planned the South Polar flight together. "Fate had sidetracked him," Byrd wrote, "but he was not forgotten." Byrd brought a stone from Bennett's grave with which he would weight the American flag he planned to drop at the Pole.

With the *Floyd Bennett*'s engines running hot and smooth, the cumbersome, heavily loaded, trimotor wallowed on its skis across the packed snow, gradually picking up speed until it fought its way into the air.

Byrd later said in his story published in *Great Adventures,*

> As the skis left the snow I saw my shipmates on the ground jumping, shouting, throwing their hats in the air, wild with joy that we were off for the Pole. We circled, emerged from clouds into sunshine....Snow-covered peaks 100 miles away glittered like fire in the sun's reflection.
>
> What we faced far surpassed the demands of a simple flight of 800 miles to the Pole. For hundreds of miles we would fly over a barren, rolling surface, then climb a mountain rampart thousands of feet high and continue across a 10,000-foot plateau....Now we began to climb. Before us, beyond the great mountains, lay uncertainty.
>
> McKinley struggled with his camera, I navigated, June sent radio messages, fed gas from cans to tanks, and cranked his movie camera. We were heading for Heiberg Glacier, the plane near its absolute ceiling. Amundsen had reported the high point of the pass as 10,500 feet. Peaks towered on both sides. To our right stretched a wider glacier. Should we tackle Heiberg, altitude known but with air currents around those peaks that might dash us to the ice, or should we try the unknown glacier? We had to choose quickly—we were heading into the mountains at more than a mile a minute. We chose the unknown glacier.
>
> Bernt Balchen fought for altitude while air currents tossed the plane about like a cork in a washtub. Suddenly the wheel turned loosely in his hands; the ailerons failed to respond. Above the engines' roar Bernt shouted, "It's drop 200 or go back!"
>
> "A bag of food overboard!" I yelled. McKinley shoved a 150-pound bag of emergency rations out the two-foot trapdoor.

"The situation looked distinctly uncomfortable," Fuchs wrote after a snow bridge collapsed under his Sno-Cat's weight.

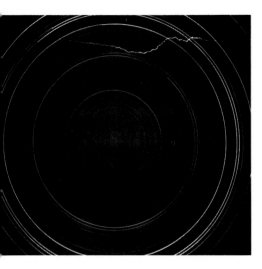

The controls responded. Slowly we climbed. But we were still too low. Again the wheel turned in Balchen's hands.

"Quick! Dump more!" he shouted.

I pointed to another bag. Mac nonchalantly shoved it through the trap-door. Again the controls took effect.

Finally we reached the pass. We ambled over—a bit to spare. Bernt yelped in relief. No mountains ahead. We faced a clear route to the Pole, dead ahead over the horizon!

Our next thought was the engines. The plateau was so high that to fly above it the plane remained near its ceiling, requiring full power for its lift. The loss of one engine meant landing in the snow.

Then the starboard engine sputtered. June rushed to the gas tank valves. Balchen manipulated the wheel. The gasoline mixture had been made too lean in our effort to conserve fuel. A quick adjustment and the engine sang again.

We had time to look around. The polar plateau at last—level, white, limitless! We passed clusters of haycocks, small domes of snow that conceal deep pits. We saw glittering sastrugi, hard, wind-formed snow ridges with knife edges. About 50 miles from the Pole, clouds approached—we would have to race them back to the pass. If we lost, our retreat would be cut off. But the big moment had come!

I handed June a message to radio to Little America: *We have reached the vicinity of the South Pole. Flying high for survey. Soon turn north.*

The temperature was 15° below zero F. Clouds obscured the horizon in several places. No mountains were in sight. The Pole lay in the center of a vast plain. We opened the trapdoor and dropped our flag.

Eighteen hours and forty-one minutes after taking off from Little America—"the loneliest city in the world"—Byrd and his crew returned. "We were deaf from the roar of the engines, tired from the strain of the flight. But we forgot all that in the tumultuous welcome of our companions."

The National Geographic, which had (as Byrd remarked) generously supplied financing and enthusiasm, devoted almost the entire August 1930 issue to Byrd's expedition.

In 1935, Lincoln Ellsworth flew his single-engine *Polar Star* across Antarctica and won the Society's Hubbard Medal. Three expeditions later and less than thirty years after his record flight to the South Pole, Richard E. Byrd, "Admiral of the Ends of the Earth," was appointed Officer in Charge of U.S. Antarctic Programs with an office in Washington, D.C. The first U.S. Navy plane landed at the South Pole in 1956 and since then planes have regularly air-supplied the Amundsen-Scott scientific station built at the Pole, where snow-bound scientists braved the long, black summer months and − 102°F temperatures. On January 19, 1958, the station crew greeted Sir Vivian Fuchs when he arrived. Fuchs and his British Trans-Antarctic Expedition's Day-Glo orange Sno-Cats were then midway across the continent in their daring 2,158-mile fulfillment of what British explorer Sir Ernest Henry Shackleton—the Hubbard Medalist who had almost reached the South Pole three years before Amundsen—had called "the last great Polar journey that can be made."

But of course, it wasn't. Great Polar journeys will continue to be reported in the *National Geographic* as long as there are men and women who are willing to attempt them "not for any hope of rewards or honor," as Peary told the Society members at that banquet in 1906, "but because the thing he has set himself to do is a part of his very being."

"The End of the Beginning"

"The year 1905 marked the turning point in the Society's fortunes, the 'end of the beginning....'"
—Gilbert H. Grosvenor

"At the outgoing of the old and the incoming of the new century," President William McKinley wrote in his 1900 Annual Message to Congress, "you begin the last session of the Fifty-sixth Congress with evidences on every hand of individual and national prosperity and with proof of the growing strength and increasing power for good of Republican institutions."

In turn-of-the-century America there were only forty-five states; the population was 75,994,575; the average wage of workers in manufacturing industries was 22 cents an hour; automobiles cost about $1,500 each; the bus and the truck had not yet been invented; there were fewer than one hundred and fifty miles of paved road in the entire nation; one out of fifty-six persons owned a telephone; there was no such thing as a radio, or an electric refrigerator; the average yearly teacher's salary in a public school was $325; sixty percent of the population was rural; more people worked on farms than in any other occupation, and, as McKinley had suggested in his annual address, life in the United States was pretty good—though short. (The life-expectancy of the average white male was 46.6 years, of the non-white male 32.5 years.)

In 1900, the Boxers rose up against the Christian missionaries in China, Colette wrote her first *Claudine* novel, Conrad wrote *Lord Jim*, Dreiser *Sister Carrie*, Sigmund Freud published his *Interpretation of Dreams*, and Nietzsche died. Picasso painted *Le Moulin de la Galette*, Cézanne his *Self Portrait with Beret*, Monet his *Pool of Water Lilies*, and Sargent *The Sitwell Family*. Puccini's opera *Tosca* premiered in Rome, the first Browning revolvers were manufactured, the American R. A. Fessenden transmitted human speech over radio waves, the first flight of a Zeppelin occurred, the cakewalk was the dance of the hour, and a revolution then occurring in American magazine publishing saw periodicals splitting into two camps, camps divided by illustration methods and price.

From the latter part of the nineteenth century through the first decade of the twentieth, of all the methods by which ideas were readily disseminated— and, at that time, those means were principally newspapers, magazines, lecture platforms, church pulpits, theater stages, and the graphic arts—none underwent so dramatic an increase in effectiveness as did magazines.

It had always been difficult to make a clear-cut distinction based on content between newspapers and magazines. And in the 1880s and 1890s, with the advent of the newspaper Sunday "supplements" and special Saturday editions—which in essence were weekly magazines containing signed literary

Preceding spread: *Delegates participating in the Eighth International Geographic Congress aboard a Chicago horse-drawn sightseeing coach. Gilbert H. Grosvenor found this 1904 Congress, meeting in the United States for the first time, "an eye-opener." The editor (hat on knee) sits on the top deck facing his wife, Elsie.*

pieces by leading writers—the difference became even more difficult to perceive. Although the introduction of the newspaper supplements had little impact on the major "quality" magazines like *Scribner's Magazine, Lippincott's, The Century,* and *Harper's,* the lesser magazines retaliated by competing directly with the newspapers by publishing breaking news stories and covering the latest controversy and events. This competition resulted in many of the magazines dropping their cover prices to 10 and 15 cents. [Information based on *A History of American Magazines* by Frank Luther Mott.]

Furthermore, the emergence of a vast, educated, ambitious middle class (generated by the increasingly sophisticated, expanding public-school systems and the easier access to colleges and universities) had created a market for an increasing number of periodicals tailored to fit its specific needs.

Among the thousands of magazines founded between 1885 and 1905—the formative years of the *National Geographic*—there were scores of specialty periodicals such as *American Anthropologist, Astrophysical Journal, Popular Astronomy, Terrestrial Magnetism, American Geologist, Journal of Geology, Physical Review, Journal of Physical Chemistry, Journal of Analytical and Applied Chemistry, Aquarium, Biological Bulletin, American Museum Journal,* and *Plant World.* Approximately 3,300 periodicals were published in America in 1885; twenty years later there were about 6,000. During those same two decades almost *11,000* different magazines had been launched [according to Mott]; and the 6,000 magazines being published at the end of that period is less a reflection of the near-doubling of the number of periodicals available than an indication of how many periodicals during that same period had failed. With this in mind, the decision of the fledgling *National Geographic's* young editor, Gilbert H. Grosvenor, to ignore—if not violate—every single recommendation made by S. S. McClure, the publisher of one of that era's most successful magazines, can be seen as either self-confidence bordering on arrogance or as devotion to the highest personal convictions. Be that as it may, at the turn of the century the *National Geographic* was only one of about 5,500 magazines being published in America—and one of the very, very few that would survive.

It survived because, during the revolution occurring in magazine publishing, brought about by the introduction of the new, cheaper halftone method of reproducing photographs, the *National Geographic's* young Assistant Editor allied himself and his Magazine with the winning side.

Initially certain publishers resisted the halftone illustrations. They felt that dot-patterned photographic reproductions looked "trashy" compared with the fine-line steel engravings. But, halftone soon took over the field and one of the major reasons for this was economic: a full-page steel engraving cost $100, whereas a halftone could be purchased for less than $20.

"In those days many publications scorned photographic illustrations," Gilbert Grosvenor recalled. "The famous and successful editor of *The Century* [Richard Watson Gilder] had gone so far as to predict in print that 'people will tire of photographic reproduction, and those magazines will find most favor which lead in original art.' This same editor once told me that he—and the public!—considered our halftone photoengravings 'vulgar' and preferred steel engravings costing $100 a plate."

Other innovative editors recognized the promise low-priced illustrations held and, like Grosvenor, ignored the criticism of editors like *The Century's* Gilder and took advantage of the new techniques that made it possible to ac-

This previously unpublished photograph of Elsie Bell Grosvenor was taken by her husband, Gilbert H. Grosvenor, in 1901, a year after their marriage.

quire a Levy-process halftone for between $7 and $8 for a full-page plate.

"It was like striking gold in my own backyard when I found Government agencies would lend me plates from their publications," Grosvenor wrote. "I illustrated numerous articles in this way."

Any magazine editor could have had these same photoengravings, but they were not interested. Grosvenor—recalling both Alexander Graham Bell's earlier advice about the material available from the government's nearby scientific agencies at no cost and the success of his father's liberally illustrated, scholarly book on Constantinople, which, with its 230 photoengravings, had been highly praised and sold well—never wavered in his confidence that photographic illustrations were a valuable asset to the Magazine.

During this period Grosvenor had been using the Society's popular and income-producing lecture series as "a barometer of public opinion concerning good geographic subject matter for The Magazine." He would place himself in Washington's 1,200-seat Columbia Theater where the lectures were presented, in a position from which he could observe both the audience and the speaker. Attendance during technical lectures on the more mundane geologic and geographic topics would average no more than twenty. In 1902, however, on the night of the lecture about the disastrous volcanic explosion of Mont Pelée on the island of Martinique, the Columbia Theater was jammed.

Mont Pelée had exploded at 7:50 A.M. on May 8, 1902, with a deafening roar heard hundreds of miles away. In less than three minutes the city of St. Pierre, its neighboring homes, and seventeen out of eighteen ships in the harbor had been destroyed by the superheated steam and flaming gases from the volcano's split belly, and 30,000 men, women, and children were killed.

Second Engineer Charles Evans of the *Roraima*, anchored off St. Pierre and the only ship to survive, was to report:

> "I was standing at the ship's rail, looking at the mountain. Suddenly it seemed to explode. An immense dark cloud shot up, blacking out the sky. Then a second cloud, even larger than the first, burst from the side of the volcano and rushed down on the city. Its speed was unbelievable. Flames spurted up wherever it touched. A huge wall of hot ashes filled the air, coming toward us *against* the wind....
>
> "Rope and bedding on the *Roraima* caught fire. I turned to run below and was burned on the back. The shock of the explosion hit like a hurricane; I could get no air to breathe. Then an enormous wave struck our port quarter. We keeled over so far the bridge went under and water poured into the hold through the fiddlers."
>
> —From *Great Adventures with National Geographic*

Although the lecturer had photographs of the Mont Pelée disaster, he had thought them too grisly to show—or perhaps he was mindful of Associate Editor W J McGee's admonition that "the excessive use of picture and anecdote is discouraged." In any case, it was precisely the *lack* of pictorial illustration during this particular Society lecture that convinced Grosvenor that McGee's thinking was in error, for in the course of the lecture Grosvenor overheard two young women behind him complain, "Why doesn't he show exciting photographs? That's what we want."

The two young ladies' frustration over the lack of photographs, Grosvenor later recalled, "helped to confirm my belief that The Magazine should use pictures—realistic ones replete with human interest...the character the Geo-

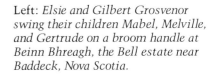

Left: *Elsie and Gilbert Grosvenor swing their children Mabel, Melville, and Gertrude on a broom handle at Beinn Bhreagh, the Bell estate near Baddeck, Nova Scotia.*

Bottom left: *Summers at Baddeck were not all play; in a tent pitched in a sylvan glade Grosvenor diligently edited the Magazine's articles—despite occasional interruptions from his children (Gertrude and Melville, in this instance) and his inventor father-in-law.*

Above: *The joy young Melville Bell Grosvenor felt in his grandfather's company glows from this photograph. "I would look forward to the trip to Nova Scotia all year long," Melville later recalled. "[My grandfather] was such a vibrant person. He had a wonderful way with children."*

graphic should take gradually became clear: lucid, concise writing; material of general, not academic, interest; an abundance of pictures. Yet pictures were hard to find, and so were good articles....in those days we had to solicit [material]. So I spent much time seeking contributors in Government departments and also in the Cosmos Club, which had then, as now, a large membership among scientists."

Most of the *Geographic*'s contributors up to that point had been connected with the various government bureaus and departments. And although the Magazine had begun to publish its now familiar "travelogue" pieces—S. P. Langley's "Diary of a Voyage from San Francisco to Tahiti and Return..." had appeared in December 1901; Congressman Ebenezer J. Hill's "A Trip through Siberia" ran in February 1902—the Magazine continued to reflect a markedly Washingtonian point of view. The Pelée eruption, however, provided Grosvenor with a deluge of material of more national interest.

On May 14, 1902, six days after the eruption, the National Geographic Society dispatched three of its members "on a special expedition to Martinique and St. Vincent to investigate the volcanic conditions of the West Indian regions." They were Robert T. Hill, of the U.S. Geological Survey; Israel C. Russell, who twelve years earlier had led the Society's Mount St. Elias expeditions; and Commander C. E. Borchgrevink, an Antarctic explorer who had studied Erebus and Terror, the southernmost volcanoes of this earth. The June 1902 *Geographic* reported:

> The expedition is the most important and best equipped commission ever sent out to study actual volcanic action. Results of great scientific and practical consequence may be expected to flow from their work. On their return to the United States they will report the results of their observations to the Society. This report, forming a series of illustrated articles, will be published in full in the journal of the Society, the *National Geographic Magazine.*

The June issue, with its announcement of the expedition, an article on the 1883 eruption of Krakatoa excerpted from a British author's book, a short essay by Grosvenor on volcanoes in general with a map showing their distribution, a report on "Magnetic Disturbance Caused by the Explosion of Mont Pelée," a map on the volcanic islands of the West Indies, and a piece by the noted author Lafcadio Hearn on "The Island and People of Martinique," also carried a preliminary report by the expedition that contained the following Associated Press account of Professor Angelo Heilprin's* ascent of Mont Pelée:

> "Saturday morning [May 31] Professor Heilprin determined to attempt the ascent to the top of the crater....The volcano was very active, but amid a thousand dangers Professor Heilprin reached the summit and looked down into the huge crater....He saw a huge cinder cone in the center of the crater. The opening of the crater itself is a vast crevice 500 feet long and 150 feet wide.
> "While Professor Heilprin was on the summit of the volcano several violent explosions of steam and cinder-laden vapor took place, and again and again his life was in danger. Ashes fell about him in such quantities at times as to completely obscure his vision. One particularly violent explosion of mud covered the Professor from head to foot with the hideous viscid

"IT IS NO EASY MATTER TO BRING DOWN ONE OF THESE STURDY FLYERS WHEN THE WIND IS BLOWING HARD," the Geographic *noted beneath this photograph of Melville (right) at Baddeck helping Alexander Graham Bell and others haul to earth one of the inventor's giant tetrahedral kites.*

* Heilprin, who became a member of the National Geographic Society's Board of Managers, was also President of the Geographical Society of Philadelphia.

"...I have been continuously at work upon experiments relating to kites. Why, I do not know, excepting perhaps because of the intimate connection of the subject with the flying-machine problem," Bell reported in a 1903 address to the National Academy of Sciences in Washington, D.C. In the course of his pioneer studies into developing a stable flying machine capable of taking off at slow speeds, Bell experimented with a multitude of kite shapes and styles.

and semi-solid matter. He still persisted in his study and observations, however, and twice more was showered with mud. He learned, as had been suspected, that there were three separate vents through which steam issued."

The June issue was simply a warm-up for the July 1902 issue, which carried on its cover the title "MARTINIQUE NUMBER."

This issue contained a forty-four-page report by Robert T. Hill on the explosion of Mont Pelée on Martinique, ten pages on "Volcanic Rocks of Martinique and St Vincent," three pages of "Chemical Discussion of Analyses of Volcanic Ejecta from Martinique and St Vincent," and two pages on "Reports of Vessels as to the Range of Volcanic Dust." This last included the following account:

> From the log of the steamship *Louisianian*, Captain D. Edwards, Liverpool to Trinidad, April 25 to May 9, 1902:
>
> "Arrived in Carlisle Bay, Barbados, May 7, at 11 a.m. (Martinique N.W., 140 miles), the weather being fine and clear. Between 1 and 3 p.m. reports as of heavy artillery firing were heard, and shortly afterward a dense black cloud appeared in the west, in the direction of St Vincent (W. 100 miles), and gradually moved E.S.E. At 4 p.m. the whole sky was overcast, except a low arch to the northward. At 4.30 light showers of dust began to fall, and it was so dark that lights had to be burned on the ships and ashore. At 5.30 we departed for Trinidad (Port of Spain), the weather being so dark that we could not distinguish a large mooring buoy at a distance of 40 yards. At this time the dust was pouring down and speedily covered the decks to the depth of a quarter of an inch...."
>
> Log of ship *Lena*, Nibbs, master, Barbados to New York:
>
> "While at Barbados a heavy rain of volcanic dust fell from Mount Soufrière (W., 100 miles) on the decks and awnings of the vessel. Seven tons of same were thrown into the hold for ballast...."

Most important of all, the July 1902 issue carried a dozen photographs graphically depicting the destruction caused by the volcano's eruptions.

In February 1903, Gilbert H. Grosvenor was given the title Editor, with full authority to try his own ideas for the development of the Magazine. (He was also appointed the Society's Director, with control of all business and membership affairs.) Not long after, Grosvenor sought Bell's counsel about publishing photographs of Filipino women working in the fields, naked from the waist up. Since, as Grosvenor later recalled, "the women dressed, or perhaps I should say undressed, in this fashion...the pictures were a true reflection of the customs of the times in those islands."

Bell advised the young editor to print them and agreed with Grosvenor that prudery should not influence the decision.

The photograph, one of four illustrations for a May 1903 article on "American Development of the Philippines" was captioned "Primitive Agriculture. Tagbanua Women Harvesting Rice, Calaminanes Islands." As Tom Buckley, in his *New York Times Magazine* article "With the National Geographic On Its Endless, Cloudless Voyage," said of the response to the *Geographic*'s publication of that photograph, " So unquestionably genteel was the magazine, and so patently pure its anthropological interest, that even at a time when nice people called a leg a limb it never occurred to anyone to accuse Grosvenor of impropriety."

Above: *"PRIMITIVE AGRICULTURE. TAGBANUA WOMEN HARVESTING RICE, CALAMINANES ISLANDS,"* was the caption for this photograph in the May 1903 Magazine.

When the young editor sought the inventor's counsel about publishing photographs of half-naked Philippine Island natives, Alexander Graham Bell advised Grosvenor that prudery should not influence the decision. Grosvenor agreed and later defended publication on the grounds that "the pictures were a true reflection of the ... times in those islands."

Opposite, top: *A tranquil street scene in St. Pierre, Martinique, before the devastating 1902 eruption. Mont Pelée is seen looming in the background.*

Opposite, bottom: *"7:50 A.M., May 8, 1902: Blotting out the sky, churning down Pelée's slope at 100 miles an hour, superheated steam and ash from the volcano's belly smash the city, snuff out its life, roar across ships in the harbor where terrified seamen leap for cover. Only one vessel in 18 stays afloat as the wall of death engulfs them."* This caption accompanied Paul Calle's painting for "Mont Pelée: Volcano that Killed a City" (a rewrite for the Society's 1963 book Great Adventures *of Robert T. Hill's 1902 article "Volcanic Disturbances in the West Indies").*

It would have been about this time, too, that Grosvenor's seven guiding principles for editing the Magazine began to evolve, principles that reflected, in current *Geographic* Editor Wilbur E. Garrett's words, Grosvenor's "own cultured background, thoughtful personality, and Victorian courtesy to create a different, but highly successful, form of journalism." It would be these principles that would shape both the strengths and the weaknesses of the Magazine over the next fifty years. In a report subsequently made to his Board of Trustees (successors to the Society's Board of Managers), Grosvenor stated these principles as follows:

1. The first principle is absolute accuracy. Nothing must be printed which is not strictly according to fact....
2. Abundance of beautiful, instructive, and artistic illustrations.
3. Everything printed in the Magazine must have permanent value....
4. All personalities and notes of a trivial character are avoided.
5. Nothing of a partisan or controversial character is printed.
6. Only what is of a kindly nature is printed about any country or people, everything unpleasant or unduly critical being avoided.
7. The contents of each number is planned with a view of being timely....

"Fortunately," wrote Frederick G. Vosburgh, Grosvenor's associate of many years and Editor of the Magazine from 1967 through 1970, "some of these points—notably numbers 5 and 6—had a certain built-in elasticity, and the Editor determined the amount of stretch. There was never any doubt about that, or his editorial courage."

Grosvenor's editorial courage was strengthened by an incident that occurred early in his editorship. A paper was submitted to the Magazine by the distinguished Harvard geographer William Morris Davis, which Grosvenor found "exceedingly hard to digest." Nevertheless he passed it on to Alexander Graham Bell, who admitted that he, too, thought the paper difficult to understand, but because of the high academic stature of its author, Bell suggested Grosvenor go ahead and print it.

"With many misgivings, I printed the article," Grosvenor later wrote. "Soon letters of protest from educators deluged us, among them a letter from G. Stanley Hall, President of Clark University, one of the most ardent supporters of our project, who swore that if that article was to be the kind of geography we published, we had better discontinue our efforts.

"From that day," Grosvenor continued, "no sentence has found space in the National Geographic that could not be readily understood."

Recalling the sort of editor Gilbert Hovey Grosvenor was, Vosburgh wrote:

Gilbert Grosvenor had no patience with murky thinking. "What does this mean?" he would pencil sternly beside a paragraph long on pretentious words but short on clarity of thought.

Bombast bored him. "Come down off your soapbox," he said firmly to writers carried away by their opinions to the point of speechifying in print. "Stick to the facts. Our readers can be trusted to form their own opinions."

"People like to learn," he once observed, "but dislike the feeling of being taught."

Use a long word when a short one would do as well, and the manuscript would come back from his paper-piled office with the offending word circled and labeled: "This is a jawbreaker."

Qualify a statement with the lazy phrase "is said to" and you would be called into his presence and informed: "The *National Geographic Magazine* does not publish hearsay. Either it is a fact or it is not. Find out."

If you came to consider yourself an expert on a field of science or a country and affected esoteric terms or high flown language, Gilbert Grosvenor would puncture your inflated ego and bring you down to earth. Snobbish use of foreign words without translation was, and is, similarly taboo, on the grounds that no one knows every language and it is our business to make ourselves understood.

Once he spelled it out for me patiently: "If you don't tell the people, they won't know."

This mild-seeming gentleman of courtly manners had the soul and heart of a fighter. Those who challenged his plans and principles struck steel beneath the velvet....

Hubbard Memorial Hall with a bust of Gardiner Greene Hubbard in the foreground.

On April 26, 1902, the cornerstone of the Society's new headquarters, Hubbard Memorial Hall, had been laid at the corner of 16th and M Streets in Washington. By October 1903, the National Geographic Society had moved out of its rented office in the Corcoran Building and into its new headquarters paid for by the Hubbard and Bell families. By then Dr. Bell had already resigned the presidency of the Society—after telling his son-in-law, "Bert, you are competent to paddle your own canoe." Bell wanted to devote himself to his scientific research, which, at that time, was concentrated upon tetrahedral kites.

In June 1903, the *National Geographic Magazine* had run thirty-two pages of text and seventy-nine photographs of Bell's experimental kites in an issue devoted primarily to "The Tetrahedral Principle in Kite Structure." Four years earlier, in April 1899, Bell had addressed the National Academy of Sciences in Washington on the subject of "Kites with Radial Wings." Since that time, Bell had written for the *Geographic*, he had been "continuously at work upon experiments relating to kites. Why, I do not know, excepting perhaps because of the intimate connection of the subject with the flying-machine problem."

Bell's piece continued:

> We are all of us interested in aerial locomotion; and I am sure that no one who has observed with attention the flight of birds can doubt for one moment the possibility of aerial flight by bodies specifically heavier than the air. In the words of an old writer, "We cannot consider as impossible that which has already been accomplished."
>
> I have had the feeling that a properly constructed flying-machine should be capable of being flown as a kite; and, conversely, that a properly constructed kite should be capable of use as a flying-machine when driven by its own propellers. I am not so sure, however, of the truth of the former proposition as I am of the latter.

On December 17, 1903, six months after Bell's article appeared in the *National Geographic*, Wilbur and Orville Wright realized one of man's oldest dreams when they achieved the world's first successful controlled, powered, manned, heavier-than-air flight, with Orville at the controls of their Wright Flyer. This epoch-making 120-foot flight lasted only twelve seconds, but aviation was born.

The new Society headquarters in Hubbard Hall "bulked as large as a palace" in young Gilbert Grosvenor's mind, and he "resolved that we would prove worthy of so splendid a headquarters. Zestfully I attacked the myriad problems that beset a lone editor. Sometimes, however, those problems seemed insurmountable...."

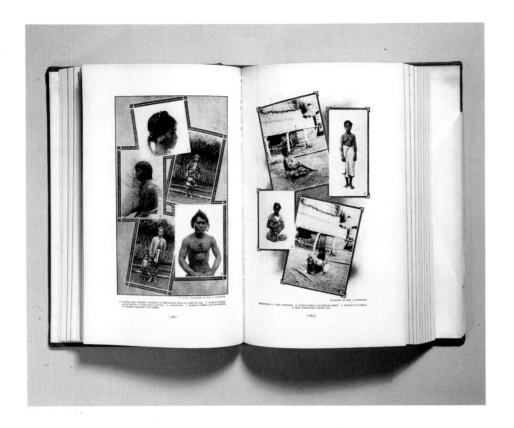

One such problem occurred in December 1904, when the printer called Grosvenor to say he needed eleven more pages for the January issue and the young Editor did not have a decent manuscript available, nor did he feel in the mood to come up with eleven pages of copy himself. By coincidence, on his desk lay that day's mail, and on top was a large, bulky envelope bearing foreign stamps. Still brooding about how to fill the January issue, Grosvenor listlessly opened the envelope then looked "with mounting excitement at the enclosures that tumbled out."

Before him lay fifty photographs of the mysterious city of Lhasa in Tibet, taken by the explorers Tsybikoff and Norzunoff for the Imperial Russian Geographical Society, which was offering them free to the *National Geographic.* These photographs—among the first ever taken of the Tibetan capital—were so extraordinary that Grosvenor quickly chose eleven of them and told the printer to fill the Magazine's empty pages with the photographs of Tibet. Then, expecting to be fired for stuffing the pages of a serious magazine with scenic photographs accompanied only by captions, Grosvenor went to the club "for a holiday," and later went home and told his wife what he had done.

Even in that era of trumpeted "illustrated" magazines, for Grosvenor to have used that many nontechnical photographs to support a basically textless piece was more than daring, it was heresy. But, as Grosvenor told Elsie, it wasn't just using the photographs that worried him, it was that he had paid for the halftone plates out of the Society's meager bank account.

Grosvenor's anxiety was swiftly dispelled when, upon the January issue's publication, Society members stopped him on the street to congratulate him on the Magazine and to say how much they had enjoyed "the first photographs of romantic Lhasa."

The National Geographic Society flag with its three stripes—blue representing the sky, brown the earth, and green the ocean—was designed by Gilbert Grosvenor's wife, Elsie, for the 1903 Ziegler Polar Expedition.

As Grosvenor later recalled, Elsie "asked me...what my idea was, and I said I would like a flag that was recognizable. In Washington we'd been watching parade after parade with notables and nine-tenths of the flags were so complicated, having a seal with the motto around it, that the bystander from the sidewalk couldn't tell what it was—what state.

"I just asked Mrs. Grosvenor if she would please design something that was readable from the sidewalk; and so she hit upon these three stripes, and then she thought she'd put 'National Geographic Society' in big letters on it so that everybody could read what's on the flag. You see my idea: a flag that is not readable fails in the purpose for which flags exist. A flag is supposed to tell the observer who is assembled with it."

"Our members liked the pictures better than the manuscripts we had been feeding them," Grosvenor recalled in 1951. "They showed their pleasure by sending in many new members. The incident taught me that what I liked, the average man would like, that is that I am a common average American. That was how the *National Geographic Magazine* started on its rather remarkable career of printing many pictures, and later more *color* pictures, than any other publication in the world."

The success of the Tibet photographs lulled Grosvenor into thinking his role as Editor would be easier; all he needed to do was to "fill our pages with pictures." But it was as difficult to come up with good photographs of geographic interest as it was to find good manuscripts. "The only photographs that seemed obtainable then were of mountains and scenery," he later complained. But luck, again, was with him.

The following month, in February 1905, Grosvenor's cousin William Howard Taft, then Secretary of War, told him that in April the War Department would be publishing the first "Census Report of the Philippines" and that the report would be illustrated with numerous photographs. If Grosvenor would inform Society members of this interesting publication, Taft told the young Editor, the National Geographic Society could help both the government and the people of the Philippines.

Grosvenor's interest had been pricked the moment his cousin had mentioned "photographs." "That word had become as musical to my ear as the jingle of a cash register to a businessman," Grosvenor wrote.

Taft had the Census Director lend Grosvenor the copper plates of any of the photographs contained in the report that Grosvenor wished to reprint, and Grosvenor chose thirty-two full-page plates containing 138 photographs in all for the April 1905 issue.

To have obtained such photographs on its own, the Society would have had to pay several thousand dollars, plus at least another $350 to engrave that many photographic plates, and, as Grosvenor later pointed out, it was "money The Society did not have at the time."

The April 1905 "A Revelation of the Filipinos" issue was such a success and brought in so many new members that Grosvenor had to reprint the issue.

The time seemed right for a determined membership campaign. Grosvenor hired additional clerical help and boldly began spending money in an all-out membership drive. By September, membership in the Society had soared to 10,000, triple the membership of the year before. The Magazine was doing so well that that same month Grosvenor was able to inform the Board of Managers that the Society would finish the year not only in the black but with a surplus of $3,500. In recognition of this milestone, Grosvenor introduced the following resolution to the Board:

> *Resolved:* That the National Geographic Society, through its Board of Managers, thank Dr. Bell for his generous subscription to the work of The Society from 1899 to 1904, and inform Dr. Bell that The Society is now on such a substantial basis that it can relieve him of his subscription for 1905.

Now, for the first time since April 1, 1899, Bell would no longer be called upon to donate the $100 per month toward Gilbert Grosvenor's salary. "Elsie and I were jubilant," Grosvenor would later recall. "At last her generous father had seen his dream realized: a geographic magazine that would support the

Society. His total gift to establish the magazine was $6,900 instead of the $87,000 he had offered in 1899."

That same month, membership nominations were coming in in such numbers that Grosvenor asked if Dr. Bell's secretary might be able to come into the Society offices to help. As the two young men were working together, Grosvenor asked the secretary if he knew of anyone who might like to work at the Geographic permanently, and "one morning," Grosvenor later recalled, "this splendid looking young man—very husky—came in. He handed me his card and I looked at it—John Oliver La Gorce." He continued:

> The word "John" appealed to me very much. It was a most honored name in the Bible. The name "Oliver" was one of the magic names of chivalry. I didn't know at the time what "La Gorce" stood for, but I soon found that his father was a colonel of artillery in a very distinguished French family. I also learned that his mother was Irish, and I thought the combination of Irish and French would be very suitable. He was sure to be a good mixer, and practical.
>
> So I asked Mr. La Gorce if he would like the job, explaining—as Dr. Bell had explained to me in offering a job—that it might be a steppingstone to something better. He said, yes, he would like it. We made no contract. I never asked him for a reference.
>
> A few days after, Dr. Alexander Graham Bell came down to call. He was received by Mr. La Gorce, and he came into my office and asked, "Who is that fine young fellow you have out there? What a gracious manner he has." I introduced them, and it was a mutual admiration society.

La Gorce remained at the Society for the next fifty-two years and retired as President and Editor on January 8, 1957, two years after having been awarded the Grosvenor Medal "for Outstanding Service to the Increase and Diffusion of Geographic Knowledge for Fifty Years, 1905–1955."

The year Grosvenor hired La Gorce, he also decided that the Society should hold an annual banquet. Grosvenor knew that not only would Washington members enjoy such a function, but if he were able to convince a nationally known figure to serve as the banquet's principal speaker, the Society's reputation would be enhanced. Grosvenor felt an ideal choice would be the Secretary of War, his cousin William Howard Taft. Although Taft's bulk was a clear indication of his love for eating, Grosvenor could not be sure that banquet fare and the tug of familial kinship would provide Taft reason enough to accept, so the young Editor was not beyond exercising a little manipulation. Knowing that Taft's wife enjoyed social evenings, Grosvenor made sure to invite the Secretary in the presence of Mrs. Taft.

"Why, what a nice suggestion!" Mrs. Taft responded. "We would be delighted to attend."

Grosvenor had his banquet speaker; Mrs. Taft had her social evening; the Secretary of War came up with a commendable address; and the Society's first annual banquet was a success.

By the end of that year, membership in the Society had grown from 3,662 to 11,479, and "the year 1905," Grosvenor would later recall, "marked the turning point in the Society's fortunes, the 'end of the beginning.' " The National Geographic Society's paid staff numbered nine people: Gilbert Hovey Grosvenor, Editor and Director; John Oliver La Gorce, Assistant Secretary; and an Assistant Treasurer, a Librarian, four clerks, and a janitor.

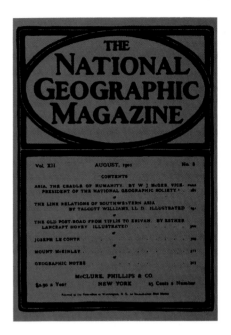

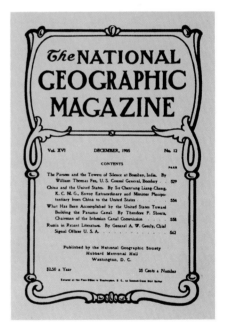

In 1901, the design of the National Geographic Magazine's *cover changed a fourth time to the style shown at top. That design remained until 1904, when the heavy, dark frame and oval border around the contents and title were discarded in favor of a lighter "art nouveau" look below. This fifth design lasted until 1910.*

At the end of 1905, the circulation of the Magazine was 11,479.

"Lord, that's enough now. Please stop it."

"Lord, that's enough now. Please stop it."
—a tearful 16-year-old girl standing with her mother in their violently heaving front yard during Alaska's 8.5 magnitude earthquake on Good Friday, March 27, 1964

"VANCOUVER! VANCOUVER! This is it...."
—the last words of geologist David Johnston, at 8:32 A.M. on May 18, 1980, six miles from Mount St. Helens

For several weeks early in that spring of 1980, *National Geographic* Assistant Editor Rowe Findley had been working on a prospective Magazine article about the national forests. On March 21, a Forest Service friend telephoned Findley at the Society's headquarters to tell him that earthquakes were shaking Washington State's 9,677-foot-high Mount St. Helens, the restless Cascade Range volcano that dominated Gifford Pinchot National Forest. When Findley began researching the mountain, he came upon Drs. Dwight Crandell and Donal Mullineaux's paper "Potential Hazards From Future Eruptions of Mount St. Helens." Mount St. Helens, Findley read, had been the most active of the Cascade volcanoes during the mid-nineteenth century, and, the authors predicted, was due for another eruption before the end of this one.

By March 26, the quakes had grown in strength and force and, as Findley later wrote, "the dormant volcano was stirring and stretching." He booked a flight out early Friday, March 28. By the time he arrived, intermittent plumes of smoke and ash were rising two miles above the peak, dusting its northeast slope sooty gray. It was, Findley noted, "the start of a geologic event—the first volcanic eruption in the contiguous 48 states since California's Lassen, another Cascade peak, shut down in 1917 after a three-year run."

Journalists, geologists, crackpots, and the curious, drawn by the restive volcano, poured into nearby communities eager to bear witness to the potentially cataclysmic drama. The volcano, however, like a supremely confident old performer, refused to heed any curtain time but its own.

Through the remainder of March and all of April the mountain, Findley wrote, still did not "fully awake. A second crater appeared beside the first, then the two merged into a single bowl 1,700 feet across and 850 feet deep. But the eruption level, geologists said, remained 'low-energy mode.' "

Findley spent time during the week between May 11 and May 18 with his new friends Reid Blackburn, a twenty-seven-year-old photographer on loan to the *National Geographic* from the Vancouver *Columbian,* and eighty-four-year-old Harry Truman, whose refusal to leave his Spirit Lake lodge five miles from the volcano's peak had captured media attention.

On May 15 and 16, clouds closed in, making the mountain nearly invisible. On Saturday, May 17, the sun shone but the mountain, Findley wrote, "drowses

Preceding spread: *On January 23, 1973, a volcano was born when a fissure opened in the earth on the outskirts of Vestmannaeyjar, the only town and port of Iceland's Heimaey Island. Within twenty-four hours, ships and planes evacuated most of the island's 5,300 people; and over the next weeks the deep advancing lava and falling ash buried much of the port city while adding about one square mile of new territory to the island. Today Eldfell, as the new volcano was christened, looms over Vestmannaeyjar and, as the Society reported in its 1983 Book Service volume* Exploring Our Living Planet, *"local teenagers are said to find Eldfell's warm crater an ideal trysting place."*

Opposite: *St. Pierre, Martinique, shortly after the May 8, 1902, eruption of Mont Pelée, whose explosion, heard hundreds of miles away, released a cloud of superheated steam and gases that destroyed the city in less than three minutes and left 30,000 dead in its ruins.*

*Sunday, May 18, 1980: Photographer
Gary Rosenquist, in the right place at
the wrong time, snapped this extraordinary sequence of Washington's
Mount St. Helens blowing the top
1,300 feet of its crest in a massive
eruption that left scores of people dead
or missing.*

*"Some two million birds, fish, and
animals perished in the May 18 explosion," noted the* Geographic *in its
January 1981 "Eruption of Mount St.
Helens" report, "and more than 150
miles of trout and salmon streams
and 26 lakes were destroyed. The forests near the mountain were simply
removed—uprooted, shredded, or
blown away by sandblasting winds of
hurricane force...."*

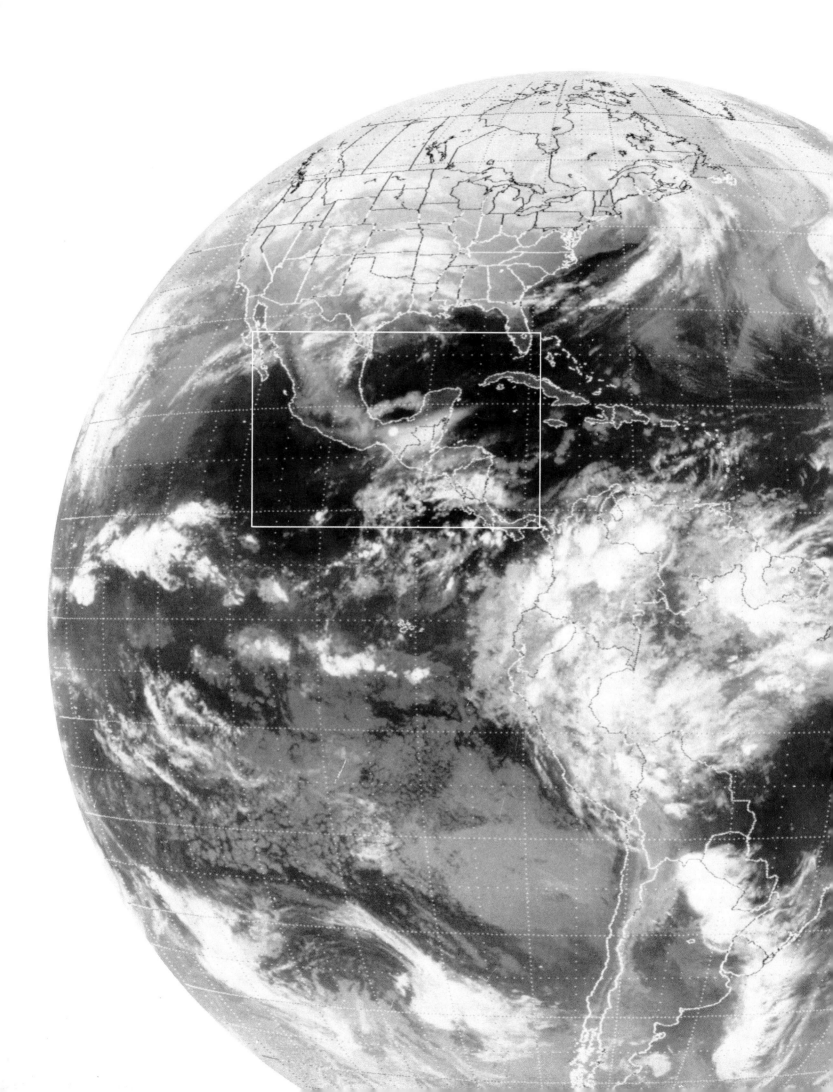

on. The north-face bulge continues—swelling five feet a day; other signs say that nothing is about to happen." Findley drove to Cougar, a timber-industry settlement twelve miles southwest of Mount St. Helens.

"Sunday, May 18. First sun finds the mountain still drowsing," wrote Findley (in his January 1981 article titled "St. Helens: Mountain With a Deathwish"):

Because it is drowsing, I decide not to watch it today, a decision that soon will seem like the quintessence of wisdom. Because it is drowsing, others—campers, hikers, photographers, a few timber cutters—will be drawn in, or at least feel no need to hurry out. Their regrets will soon be compressed into a few terrible seconds before oblivion.

Ten megatons of TNT. More than 5,000 times the amount dropped in the great raid on Dresden, Germany, in 1945. Made up mostly of carbon dioxide and water vapor, innocuous except when under the terrible pressure and heat of a volcano's insides and then suddenly released.

That 5.0 quake does it. The entire mountainside falls as the gases explode out with a roar heard 200 miles away. The incredible blast rolls north, northwest, and northeast at aircraft speeds. In one continuous thunderous sweep, it scythes down giants of the forest, clear-cutting 200 square miles in all....Within three miles of the summit, the trees simply vanish—transported through the air for unknown distances.

Then comes the ash—fiery, hot, blanketing, suffocating—and a hail of boulders and ice. The multichrome, three-dimensional world of trees, hill, and sky becomes a monotone of powdery gray ash, heating downed logs and automobile tires till they smolder and blaze, blotting out horizons and perceptions of depth. Rolling in the wake, the abrasive, searing dust in mere minutes clouds over the same 200 square miles and beyond, falling on the earth by inches and then by feet.

The failed north wall of the mountain has become a massive sled of earth, crashing irresistibly downslope until it banks up against the steep far wall of the North Toutle Valley. This is the moment of burial for Harry Truman and his lodge, as well as for some twenty summer homes at a site called the Village, a mile down the valley.

The eruption's main force now nozzles upward, and the light-eating pillar of ash quickly carries to 30,000 feet, to 40,000, to 50,000, to 60,000. ...The top curls over and anvils out and flares and streams broadly eastward on the winds.

The shining Sunday morning turns forebodingly gray and to a blackness in which a hand cannot be seen in front of an eye.

In the eerie gray and black, relieved only by jabs of lightning, filled with thunder and abrading winds, a thousand desperate acts of search and salvation are under way.

Psychological shock waves of unbelief quickly roll across the Pacific Northwest. In Vancouver and Portland, in Kelso and Longview, and in a hundred other cities and towns, the towering dark cloud is ominously visible.

A phone call...sends me outside to gaze at the spectacle. As soon as I can, I get airborne for a better look and recoil from accepting what I see.

The whole top of the mountain is gone.

Lofty, near-symmetrical Mount St. Helens is no more. Where it had towered, there now squats an ugly, flat-topped, truncated abomination. From its center rises a broad unremitting explosion of ash, turning blue-gray in the overspreading shadow of its ever-widening cloud. In the far deepening gloom, orange lightning flashes like the flicking of serpents' tongues. From the foot of the awesome mountain there spreads a ground-veiling pall.

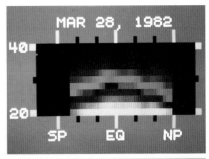

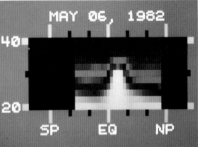

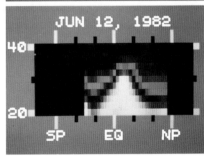

Above: "Comparable in ash output, the Mexican and U.S. volcanoes had very different impacts. Mount St. Helens' predominantly lateral blast laid waste an area of...approximately 232 square miles. El Chichón's scorched earth was only a quarter that, but...atmospheric conditions together with the force of the blast enabled it to penetrate that cold-air barrier and enter the stratosphere....Computer profiles of satellite data taken over Europe document the cloud's spread through the stratosphere...."

Opposite: "Colossal plume of natural air pollutants, the largest in the Northern Hemisphere since the eruption of Alaska's Mount Katmai in 1912, El Chichón's April 4 [1982] explosion is seen as a bright circular puff over Mexico's isthmus in this weather-satellite image."

Overleaf: Masked against El Chichón's abrasive ash, residents of San Cristóbal de las Casas, nearly sixty miles southeast of the volcano, clear their streets at noon.

Somewhere down there lies Coldwater I [a U.S. Geological Survey Camp], above the rushing waters of Coldwater Creek and the valley I had left in verdant beauty only 40 hours before.

Reid Blackburn at Coldwater I, eight miles from the blast, had time to fire four frames of film and dash for his car before being caught in the burning, suffocating ash. Harry Truman and his lodge, buried under hundreds of feet of ash and debris, became part of the mountain he had always said was a part of him. They were two of the dozens who lost their lives when Mount St. Helens erupted.

The whole top of the mountain, as Findley had noted, was indeed gone; the explosion blew away almost a cubic mile of its summit, reducing its elevation from 9,677 feet to 8,364 feet.

Two years later, in March and April 1982, a small, 4,134-foot-high volcano, El Chichón in the Mexican state of Chiapas, was blown apart in a series of violent eruptions that hurled 500 million metric tons of ashes, dust, and gases into the atmosphere. El Chichón, half the size of Mount St. Helens, filled the air with at least ten times the amount of debris, and, although the official death toll was 187, unofficial estimates of the number of victims ran into the thousands.

And on November 13, 1985—two months after a massive offshore earthquake sent seismic waves racing more than 350 kilometers east toward Mexico City, destroying buildings and killing possibly as many as 30,000 people—a minor eruption of the 17,822-foot-high volcano Nevado del Ruiz, 3,200 kilometers to the south, melted part of its ice cap, triggering a murderous mudslide that killed some 20,000 persons in Armero, thirty miles from its peak. There, in that once-busy agricultural center, the mudslides struck after 11 P.M. when most were asleep, entombing them in mud nearly ten feet deep.

Mexico City's quake occurred on September 19, 1985, at 7:18 A.M., when a huge slab of the earth's crust moved twelve miles beneath the ocean's surface.

"Seismic waves a thousand times more powerful than the atomic bomb that leveled Hiroshima fanned out toward Mexico City, in pulses traveling 25,000 kilometers an hour (15,500 mph) and registering 8.1 on the Richter scale," wrote *Geographic* senior editorial staff member Allen A. Boraiko. "So violent was the earthquake that tall buildings trembled in Texas, water sloshed in Colorado pools, and the entire Earth vibrated like a struck bell."

When we look at photographs taken of the earth by the astronauts and see our lovely, cloud-traced, blue sphere floating serenely through the star-studded, black velvet canopy of space, the image our planet projects is one of deceptive tranquillity. Within those clouds lurk violent, thunderous forces unleashing havoc upon an earth whose fragile crust is constantly grinding and trembling and threatening to explode beneath our feet.

As Kendrick Frazier, a former editor of *Science News* magazine, pointed out in his book *The Violent Face of Nature*, at any given time 1,800 thunder-

---

Workers in front of the fallen wooden cupola of the Santa Rosa Courthouse, damaged in the San Francisco earthquake of April 18–19, 1906. Some fifty miles north of San Francisco, Santa Rosa suffered greater earthquake damage for its size than any other town in the area.

storms are churning across the expanse of our planet; 100 lightning bolts strike its surface each second; in late summer, of the fifty or so hurricanes and typhoons that are spawned each year, one—and maybe more—is moving toward a populated coastline. Half a billion people live on floodplains that supply food for a third of the world's population. Rampaging waters somewhere are inundating peoples' homes and croplands.

More than fourteen thousand earth tremors, strong enough to jiggle seismograph needles, shake our planet each week. Twice a day an earthquake will somewhere damage buildings; from fifteen to twenty times a year an earthquake will occur with enough force to result in widespread death and destruction. "And all the while," Frazier adds, "there are 516 active volcanoes waiting to spring loose their violence. An eruption begins somewhere every 15 days."

In the summer of 1927, when Secretary of Commerce Herbert Hoover, in his capacity as director of relief operations, looked upon the devastation caused by the overflowing Mississippi River, he said, "We are humble before such an outburst of the forces of Nature and the futility of man in their control." Hoover called it "the greatest peace-time disaster in our history."

The *National Geographic* was there to report on "The Great Mississippi Flood of 1927," just as it has been there to cover nearly all the other great disasters that have shaken our planet since the Society's founding. In fact, starting with its first issue's report on The Great Storm of March 11–14, 1888, accounts of floods, blizzards, famines, droughts, hurricanes, pestilences, avalanches, earthquakes, and volcanic eruptions, photographed in vivid detail and written up with gusto, have figured prominently in the Magazine's pages.

Nowhere is the *Geographic's* reporting more moving than in the accounts of starving children. And famine cares not where it takes its victims, whether in drought-stricken Bangladesh as the Magazine's Steve Raymer reported in 1975:

> The scene is chilling, horrifying. Several thousand starving Bengalis wait patiently, it seems, to die in a refugee camp in Rangpur, one of the remotest districts in the poor, desperate land of Bangladesh.
>
> There is only enough food on hand from the United States and Canada to provide each with a daily cup of flour, but no powdered milk. Many are too weak to eat, or to swat at the flies swarming around the kitchen.
>
> I see a child—a naked skeleton—waiting for his meager ration; his withered body bears the telltale signs of advanced malnutrition. Others like him sit almost lifeless in their filth. A woman clad in rags clutches an infant so thin his ribs look like a birdcage beneath his peeling skin. I see a tear in the mother's eye.

Or, as Melville Chater wrote in the Magazine in 1919, the famine finds its victims in Armenia:

> As we neared Igdir our interpreter, a cheery, affable young Armenian... turned to us from the front seat and inquired with just a trace of the showman's manner:
> "What you like to see, gentlemens?"

Alaska's March 27, 1964, Good Friday earthquake dropped buildings and pavements as much as thirty feet in downtown Anchorage. Two days later, troops patrol the area and survey the devastation caused by the strongest quake in North America since 1899.

"Conditions," snapped the doctor.

"You like best conditions of dyings or deads? Dyings is easy to see everywhere in the streets. But I know where many deads are, too—in what houses—if you like."

"Drive on," I said hastily. "We'll decide later."

The town of Igdir, with its local and near-by populations of 30,000 Armenians, 20,000 Tatars and 15,000 Yezidis, revealed some squalid streets with but a few people seated disconsolately here and there as we drove in. Throughout those tortuous, sun-beaten byways no children played and no animals roamed. The air was heavy with dreadful silence, such as hangs over plague-smitten communities.

We found the children, such as they were, inhabiting an orphanage wherein one sickened at putridity's horrible odor, and were informed that there were neither medicines nor disinfectants wherewith to allay the condition of the many little sick-beds.

Sick? Say, rather, the bed-ridden—a word which more justly describes those tiny, withered up, crone-like creatures, upon whose faces the skin stretched to a drumhead's tightness: whose peering eyes show terror and anguish, as if Death's presence were already perceptible to them, and who lay there at Famine's climax of physical exhaustion....

"They'll all die," was the brusque observation of the doctor, who had taken one glimpse and gone out. "We can't do them any good. Silly business, anyway, to come out here in a broken-down car."

"We will now see conditions of the deads?" inquired our interpreter, sweetly....

Further on in his Armenia piece Chater tells of coming upon "fifty wizened children [sitting] about a long board" in an open space "eating the American Committee's daily dole of boiled rice. This was accomplished at a gulp," he reported. Then, he added:

> ...the children scattered, searching the ground as I had seen others do beside our car at Alexandropol. Soon one was chewing a straw, another the paring of a horse's hoof, a third a captured beetle.
>
> One seven-year-old girl crouched by herself, cracking something between two stones and licking her fingers. The doctor bent over, examining the object. He asked with peculiar sharpness, "Where did she get that—that bone?"
>
> The child looked up with a scared, guilty glance; then her answer came through the interpreter, who said in a low voice, "Yonder in the graveyard."
>
> I am not sure that we preserved our composure.

If the weather or political oppression doesn't cause the drought that creates the famines, there are the locust plagues such as the one Ethiopia suffered in 1958, in which the insects devoured 167,000 tons of grain—enough to feed a million people for a year.

In his stunning and disturbing July 1977 article, "The Rat, Lapdog of the Devil," Thomas Y. Canby noted that rats had caused the Bombay plague epidemic of 1898 which killed 12½ million Indians, and that during the Black Plague of the Middle Ages, the rat, host for the plague-carrying flea, had caused an estimated 25 million deaths. "In a world haunted by threat of famine," Canby continued, "[rats] will destroy approximately a fifth of all food crops planted. In India their depredations will deprive a hungry people of enough grain to fill a freight train stretching more than 3,000 miles."

The average rat, Canby pointed out, can

wriggle through a hole no larger than a quarter;
scale a brick wall as though it had rungs;
swim half a mile and tread water for three days;
gnaw through lead pipes and cinder blocks with chisel teeth that exert
   an incredible 24,000 pounds per square inch;
survive being flushed down a toilet, and enter buildings by the same
   route;
multiply so rapidly that a pair could have 15,000 descendants in a year's
   life span;
plummet five stories to the ground and scurry off unharmed.

"This year in the United States alone," Canby continued, "rats will bite thousands of humans, inflicting disease, despair, terror. They will destroy perhaps a billion dollars worth of property, excluding innumerable 'fires of undetermined origin' they will cause by gnawing insulation from electrical wiring."

That's the bad news. The good news is that "in laboratories the rat has contributed more to the cure of human illness than any other animal." Furthermore, the author pointed out, the rat, when deep-fried in coconut oil, has "the pleasing gaminess of squirrel or rabbit."

Of all the natural disasters that can befall man none are more dramatic than earthquakes—themselves, in many cases, the cause of volcanic explosions. The magnitude of an earthquake is measured on the scale developed in the 1930s and 1940s by Charles F. Richter and Beno Gutenberg of the California Institute of Technology. The Richter scale is not geometric but *logarithmic.* Each increase of 1 in the scale represents an increase of 32 times in the force of the tremor, and ten times in the size of the seismic wave measured by the seismograph. Therefore, an earthquake such as the one that struck San Francisco in 1906, which measured 8.3 on the Richter Scale, is not twice as powerful as one of, say, a 4.1 magnitude, but a *million* times more powerful.

Between 1970 and 1980, twenty-one earthquakes of a 6.3 magnitude or higher occurred, killing close to 960,000 people.

On March 27, 1964, an earthquake measuring 8.5 struck Alaska—the strongest quake to hit the North American continent in this century. Bill Graves reported on the damage for the July 1964 Magazine:

> It began years before that fatal Good Friday of March 27, 1964. Deep in the earth, perhaps 12 miles beneath the region north of Prince William Sound, fearful and little-understood forces were at work on the earth's crust, twisting and straining the great layers of rock as a truck strains its laminated springs going over a bump. Eventually, at a point called the focus, the rock gave way, snapping and shifting in an instant with the force of 12,000 Hiroshima-size atomic explosions.
>
> The devastation spread with terrible speed in an arc 500 miles long.... Crackling through the earth at thousands of miles an hour, the shock wave sliced, churned, and ruptured the land like some enormous disk harrow drawn over the surface. Highways billowed with the upward thrust of the shock, great concrete slabs overlapping one another like shingles set awry.

◀ ──────────────────────────────

After leaving 250 dead in the Caribbean, Hurricane Allen's 190-mph winds and ferocious seas lashed Corpus Christi, Texas, in August 1980.

Rail yards heaved and buckled, twisting tracks into bright curls of steel. Serene, snow-capped mountains shuddered, loosing cascades of ice and rock that sheared slopes razor clean of brush and trees....

The shock wave struck and raced on, but in passing it stirred other, sequel forces. Somewhere off the crescent of Alaska's southern coast, the sea bottom had heaved and plunged violently, setting millions of tons of water in motion. It was the motion of a tsunami, a seismic sea wave, whose effect onshore can be that of a battering ram. The time was 5:36.

In downtown Anchorage, pavements "rippled like an ocean wave" and dropped as much as thirty feet. "In the split second it took the earthquake to flicker the length of 4th Avenue, the shock scythed the ground out from under a score of buildings on the northern, older side of the street....So neat was the job on the Denali Theater that the building dropped 10 feet below sidewalk level without popping a single lightbulb on its old-fashioned marquee. On the south side of the street the shops and offices quivered, lost a window or two, and stood firm."

Kodiak was struck by "a cresting, thirty-foot-high wall of water that thundered up the channel, lifting 100-ton crab boats on its shoulder and flinging them like empty peanut shells over the harbor's stone jetty, and sometimes two or three blocks into town." Valdez residents, too, remember a mountainous wave and how " in the trip-hammer impact...and its terrible backwash, the pier at Valdez, with 28 stevedores and onlookers, vanished forever."

Hurricanes, too, can spawn storm tides, or surges, walls of water that sweep ashore with the high winds. Since 1900 more than 13,000 people have been killed by hurricanes in the United States alone; on September 8, 1900, 6,000 lost their lives when a twenty-foot storm tide inundated Galveston Island, Texas. Drowning claims nine out of ten victims of hurricanes. In 1970, 500,000 were killed when a cyclone—the western Pacific and Indian Ocean term for a hurricane—struck the coastal islands of Bangladesh.

Since the Society's founding the *Geographic* has published ninety-two articles directly related to destruction caused by natural disasters: of these there have been thirty-two articles on volcanic eruptions, nineteen on earthquakes, eleven on floods, seven on hurricanes, five on droughts, two on blizzards, and one each on monsoons, avalanches, and typhoons.

Why has the *Geographic* devoted such attention to the devastations caused?

"Disasters are the violent disruptions that rip the fabric of our society," James Cornell wrote in *The Great Disaster Book*. "They are the sudden revelations of life's fragility....our nightmares come true." Coverage of natural disasters is nothing more than the diffusion of geographic knowledge about the most elemental forces of nature; and, as with the young women attending the Society's 1903 lecture on Mont Pelée's eruption—whom Gilbert H. Grosvenor overheard complain about the lack of "exciting" photographs—an elemental force of human nature when confronted by natural events beyond control and comprehension is to want to *see*.

Bangladesh, 1974: "My God, these people were starving!...I'll never forget the agony on the desperate faces of Bangladesh's starving thousands that *day."*—National Geographic *photographer Steve Raymer writing about his assignment "The Nightmare of Famine" for the July 1975 issue.*

The Language of the Photograph

As Gilbert Hovey Grosvenor's editorial policies evolved with the new century, the *National Geographic* evolved too, moving increasingly toward its maturity as a popular magazine of catholic interests, reflecting Grosvenor's confidence that "what I liked, the average man would like..."

Bell's articles on the construction of kites were followed by articles on topics as varied as geographical distribution of insanity in the United States, immigration, the "Proportion of Children in the United States," the San Francisco earthquake, Polar exploration, eugenics, American industries, "Queer Methods of Travel in Curious Corners of the World," "Children of the World," and so on.

Two articles, indicative of the enlarging sphere of interest, appeared in the July 1905 issue and are worth noting: the first, "The Purple Veil: A Romance of the Sea," was signed "H. A. L."—the last initial standing for, readers learned the following year, "Largelamb." H.A. Largelamb's "The Purple Veil" opens:

> Off the New England coast a curious object is often found floating on the water, somewhat resembling a lady's veil of gigantic size and of a violet or purple color....In 1871 the late Prof. Spencer F. Baird had the opportunity of examining one of these objects at sea, and he found it to present the appearance "of a continuous sheet of a purplish-brown color, 20 or 30 feet in length and 4 or 5 feet in width, composed of a mucous substance, which was perfectly transparent...." On examining the substance with a magnifying glass...it was obvious that the purple veil, as a whole, was the egg-mass of a fish.

What fish? The *Lophius piscatorius*, the author tells us, "variously known as the 'Goose-fish,' the 'All-mouth,' or the 'Angler,' one of the most remarkable fishes in existence." And one of God's ugliest, too, a fact irrelevant to Largelamb, who explains the "Goose-fish" name as having derived from its

> "having been known to swallow live geese," a statement almost incredible; but a reputable fisherman told the late G. Brown Goode that "he once saw a struggle in the water, and found that a Goose-fish had swallowed the head and neck of a large loon, which had pulled it to the surface and was trying to escape."

Largelamb provides additional accounts of goosefish stomachs having been opened to reveal within "seven wild ducks," "large gulls," and on a third

Preceding spread: *Answering President Woodrow Wilson's call for a show of support to alert the nation to defense needs before the U.S. entered World War One, 150 employees of the National Geographic—the women in white, the men in straw hats—march up Pennsylvania Avenue, Washington, D.C., behind Society President Gilbert H. Grosvenor, his wife, Elsie, and Society Assistant Secretary George W. Hutchison in the Preparedness Parade organized for Flag Day, June 14, 1916.*

occasion "six coots in a fresh condition," all of which leads H.A.L. to conclude: "The fish is a most voracious, carnivorous animal—indeed omnivorous—and quite indiscriminate in its diet."

"The Purple Veil" marked H. A. Largelamb's first appearance in the *Geographic*, but not the first appearance of "The Purple Veil"'s famous author; the byline "H. A. Largelamb" is an anagram: unscramble the letters and you get none other than the ubiquitous A. Graham Bell.

The second article of note that July 1905 was written by the Honorable Eki Hioki, First Secretary of the Japanese Legation, one of the earliest publications of a foreign power's point of view in the *National Geographic*. In "The Purpose of the Anglo-Japanese Alliance," Hioki concluded:

> Instead of Japan coveting the possessions of the United States in the Pacific, Japan welcomes the United States as a neighbor....
>
> For the same reason that Japan does not menace the United States politically Japan does not threaten the United States commercially....There never has been, is not now, or ever will be a strong commercial rivalry between Japan and the United States....The United States will not be swamped by the products of the loom and the forge of Japan....

Although Grosvenor had recognized early in his editorship the value of "beautiful, and instructive" illustrations—the eruption of Mont Pelée, the Lhasa, Tibet, and Philippine Census Report photographs, for example, had been resounding successes—those early illustrations still tended to be of *places* and *things* rather than people; on those occasions when people were shown, they were usually posed in native costume. However, the January 1905 *Geographic* article "Our Immigration During 1904" contained photographs that were eloquent despite the poses. The article was, a member of the Society's Book Service would write in 1981 in *Images of the World*, "illustrated by lineups at Ellis Island: weary old men, a young black man, a barefoot woman wearing a babushka. In photos they are what they would become in poetry— the tired, the poor, the huddled masses yearning to breathe free."

There was still the sense, however, that Grosvenor's recognition of such eloquence might have been more accidental than intentional. But just as the *Geographic* began to move toward "exotic natives in colorful costumes," it began to move away from the "Marine Hydrographic Surveys of the Coasts of the World" type of article. In its place came the first-person-narrated "adventure," as in this excerpt from the March 1906 article "Morocco, 'the Land of the Extreme West'":

> We had gathered in the drawing-room directly after dinner, when we were startled by loud screams from the servants' quarters. Followed by my stepson, Mr. Cromwell Varley, whose wife and two daughters, just home from school at Geneva, completed, with [my wife] Mrs. Perdicaris, our family circle, I rushed down a passage leading to the servants' hall, where I came upon a crowd of armed natives.
>
> Even then we did not realize our danger, but thought...that they, like ourselves, had rushed in to learn the cause of the uproar....
>
> As I turned to inquire of these natives who crowded about me as to what had occurred, I saw some of our European servants already bound and helpless and, at the same moment, we ourselves were assailed by these intruders, who struck us with their rifles. At the same instant our hands were roughly twisted and bound behind our backs with stout palmetto cords that cut like knives.

Author Ion Perdicaris relates how he, his family, and servants are forced out of the house where they are held under guard by the rifle-toting invaders, while

a little apart stood their leader, a man of fine presence, attired in the handsome dress worn by the native gentry. One of my men was reproaching this personage bitterly for this unprovoked aggression.

The leader of the mountaineers raised his hand and, in low but emphatic tones, declared that if no rescue were attempted nor any disturbance made, no harm would befall us and in a few weeks we should be safely back among our people, adding, "I am Raisuli! the Raisuli!"—this...being his clan appellation, since this chereef, or native nobleman, is known among his own followers as Mulai Ahmed ben Mohammed, the Raisuli.

On hearing him declare his name I felt at once that the affair was possibly more serious than I had hitherto anticipated....

Raisuli had indeed been reported to be on the warpath for some time past, but...no one imagined he would attack any one in the immediate neighborhood of Tangier....

Approaching him, bound as I was and in evening dress, I said to him in Arabic, "I know you by name, Raisuli, and I accept your safe conduct, but we cannot go with you thus. We must have our overcoats, hats, and boots."

"Which of your servants shall I have released to return to the house for what you require?" replied Raisuli.

I selected Bourzin, the younger of the guards, on duty that evening....

The tame photographs accompanying Perdicaris' article—"A Group of Camels Passed on the Way to Tsarradan. The Site of Our Captivity," "A Waterwheel Driven by a Donkey on the Road to Tsarradan. Raisuli's Stronghold. Notice the earthen jars and the blindfolded animal," "Moorish Women at the Spinning-wheel Waited on by a Slave"—gave no hint of the international furor the Perdicaris incident had caused. Text and illustration were still separate entities; all they had in common was locale. But the captivity of Perdicaris was serious enough to prompt President Theodore Roosevelt to instruct his Secretary of State to cable, "WE WANT PERDICARIS ALIVE OR RAISULI DEAD." Perdicaris had been freed by the time his article appeared and, four months later, in July 1906, the *Geographic*'s cover proclaimed:

"Photographing Wild Game with Flashlight and Camera"

By Hon. George Shiras, Third
Member of Congress 1903–1905

With a Series of 70 Illustrations of Wild Game—
Deer, Elk, Bull Moose, Raccoon, Porcupine
Wild Cat, Herons, Ducks, Snowy Owl,
Pelicans, Birds in Flight, etc.

"In 1906 George Shiras 3rd, a member of Congress and scion of a prominent family, walked into my office with a box full of extraordinary flashlight photographs of wild animals," Grosvenor later wrote. "He had invented the technique for making such pictures, and an exhibit of his work had won gold medals at Paris and St. Louis Expositions."

"With mounting excitement I sorted the photographs into two piles, one towering, the other small," Grosvenor continued. "Mr. Shiras had been able to interest a leading New York publication in only three of his pictures, so he was astounded when told I intended to print every photograph in the large pile. And I did—74 of them on 50 pages with only four pages of text, in the July issue." Shiras' text contained a description of how he had mounted two cameras, focused at between thirty and forty feet on the bow of a light fourteen-foot boat. Above the cameras was a powerful lamp. Shiras would float down along the edge of the streams or lake "jacklighting" the wildlife, in order to get close enough to photograph them. "There is no sound or sign of life," he wrote, "only the slowly gaining light...."

"When these extraordinary pictures of wildlife appeared, letters poured in demanding more natural history," Grosvenor reported. "But two distinguished geographers on the Board resigned, stating emphatically that 'wandering off into nature is not geography.' They also criticized me for 'turning the magazine into a picture book.' "

By the end of 1906, the Magazine's circulation had risen to 19,237, a figure Grosvenor felt warranted a campaign for additional advertising, "a task I assigned to John Oliver La Gorce," Grosvenor later recalled. "I gave him a free hand except for policy rules that I imposed, among them a firm ban on liquor, beer, wine, and patent medicine advertisements. The magazine even then was widely used in schools. Also, I decided it would benefit both advertisers and readers if ads were printed separately from pictures and text."

By 1908, more than half of the *National Geographic*'s pages were photographs. Referring to one particular article, Thomas Barbour's August 1908 "Further Notes on Dutch New Guinea," a Society editor noted in *Images of the World* how the Magazine had begun to show not only how people lived,

"but also how they looked....Barbour laid aside his butterfly nets and picked up his camera. One of his photographs shows, at first glance, a typical lineup of natives in loincloths or less. These men of Tobadi Village 'wear boars' tusks in their noses, feathers in their hair, and in their ears almost anything.' But the fierce-looking men are not portrayed as savages. Shy young boys stand amid them. One man holds a boy's hand; another man paternally clasps a boy's shoulder. There is humanity here." *

On April 6, 1909, Commander Robert E. Peary reached the North Pole—or as close to the North Pole as he could determine. Gilbert and Elsie Grosvenor had traveled from Baddeck down to Sydney, Nova Scotia, to meet Peary's ship, the *Roosevelt*, upon his return. But Peary was delayed by storms. While the Grosvenors were waiting, the announcement was made that Dr. Frederick Cook was claiming discovery of the Pole. Grosvenor later recalled in a 1962 interview:

> I didn't for the world believe it; I didn't believe Cook because I had known him well. And we waited from day to day there in Sydney, and finally I got a hard telegram from Mr. Gannett, the Vice President of the Society, who said, "Better come to Washington to help me...." The President of the Society at the time, Willis L. Moore, was about to back Cook and Mr. Gannett said that he wanted my help to prevent the situation. So I hurried back [to Washington] and, of course, was very strongly and firmly convinced of Peary's claim....

Grosvenor's disputes with Society President Willis Moore were nothing new. "Moore wanted to assume, as President, authority over me which had not been given him," Grosvenor recalled. Moore had tried various tactics: two years earlier he had attempted to get La Gorce fired because Grosvenor's assistant had accepted a $2.50 payment for a newspaper article; Moore asked for a revision of the bylaws to give the President authority over all Society business; when that failed, he tried to get authority over all financial matters.

Opposite and above: "*Straight toward the mark of the shining eyes the canoe is sent with firm, silent strokes. The distance is [only]...Twenty-five yards now, and the question is, Will he stand a moment longer! The flashlight apparatus has been raised well above any obstructions in the front of the boat, the powder lies in the pan ready to ignite at the pull of a trigger. ...Fifteen yards now, and the tension is becoming great. Suddenly there is a click, and a white wave of light breaks out from the bow of the boat—deer, hills, trees, everything stands out for a moment in the white glare of noonday. A dull report, and then a veil of inky darkness descends. Just a twenty-fifth of a second has elapsed, but it has been long enough to trace the picture of the deer on the plates of the cameras....*" So wrote George Shiras, 3rd, in the July 1906 Geographic *of his innovative technique for photographing animals at night.*

Grosvenor's decision to publish seventy-four Shiras photographs in that issue resulted in the resignations of two Board members, who felt that "wandering off into nature is not geography." *A white owl perched above a Michigan river was also displeased; the flash so terrified the owl, Shiras wrote, that it* "fell 15 feet into the water, swore like a trooper, and waded ashore."

---

*There was also, as the editor pointed out, danger: "Barbour visited villages, asked questions, and produced a vivid report about native life. He took chances, as did his wife, the first white woman many of the New Guineans had ever seen. One day the Barbours approached a temple. 'When we tried here, several times, to persuade the crowd to admit Mrs Barbour,' he wrote, 'a single gesture gave a final answer; that gesture was the swift passing of the hand across the throat.' "

In each case, Grosvenor, with the full support of the Board of Managers, was able to hold his power intact. In 1910, upon Moore's retirement, Henry Gannett was elected President of the Society. "Mr. Gannett was a foremost geographer of the time, and everybody had great regard for him," Grosvenor recalled. "One member of the Board got up after Mr. Gannett was elected, and said, 'Thank the Lord we now have as the head of The Society a *real* scientist.' Mr. Moore...had been chief of the Weather Bureau for a long time and considered himself a very great scientist. But he was a politician...and was not regarded as a scientist by the meteorologists.* When this was said of the new president, Mr. Moore got his hat and got up immediately and he left the meeting. He resigned a few days later from the Board. One member of the Board thought [Moore behaved the way he did because it was his desire] to get on the payroll of the National Geographic. By that time the Society was 'roaring' we were getting members so fast."

The first series of hand-tinted illustrations appeared in the November 1910 issue of the *Geographic*—"color paintings," Grosvenor called them: twenty-four hand-tinted color pictures made from black-and-white photographs of "Scenes in Korea and China" taken by wealthy New Yorker William W. Chapin, who had made his fortune investing in Kodak during its early days. Color film had not yet been perfected, and Chapin had a Japanese artist color his photographs based on careful notes he had kept on the costumes and backgrounds he had photographed.

One photograph showed a rear-view of a Korean farmer wearing a large woven straw hat and a loose-hay cape to ward off rain. "This and other photographs in a twenty-four-page spread caused a sensation when they appeared," readers of a *Geographic* history later learned. A reader today would be less impressed by the photographs of heavily laden coolies, Buddhist nuns, shackled prisoners, and shy Manchu women strolling along a Peking street—especially since the reader would want more information about the subjects than captions such as "Manchu Women: Peking, China" provided. Even the text seemed distant, removed, with lines like: "The street scenes of Seoul offer great variety for the kodak, the burden-bearers of both sexes furnishing a constant change of scene; most of them being willing victims, entirely satisfied with a small tip."

Each color page cost four times the amount of a black-and-white page to reproduce and Grosvenor had been tempted to publish only a few color photographs at a time throughout the year to spread the expense; but he decided that doing so would spoil the effect of the series. The membership's excited response to Chapin's hand-tinted photographs and the leap in advertising revenue convinced Grosvenor that the impact of color was worth the expense. The problem, however, was that the development of color photography still lagged behind Grosvenor's enthusiasm for printing it.

"By 1910," Grosvenor was later to write, "income was sufficient for me to start a photographic laboratory. Soon the *National Geographic* began pioneering in the use of color photographs made by the Lumière Autochrome process, and later by other processes—Agfacolor, Finlay, Dufay. As each improvement became available, the magazine put it to superlative use."

---

*Nor, presumably, was Moore regarded as a scientist by the politicians. The day before William Howard Taft's March 4, 1909, inauguration as President, Moore sent him a telegram assuring Taft of fair weather. The morning of the inauguration enough snow fell on the capital to interrupt all traffic.

In 1910, Queen Victoria's son, King Edward VII, died and was succeeded by George V; Japan annexed Korea; China abolished slavery and Congress passed the Mann Act (prohibiting the transportation of women across state lines for immoral purposes); Halley's Comet was observed; and Mark Twain, Leo Tolstoy, and Julia Ward Howe died. Cézanne, Van Gogh, and Matisse had paintings shown in Roger Fry's Post-Impressionist exhibition in London. Prince Albert I founded the Musée Océanographique in Monaco; the ballet for Stravinsky's "The Firebird" was performed in Paris; and the dance craze sweeping Europe and America was the South American tango. Barney Oldfield drove a Benz racer 133 mph at Daytona; Henri Farman flew approximately 300 miles in eight and a quarter hours; the "weekend" was becoming popular in the United States; on Broadway Marie Dressler was singing "Heaven Will Protect the Working Girl" (someone had to; by 1910 there were 7.5 million women and girls employed outside the home); and Society membership would reach 74,018.

The January 1910 issue of the *Geographic* contained a report on the Society's annual banquet, on December 15, 1909, and the presentation of a special gold medal to Commander Robert E. Peary for his discovery of the North Pole.

Beginning with its February issue, the Society adopted the cover design longtime readers of the Magazine remember with affection to this day: the yellow border with the four globes representing views of the four hemispheres imbedded in a gray "grosgrain ribbon," with acorns and oak and laurel leaves.

One of the most powerful pieces from this era was "Race Prejudice in the Far East" by Melville E. Stone, General Manager of the Associated Press, which appeared in the December 1910 issue. Stone argued that rather than accept that the Oriental mind was "unfathomable," he still had "some respect for Cicero's idea that there is a 'common bond' uniting all of the children of men. And whatever our ignorance of, or indifference for, the Orientals in the past, it is well to note that conditions, both for us and for them, have entirely changed within the last decade."

Stone provided examples from his own personal observation of the failure of Europeans to accept the Asian:

> At the Bengal Club at Calcutta last year a member in perfectly good standing innocently invited a Eurasian gentleman...to dine with him. It became known that the invitation had been extended, and a storm of opposition broke among the members. The matter was finally adjusted by setting aside the ladies' department of the club, and there the offending member and his unfortunate guest dined alone. The next day the member was called before the board of governors and notified that another like breach of the rules would result in his expulsion.

At a Government House ball in Calcutta, white men dance with native princesses, but according to Lady Minto, wife of the Viceroy of India, Stone reported, "No white woman would think of dancing with a native; it would certainly result in ostracism." But the most horrendous example given Stone was provided by a Japanese Harvard graduate, then a minister of the Japanese crown:

> "When [Commodore] Perry came here [in 1853] and Townsend Harris (of blessed memory) followed him and made the first treaty with Japan, it was stipulated that we (the Japanese) should give them ground for their legation and their consulates', compounds. We did so. Yokohama was then an unimportant place, a native fishing village. It was the natural port of Tokio,

Thomas Barbour's photograph, published in 1908, "MEN OF TOBADI VILLAGE, HUMBOLDT BAY [NEW GUINEA]" was captioned: "Fond of ornaments, they wear boars' tusks in their noses, feathers in their hair, and in their ears almost anything. The boys, who are not yet full members of the tribe, have their hair cut as the picture shows. This is done by scraping the head with a splinter of shell from the giant clam (Tridacna). It is indeed a bloody operation."—From "Further Notes on Dutch New Guinea," August 1908 issue

The National Geographic's first color series: William W. Chapin's photographs—"Peasant in Rain-coat and Hat: Seoul, Korea" (above) and "The Manchu Family Airing: Peking, China" (opposite, bottom)—published in November 1910, "marked a turning point for the magazine," Gilbert H. Grosvenor wrote in 1963. "Returning from the Orient, Mr. Chapin offered me his entire collection of black and whites, most of which a Japanese artist had tinted by hand. Determined to introduce new features into National Geographic, I printed 24 pages of color in one number; no editor had ever run so much before. The issue created a sensation and brought in hundreds of new members."

"Dancing Girls" (opposite, top) appeared in November 1911 in a second color series titled "Glimpses of Japan."

but as we had no foreign trade that meant nothing. We gave them ground in Yokohama for their consulate. Merchants and traders followed, and we gave them ground also for their shops. The British and the Russians and other European nations came in and we gave them like concessions....

"Well, as time went on the village grew into a city....Sir Harry Parks, the British minister, asked for ground in Yokohama for a race-track. We cautiously suggested that horse-racing was said to be wicked by the European missionaries. But he insisted and we gave him the ground. Then we were asked for ground for a social club for the foreigners, and we gave them a plot on the sea front, the finest piece of land in the city.

"Later they wanted to play cricket and football, and finally golf. Well, we gave them ground for this. As the city grew, this cricket-field was so surrounded by buildings that it was practically in the center of town. Understand, all of this ground was donated. Last year we suggested that we could use the cricket-field, and we offered to give in place of it a field in the suburbs. As railways had been built meanwhile, the new field would be even more accessible than the old one was when we gave it. The foreigners demurred, and proposed that we buy the old field and with the purchase-money they would secure a new one. Finally, we compromised by paying for their improvements and furnishing them a new field with like improvements free of cost.

"The question of taxation arose. Yokohama had grown to be a city of 300,000 inhabitants, with millions of dollars invested in buildings owned by foreigners. We asked no taxes on the ground we had donated to them, but we did think it fair that they should pay taxes on their buildings. They said no, that everywhere in the West the buildings went with the ground. We submitted the question to the Americans, but they dodged the issue, saying they would do whatever the others did. Then, under the law of extra-territoriality, we were compelled to leave the decision to the British consul, and he decided against us. The case has now gone to The Hague Court.

"Finally, when I tell you that in the light of this history no native Japanese gentleman has ever been permitted to enter the club-house or the grand-stand of the race-track or to play upon the cricket-field, perhaps you will understand why there is some feeling against foreigners in Yokohama."

In June 1911, as a supplement to its regular issue, the *Geographic* offered an eight-foot-long, seven-inch-wide, panoramic photographic view of the Canadian Rockies. "It remains to this day," Grosvenor recalled in 1957, "one of the most marvelous mountain views ever photographed. Twenty peaks, passes, canyons and other features were distinctly captioned on the supplement." On March 29, 1912, Robert Falcon Scott, leader of the British Antarctic Expedition, made his last diary entry and died on his return from the South Pole. In September 1912, "Head-Hunters of Northern Luzon," a piece by Secretary of the Interior of the Philippine Islands Dean C. Worcester, was accompanied by Charles Martin's photographs—one of which was a grisly full-page picture of "An Unlucky Ifugao Head-Hunter Who Lost His Own Head and Thereby Brought Disgrace Upon His Family and Village." The headless victim, trussed like a turkey with his wings folded over his chest, is being carried off on some men's shoulders. The text, with its description of the ceremony surrounding the triumphant warriors' return with the victim's head, was equally grim.

In 1913, construction was begun on the Society's new administration building next to Hubbard Hall; that year Gilbert H. Grosvenor bought a hundred-acre farm in Bethesda, Maryland, and discovered his love for birds.

The June 1913 issue contained fifty color plates of "Fifty Common Birds of Farm and Orchard."

Charles Martin (above at left) was an Army sergeant assigned to the staff of Dean C. Worcester, Secretary of the Interior of the Philippine Islands, when his photographs caught Grosvenor's eye.

Martin's grisly photograph (opposite, bottom right) accompanying Worcester's 1912 article "Head Hunters of Northern Luzon" was titled "An Unlucky Ifugao Head-Hunter Who Lost His Own Head and Thereby Brought Disgrace Upon His Family and Village." The two other arresting Martin photographs from that era, Kalinga Province chiefs listening to a phonograph (opposite, top) and Luzon natives smoking tobacco (opposite, bottom left) had not previously been published.

In 1915 Charles Martin was hired to become the Society's photographic laboratory chief.

Gilbert H. Grosvenor later recalled, "My twin brother, Edwin Prescott Grosvenor, had told me that the U.S. Department of Agriculture's edition of 100,000 copies of a circular, *Fifty Common Birds of Farm and Orchard*, had been exhausted in two weeks and that the Department was not able to supply thousands of applications for it. I borrowed the color plates and republished the bulletin, with credit to the authors, in the *National Geographic Magazine*. Our members liked it immensely."

The first Lumière Autochrome appeared in July 1914, a single photograph of "A Ghent Flower Garden." The caption explained that the photograph had been taken at the last exhibition held in the Horticultural Hall at the World's Fair Grounds in Ghent, the "Flower Garden of Belgium," and went on to conclude, "The picture makes one wonder which the more to admire—the beauty of the flowers or the power of the camera to interpret the luxuriant colors so faithfully." Lumière Autochrome, the first commercial process in color photography, was developed in France and marketed beginning in 1907. The Ghent flower garden color photographic reproduction was significant for its technological achievement, not its subject matter. It would still be years before the novelty of color photography in a popular magazine would fade.

On June 28, 1914, Archduke Franz Ferdinand, heir to the throne of Franz Josef of Austria-Hungary, was assassinated by a Serbian nationalist at Sarajevo; on July 28, Austria-Hungary declared war on Serbia; Russia ordered full mobilization; on August 1, Germany declared war on Russia; on August 3, Germany declared war on France and invaded Belgium and Luxembourg; on August 4, England declared war on Germany; and within weeks Montenegro and Japan joined the Allies (England, France, Belgium, Russia, Serbia), and the Ottoman Empire, or what little was left of it, joined the Central Powers (Germany and Austria-Hungary).

While non-*National Geographic* readers were scurrying through sadly out-of-date atlases trying to follow the action, Society members received their August *Geographic*s containing a "Map of the New Balkan States and Central Europe" and the explanation:

> The eyes of the civilized world are now focused upon Europe and the stupendous war there beginning. Therefore the map accompanying this issue of the *National Geographic Magazine* is particularly timely. It contains the most complete and up-to-date data about central Europe, including the boundaries of southern Europe as reformed by the Balkan wars and as determined by the London Conference.
>
> This map will prove of much value to the members of the Society who wish to follow the series of military campaigns that it is feared will be without parallel in history.

Distribution of the map was a publishing coup arranged the year before by Gilbert Grosvenor when he and his wife had been caught in Edinburgh. "We couldn't get any money for three days—the banks were all closed because they expected a European war," Grosvenor later recalled. "We called the Secretary of the Royal Geographical Society. I asked him what he thought about a war. 'Oh,' he said, 'We'll never have a European war. Never.' That was 1913."

Grosvenor, aware of the growing tensions and the talk of war, ordered an updated map of central Europe from commercial cartographers upon his return. "We had several hundred thousand of them printed," he said, "and they were waiting in the cellar until the war started in 1914. Within a week we dis-

"My camera recorded splendors of Tsarist Russia when my wife, parents, and I travelled to Moscow in 1913," wrote Gilbert H. Grosvenor, shown (right) in 1913 by Baddeck Bay in Nova Scotia, examining his Speed Graphic camera, newly purchased that same year in London. Its shutter speed of 1/500 of a second was a far cry from that of the 4A Folding Kodak with which he recorded the Moscow scenes "The Cathedral of St. Basil's" (opposite, top) and "A Poor Man's Funeral Procession in Moscow" (opposite, bottom). Grosvenor had his St. Basil's photograph hand-tinted by his own instructions for inclusion in his November 1914 article "Young Russia: The Land of Unlimited Possibilities."

tributed that map of central Europe to our members. Of course, it helped very much to get more members."

The United States may, at that time, have been militarily neutral, but the *Geographic* made its sympathies clear with articles such as Arthur Stanley Riggs' long (100 pages), copiously illustrated (seventy-three black-and-white and sixteen color photographs) "The Beauties of France" (November 1915) that ran along with an editorial, "The World's Debt to France." The September 1914 issue's "Belgium: The Innocent Bystander," with its photograph of slain horses in a shelled-out village street, was followed by "Belgium's Plight" in May 1917, the month after the United States entered the war.

The November 1914 *Geographic* carried Gilbert H. Grosvenor's own "Young Russia: The Land of Unlimited Possibilities," for which he had provided both the stunning, "candid" hand-tinted photographs and a text that decried the illiteracy rate and censorship. Grosvenor reported:

> There are conditions in Russia which a visitor from the land of free schools, free speech, and a free press finds it difficult to understand; the deplorable rarity of good schools, making it a sore trial for a poor man to get his son educated; the arrival of his American newspaper, with often half a page stamped out by the censor in ink so black that it is impossible to decipher a single letter; the timidity, nay fear, of some people of being overheard when talking frankly on political subjects....

He also pointed out the "devoutness of the people," the nation's resources, that Russia's siding with the Union during our Civil War had perhaps prevented England's intervention in behalf of the Confederates, and that "the progress of the times...has set to work forces that inevitably will spell the doom of illiteracy and ignorance and make Russia in fact the land of unlimited possibilities."

While the war raged in Europe, the *National Geographic* published articles on "The France of Today" and "The German Nation," "Hungary: A Land of Shepherd Kings," "Glimpses of Holland" and "Partitioned Poland," "Armenia

and the Armenians" and "Roumania, the Pivotal State," "What Great Britain Is Doing" and "The Ties That Bind: Our Natural Sympathy with English Traditions, the French Republic, and the Russian Outburst for Liberty," and so on. Still, the Magazine found room for pieces like "The Wild Blueberry Tamed." It also was giving extensive coverage to Hiram Bingham's Yale University–National Geographic Society jointly sponsored archeological investigations of Machu Picchu, Peru, as well as "The Valley of Ten Thousand Smokes" series of the Society's expeditions into the distant and remote Mount Katmai volcanic region in Alaska, where, in 1912, there had occurred one of the most violent volcanic explosions ever recorded. As Robert F. Griggs reported for the January 1917 issue:

> The comparative magnitude of the [Mt. Katmai] eruption can be better realized if one should imagine a similar eruption of Vesuvius. Such an eruption would bury Naples under 15 feet of ash; Rome would be covered nearly a foot deep; the sound would be heard at Paris; dust from the crater would fall in Brussels and Berlin, and the fumes would be noticeable far beyond Christiania, Norway.

Membership topped 100,000 for the first time in 1911; two years later, in 1913, it was 234,000; in 1914, 337,000; and in 1915, membership reached 424,000, and Associate Editor John Oliver La Gorce was able to brag in a Society promotional pamphlet:

> ...the National Geographic Magazine has found a new universal language which requires no deep study—a language which takes precedence over Esperanto and one that is understood as well by the jungaleer as by the courtier; by the Eskimo as by the wild man from Borneo; by the child in the playroom as by the professor in the college; and by the woman of the household as well as by the hurried business man—in short, the Language of the Photograph!

In April 1916, accompanying Gilbert Grosvenor's appreciation of this country, "The Land of the Best," there appeared the Magazine's first natural-color series, among them twenty-three Lumière Autochromes by Franklin Price Knott (a respected painter who took color photographs as a hobby), including a picture of the then famous husband-and-wife dance team of Ruth St. Denis and Ted Shawn captioned, "The Poetry of Motion and the Charm of Color." A second series of sixteen Lumière Autochromes by Knott, photographs taken in Tunisia, Algeria, Holland, India, and the United States, appeared that September. Because of the war they would be the last color photographs printed for several years.

On April 2, 1917, President Woodrow Wilson called for a declaration of war against Germany. That month's issue of the *Geographic* carried "Do Your Bit for America: A Proclamation by President Wilson to the American People." There was also Granville Fortescue's "The Burden France Has Borne," with its descriptions of the "German armies spread across [France's] most productive provinces like a gray corroding acid," and what Fortescue referred to as the "Miracle of the Marne," where for "three days lustful Uhlan outguards pointed their blood-stained lance tips at the Eiffel Tower, saying confidently, 'Within the week and our flag will float from the highest pinnacle in France.' But the God who weaves the world's destiny in mystery heard the prayers of France. The miracle was performed. Paris, the most beautiful achievement of man on earth, was saved from sack and rapine."

Above: *"Invention of the Lumière Autochrome opened a new world in photography, and we were quick to take advantage,"* Grosvenor later wrote. Franklin Price Knott's Autochrome of dancers Ruth St. Denis and Ted Shawn appeared in the April 1916 issue.

Opposite: *"On a mission to save the giant sequoias, I went to California in 1915,"* Gilbert Grosvenor wrote. Disdaining a tent, he and friend Horace Albright instead *"pitched our sleeping bags at the foot of a tremendous sequoia."* (bottom) Moved by the *"majesty and friendliness"* of the magnificent huge trees, Grosvenor and the Society campaigned to protect the sequoias against the lumbering interests. Two years later, through Society contributions, 2,239 acres of the California sequoias were preserved. *"A TROOP OF UNITED STATES CAVALRY AND A FALLEN SEQUOIA"* (top) *appeared in the January 1917 issue.*

World War One Geographic *photographic coverage included:* (far right) "BANDAGING A WOUNDED DOG...*With a heroism that makes them akin to their masters, these gallant animals carry succor to the helpless and the dying who lie in no-man's land between the trenches. Heartless indeed must be the sharpshooter who can make a target of one of these dumb messengers of mercy*" (May 1917); (near right) *the membership-supported twenty-bed special* National Geographic Society Ward *in the American Ambulance Hospital at Neuilly, France (July 1917). It was published with the note* "Suppose He Were Your Boy!" *above a donation form that members could clip out and send in; and* (below) *this classroom scene showing the* National Geographic Magazine *being used to help foreign-born soldiers at Camp Kearny, California, learn English (August 1918).*

A photograph of munitions workers was titled "Munitions Manufacturing Is No Respecter of Age," with the caption underneath: "Many of the women of France who are doing their bit in the production of large-caliber shells for the big guns at the front have completed their allotted threescore years and ten, yet they gladly give the closing days of their lives 'for France.' In many cases their labor is all that they have left to give, for grandsons, sons, and husbands already have been sacrificed on the firing line."

Ex-President William Howard Taft contributed "A Poisoned World"—a reference to how "the minds of a great people...have been poisoned into the conviction that it is their highest duty to subordinate every consideration of humanity to the exaltation and the development of military force..." And Major General John J. Pershing wrote "Stand By The Soldier."

The October 1917 *Geographic* was a special issue containing full-color depictions of 1,197 different flags and standards of the world's nations and private organizations—including the National Geographic—resulting in what the *Geographic's* Editor called "the most expensive, instructive, and beautiful number of its magazine in the history of periodical literature."

The National Geographic Society and its now more than 625,000 members were off to war, and readers were invited to contribute to a National Geographic Ward in the American Ambulance Hospital in Neuilly, France, and treated to articles like "The Life Story of an American Airman in France," "Plain Tales from the Trenches," a new National Geographic Society "Map of the Western Theatre of War," and "Hospital Heroes Convict the 'Cootie.'"

The war ended on November 11, 1918, and the entire next month's issue was given over to Edwin A. Grosvenor's lavishly illustrated, remarkable article, "The Races of Europe: The Graphic Epitome of a Never-ceasing Human Drama. The Aspirations, Failures, Achievements, and Conflicts of the Polyglot People of the Most Densely Populated Continent."

The following year the entire March issue was devoted to another single subject: the noble dog. The lead article in that issue, "Mankind's Best Friend: Companion of His Solitude, Advance Guard in the Hunt, and Ally of the Trenches," was followed by "Our Common Dogs" and "The Sagacity and Courage of Dogs: Instances of the Remarkable Intelligence and Unselfish Devotion of Man's Best Friend among Dumb Animals."

As 1919 drew to a close, Theodore Roosevelt was dead, President Wilson presided over the first League of Nations meeting in Paris—although the United States Senate was later to vote against America's joining. The Third International was founded in Moscow, the Hapsburg dynasty exiled from Austria, the German fleet scuttled at Scapa Flow, and the peace treaty signed at Versailles. Sherwood Anderson published *Winesburg, Ohio*; James Branch Cabell, *Jurgen*; Hermann Hesse, *Demian*; W. Somerset Maugham, *The Moon and Sixpence*; and Carl Sandburg won a special Pulitzer award in poetry for *Cornhuskers*. Walter Gropius founded and built the Bauhaus in Weimar, Germany. Jazz arrived in Europe, Jack Dempsey knocked out Jess Willard to win the world heavyweight boxing championship, and the "Black Sox" scandal stunned professional baseball. The U.S. House of Representatives was considering limiting immigration, Lady Astor was elected to Parliament, American Oliver Smith perfected the mechanical rabbit, paving the way for modern greyhound racing, and National Geographic Society membership reached 668,174.

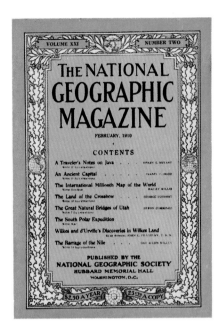

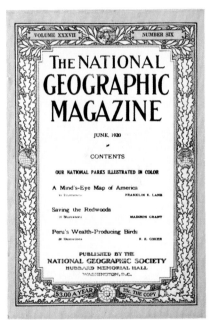

The February 1910 issue was the first to carry the distinctive acorn, oak, and laurel wreath cover. The border, however, was buff—as was the interior box containing the table of contents.

The cover design continued to change and, by 1920, the table of contents box was white.

In 1910, the Magazine's circulation was 74,018 and, by 1920, it increased nearly ten-fold to 713,208.

"Ruins, Lost Cities, and Bones"

"...I laid considerable emphasis on the fact that the map making...ought to be completed at once, while the same topographer was available and willing to undertake the work. His work is purely geographical...you replied that you were not interested in that so much as you were in the ruins, the lost cities, and the bones."

—From a letter written March 18, 1912, by Hiram Bingham to Gilbert H. Grosvenor, Esq.

Above: *Hiram Bingham in front of the main tent at his Machu Picchu expedition camp in 1912.*

Opposite: *Archeologist, historian, and statesman, Bingham, here seen helping paddle a primitive ferry in Peru, later became Governor of Connecticut and a U.S. Senator.*

Preceding spread: *The lost city of the Incas, Machu Picchu, lay hidden amid Peru's misty peaks for three centuries, until its discovery in 1911 by Hiram Bingham, who then led three National Geographic–Yale University expeditions to the site, in 1912, 1914, and 1915.*

The long, mutually beneficial relationship between the National Geographic Society and Hiram Bingham, who was to discover Machu Picchu, did not get off to a good start. In 1908, three years before Bingham found the lost city of the Incas, his wife forwarded photographs and two of his articles to *Geographic* Editor Gilbert Hovey Grosvenor, who flatly rejected them explaining, "We have on hand a number of articles on South America and therefore cannot find room for them."

When, however, Bingham's book *The Journal of an Expedition Across Venezuela and Colombia* was published the following year, 1909, Grosvenor thought it "splendid" and asked whether Bingham could send him "an article of about 4500 words describing the least known section which you traversed with as large a selection of photographs as possible." Grosvenor offered "$25 for the article and $1.50 for each photograph used."

"I scarcely know what to say," Bingham wrote back. Portions of the book and "some of the pictures...the very ones which you say you found particularly interesting," he pointed out, had already been rejected by Grosvenor nine months before. Bingham continued, "As you did not find those articles desirable (although they contained the very cream of my first trip) I do not believe that you would find anything that I might write on this last trip, worth printing or paying for."

Bingham's letter was forwarded to Grosvenor, by then summering at Baddeck, Nova Scotia. Grosvenor immediately replied, "It is very strange, but I haven't the remotest recollection of the articles you state Mrs. Bingham sent me...." He promised to "look the matter up as soon as I return to Washington." If he did look up the matter, there is no record of any action taken, and correspondence between the thirty-three-year-old *Geographic* Editor and the thirty-three-year-old Yale University Assistant Professor of Latin American History was, for the time being, dropped.

During the summer of 1911, Bingham was a few days out of Cuzco, Peru, on a winding jungle trail alongside the turbulent rapids of the Urubamba River. "Cliffs rose 2,000 feet, and above soared snow-capped Andean peaks," Bingham later recalled.

Above: *Skulls and bones littering an ancient cemetery near Lima, Peru, offer grim evidence of the looting that, beginning with the Spanish conquest and continuing up to the present, has defiled much of the Indian heritage.*

Right: *Each fall Lake Titicaca's Island of the Sun, legendary birthplace of the first Inca, provides the setting for an Inca-style ritual. Here an Aymara Indian paddles a boat with a white llama to the annual ceremony in which, to guarantee good crops, the llama's blood will be shed as a sacrificial offering to the Earth Mother.*

"Just before dark we…found ourselves at Boca San Miguel….Here…mules were rounded up for us while we [ate] a meal of rice, beans, and cold tamales.

"By the time our packs were on the mules and we ourselves were mounted, the sun had set and the quick darkness of the Tropics had descended on us," reported archeologist Matthew W. Stirling in the August 1939 National Geographic.

Dr. Stirling, discoverer of eleven massive stone Olmec heads, his wife, Marion, and their party pause for a 1940 National Geographic photograph by Richard H. Stewart on the steamy jungle trail linking the nearest place reachable by launch, Boca San Miguel, with their Tres Zapotes base camp.

Over this trail—at times so muddy that the mules sank up to their bellies—all the Stirlings' supplies were carried during their National Geographic Society–Smithsonian Institution Expeditions in southeastern Mexico in 1939 and 1940.

Above: *Dr. Matthew W. Stirling measures a giant stone head weighing over ten tons left behind by the mysterious Olmecs in the hot, humid lowlands of the Gulf coast of southeastern Veracruz about 1200 B.C.*

Without domestic animals or the aid of the wheel, Stirling wondered, "How was this great block of stone transported more than ten miles from its source...?"

Opposite: *Matthew Stirling excavated this enigmatic eight-foot-tall, twenty-four-ton stone portrait and others at La Venta, in search of clues to their puzzling past. "The Olmecs ritually 'killed'—then buried these and other large sculptures, possibly to purge them of their power," the Society later explained in its 1983 cultural atlas* Peoples and Places of the Past.

Inca terraces climbed the slopes, but these were no novelty to us. Instead we were searching for the last capital of the great Inca Empire that had extended more than the length of Peru before the Spaniards came in the 16th century. With two Yale colleagues, a Peruvian soldier, and a mule train, I had investigated every rumor of a ruin. And now, camping beside the river at the base of a mountain called Machu Picchu, we were told of another ruin on the ridge above us.

Bingham agreed to pay his informant, Melchor Arteaga ("an Indian obviously overfond of firewater"), 50 cents to lead him to the ruin; and on a drizzly July 24, 1911, the tall, gangly Bingham, a sobered Arteaga, and Sergeant Carrasco of the Peruvian Army crossed the Urubamba's seething rapids on a bridge—so flimsy, Bingham confessed, that he crawled over it—and started up the "stiff, almost precipitous slope."

On all fours we pulled ourselves up through slippery grass, digging with fingers to keep from falling. Far below, the Urubamba snarled angrily. The heat was oppressive. Arteaga moaned that there were lots of snakes—vicious fer-de-lances reputed to spring after their prey.

Calling on every reserve, I clambered through thinning jungle to where the ground leveled. Drenched with sweat, I straightened and saw a grass hut. Indians approached with gourds of spring water.

I drank in the view of steep summits between gulps of water and lungfuls of air. Our Indian hosts said there were old houses "a little farther on." With a small boy to guide and Carrasco to interpret, I set out along the ridge.

We rounded a knoll and suddenly faced tier upon tier of Inca terraces rising like giant stairs. Each one, hundreds of feet long, was banked with massive stone walls up to ten feet high.

What settlement of Incas had needed a hundred such terraces in this lofty wilderness? Enough food could be grown here to feed a city!

In my excitement I forgot my fatigue and hurried the length of a wide terrace toward the tangle of jungle beyond it. I plunged into damp undergrowth, then stopped, heart thumping. A mossy wall loomed before me, half hidden in trees. Huge stone blocks seemed glued together, but without mortar—the finest Inca construction. I traced the wall and found it to be part of a house. Beyond it stood another, and beyond that more again.

Under a ledge appeared a cave, its walls lined with niches of cut stone—a royal mausoleum. Above rose a semicircular building with sloping outer wall like that of Cusco's Temple of the Sun. Its straight courses of masonry diminished in size toward the top. The outer side of each stone was rounded to form the graceful curve of the building. No pin could penetrate between these blocks. It was the work of master craftsmen.

Stone steps led to a plaza where white granite temples stood against the sky. Here high priests in resplendent trappings had carried out rituals to the sun god. A gabled compound of beautifully built dwellings nearby must have sheltered the Inca himself. I could picture it floored with vicuña rugs and soft textiles woven by his Chosen Women—those "Virgins of the Sun" who figured so prominently in Inca religious ceremonies.

Down the slope, buildings crowded together in a bewildering array of terraced levels linked by at least 100 stairways. This beautifully preserved sanctuary had obviously never felt the tramp of a conquistador's boot. I realized that Machu Picchu might prove to be the largest and most important ruin discovered in South America since the Spaniards arrived.

—From *Great Adventures*

Moonlight bathes El Castillo, the largest of nearly a dozen restored structures in the Maya's sacred city Chichén Itzá in Yucatán, where the National Geographic Society and Mexican divers have recovered human bones and some 4,000 artifacts from its Well of Sacrifice.

"Silent are the temples, courts, and colonnades; gone the rulers, priests, and sacrificial victims; gone the artisans and builders; gone those humbler folk whose unremitting toil alone made all this pomp and pageantry possible—back to Mother Earth, enshrouded by the living green of tree and bush and flower.

"But of a moonlight night, standing on the lofty terrace before the palace of the Itzán kings, the silent city at one's feet, the temples and pyramids rising white and spectral above the dark forest, breezes whispering through the trees bring stirring tales of other days, other men, other deeds, and he who would may listen then and hear."—Sylvanus Griswold Morley on Chichén Itzá in the National Geographic Magazine, January 1925.

Bingham was sure he had found the legendary Vilcabamba, the last stronghold of the Inca ruler who, it was believed, had sought refuge there when he fled the Spanish in 1537. Bingham made a sketch map of what he had found, then continued up the Urubamba to Vitos on the Vilcabamba in search of other rumored Inca cities before returning to New Haven, Connecticut, to tell his sponsors what he had found.

On February 23, 1912, after several conversations with Gilbert Grosvenor, Bingham had written asking for a letter from Society President Henry Gannett and another from Grosvenor himself, expressing support and interest in Bingham's work. "The stronger the credentials which you can give me," Bingham wrote, "the better able I shall be to persuade…[the politicians in Lima] that our object is scientific and not commercial." Grosvenor wrote back in mid-March, "We find there is considerable feeling that the work is archaeologic and not sufficiently geographic, and I am afraid that we shall have to give up our plans for the present. I will write you again later. I am most interested in your work, and am doing all I can to put the proposition through."

Bingham was stunned.

March 18, 1912.

My dear Grosvenor:—

Your letter is very disappointing. I planned the work with especial emphasis on the archaeological side because I gathered from my conversation with you that you were particularly interested in that field. You will remember that in my first conversation with you in your office I laid considerable emphasis on the fact that the map making begun by the Yale Peruvian Expedition ought to be completed at once, while the same topographer was available and willing to undertake the work. His work is purely geographical. You remember that you replied that you were not interested in that so much as you were in the ruins, the lost cities, and the bones. Accordingly it was in an effort to meet your interest in the matter that the plans were drawn up with so much emphasis on the archaeological side of things. If you wish to present the matter to your Committee in another way, or if you would like to have me alter my plans so as to devote more time to the work of archaeological [crossed through with *geographical* written above] exploration, including geology, physiography, topography and natural history, I shall be very glad to do so….

[And]…if I can get the money for the archaeological end of things, [I] would like to do exclusive [crossed through with *extensive* written above] excavating near Cuzco, and reconnoissance [sic] excavating in the ruins discovered last time.

If you still think that it will be necessary to abandon any idea of coöperation for the present year, I shall appreciate it if you will let me know as soon as possible.

I am extremely grateful for your interest in our work and I realize that you are doing all you can to put the proposition through. I hope you will be successful.

Faithfully yours,

s/Hiram Bingham

At Gilbert H. Grosvenor's urging, the National Geographic Society finally agreed to support Bingham's return to Machu Picchu and announced in its April 1912 Magazine that it had subscribed $10,000 "to the Peruvian expedition of 1912, to which the friends of Yale University have made an equal grant." The $10,000 to Bingham was the Society's first archeological grant.

That year, Bingham and his party "spent several months at Machu Picchu clearing it from the forest and jungle and making such excavations as were necessary in order to restore it as far as possible to its original appearance...." Bingham's report on the results of that expedition, "In the Wonderland of Peru," took up the entire April 1913 issue of the Magazine; it ran 186 pages, contained eight drawings, two maps, a fold-out panorama of "The Ruins of an Ancient Inca Capital, Machu Picchu," and 234 additional photographs.

Another archeologist with a long relationship with the Society was Dr. Matthew W. Stirling, who, usually accompanied by his wife, Marion, made eight expeditions to southern Mexico between 1938 and 1946. There, over the years, they found in the jungle massive, elaborately carved stone heads—the largest weighed 24 tons—stelae with carved jaguar faces, and jade masterpieces from a civilization identified by Stirling as the Olmec, which pre-dated Columbus' discovery of the "New World" by 2,500 years. On his first expedition to Tres Zapotes in the Mexican state of Veracruz, Stirling had found a Maya monument that dated back to 32 B.C. and uncovered enough of the first huge stone head to take photographs which, when shown to the National Geographic and the Smithsonian Institution, convinced them of the archeological importance of the site. Funds were made available for a second expedition. Stirling returned to Mexico with his wife, Marion. As he later wrote for the August 1939 *Geographic*, "Thus we finally found ourselves at the little port of Alvarado on our way to the Canton of the Tuxtlas, where we were to conduct our investigations."

> The red-tiled roofs of Alvarado receded behind us as our launch crossed the shallow bay and entered the river. As we chugged between mangrove-covered banks, close on our left hand rose the great sand hills which lie behind the seacoast. Across the vast level plain to the west loomed the pale-blue range of the sierras and, towering over all, hung ghostlike the snow-covered cone of Orizaba (Citlaltepetl), principal landmark for Cortez.
>
> Our hearts beat a little faster when first we distinguished to the eastward the hazy volcanic peaks of the Tuxtla mountains, at the base of which lay our goal.
>
> What were we destined to encounter there? One thing was certain— the Colossal Head! Would we find it reposing in lonely grandeur, or would we discover the remains of a great city, worthy of such an ambitious work of art?
>
> What would be revealed by the complete excavation of the head? Was it attached to a body? If so, what would be its position—crouching, seated, or standing?

The Stirlings and their party spent the night at Tlacotalpan, then the next morning reboarded their launch for the trip that took them away from the main course of the Papaloapan River through a maze of winding channels choked with water hyacinths, channels so narrow that at times they could touch either bank with their hands, past solitary vaqueros tending cattle, large iguanas sunning themselves on vine-covered jungle walls, and shrieking parrots coursing and tumbling overhead, until at Boca San Miguel they transferred to mules for the one-and-a-half-hour ride in the dark to their Tres Zapotes camp.

The Stirlings' camp consisted of a main house to lodge the expedition's storeroom, laboratory, and "guest room" for visiting dignitaries, and two smaller structures with thatched roofs. One was the Stirlings', the other for

Opposite: *The starkly beautiful ruins of 1,000-year-old Pueblo Bonito have excited interest ever since archeologist Neil M. Judd in 1921 launched the first of seven Society expeditions there to explore and, later, stabilize and preserve New Mexico's "Beautiful Village." Today the archeological uniqueness of the parched, silent ruins of Pueblo Bonito, a center of the great Anasazi Indian pueblo culture in Chaco Canyon, New Mexico, continues to attract scientists, who ponder the many remaining unanswered questions—chief among them, Why did the Anasazi, as builders of the Southwest's most elaborate native architecture, create such a complex culture in such a harsh land and then simply disappear?*

Overleaf: *The Chacoans—Anasazi who dwelled in Chaco Canyon— were agrarian, hence dependent upon being able to "read" the harvesting and planting seasons to survive. One of the prehistoric Chacoans' most stunning accomplishments was discovered in 1977 by artist Anna Sofaer—the "Sun Dagger" by which they recorded time's passage.*
At the winter solstice, rays of sunlight fall between two huge stone slabs, neatly bracketing the spiral petroglyph on 443-foot Fajada Butte at the south entrance to Chaco Canyon. At the summer solstice a single band of light bisects the center of the spiral. The spring and fall equinoxes were heralded by an additional light that fell on the smaller petroglyph, visible to the left of the larger one.

"AMID DEATH'S SHADES THE GLINT OF GOLD...And the echo of love" opened the caption for these photographs in the May 1984 Geographic article *"The Dead Do Tell Tales at Vesuvius."*

Dr. Sara Bisel, a physical anthropologist, examined the skeleton (above) of this bejeweled forty-five-year-old Roman lady found in Herculaneum entombed by the same violent eruption of Mount Vesuvius that buried neighboring Pompeii. Among the Roman lady's treasures were two gold rings, and (opposite) twin jasper-eyed gold snake bracelets and gold earrings made for pierced ears.

Dr. and Mrs. C. W. Weiant. (Weiant was drawn to archeology while working as a chiropractor in Mexico in the mid-1920s; he had then taken a degree in anthropology at Columbia in New York.)

Because the workers changed daily, Dr. Stirling delegated to Dr. Weiant responsibility for supervision of the laborers' roll call and distribution of their pay slips. The workers would then take their pay slips to the local storekeeper, who paid them off, thereby making it unnecessary for the expedition to risk holding large sums of money in its camp. In addition, Stirling dictated all his reports to his wife, Marion, who then typed them up—on a machine which, of itself, was a source of considerable interest and entertainment to the natives. Stirling had thought his ability "in thus having all such tasks done by others" must have "pretty well impressed" the local citizens with his importance. "One day, however," he wrote, "Valvina [the maid] in a confidential mood turned to Mrs. Stirling. 'How unfortunate,' she said, 'that your husband never learned to read or write.'"

Soon after their arrival at Tres Zapotes, they began "the work of excavating the Colossal Head, which lay not more than 100 yards from our camp.... Many were the stories we heard about the Colossal Head," Dr. Stirling wrote:

> We were told that years before, treasure hunters had found much gold around it. The breast of the figure had been exposed and two alligators were reported carved upon it. In addition, we learned that it was impossible to photograph the head, because it always turned around on the film and only the back would show!
>
> Although we were rather skeptical of some of these details, nevertheless our interest was intense as we slowly cleared the great object of the heavy clay in which it was interred.
>
> With a crew of twenty men the work of exposing the monument did not take long. On the second day our queries were answered. Carved from a single massive block of basalt, the head was a head only, and it rested on a prepared foundation of unworked slabs of stone....
>
> Cleared of the surrounding earth, it presented an awe-inspiring spectacle. Despite its great size the workmanship is delicate and sure, the proportions perfect. Unique in character among aboriginal American sculptures, it is remarkable for its realistic treatment....
>
> Fully exposed to view for the first time in modern times, it still remains as great a mystery as ever, for it fits into no known aboriginal American cultural picture. Approximately 6 feet high and 18 feet in circumference, it weighs over 10 tons.
>
> How was this great block of stone transported more than ten miles from its source near the base of Mount Tuxtla? This problem, which would tax the ingenuity of an engineer with the benefit of modern machinery, included crossing the 30-foot-deep gorge of the arroyo. The ancient engineers, however, performed the feat of successfully quarrying a flawless block of basalt and transporting it in perfect condition without the aid of the wheel or domestic animals.

The Stirlings' relationship with the National Geographic spanned more than thirty years and took them, after Mexico, to sites in the jungles of Panama and Ecuador.

More recent Society-sponsored archeological digs have taken place at Aphrodisias, the great ruined Greco-Roman city in an isolated valley in southwestern Turkey, and at the former Roman seaside resort Herculaneum on the Mediterranean in Vesuvius' shadow. In addition to much in the way of

ruins that these sites have in common, they both contain those elements of romance and tragedy that embody archeology's allure.

Aphrodisias was already a thriving center during the Bronze Age and continued to be one through Greek, Roman, and Byzantine times until it was shattered by earthquakes and succumbed to the raids of twelfth- and thirteenth-century invaders. At its height of glory in the second century A.D., the city was both renowned as a sanctuary dedicated to Aphrodite and as the home of sculptors whose works were acclaimed throughout the Roman empire. Today it is the site of incomparable archeological finds. "Imagine coming upon a city of antiquity so rich in archeological treasures that choice sculptures roll out of the sides of ditches, tumble from old walls, and lie jam-packed amid colonnaded ruins," wrote Kenan T. Erim in the August 1967 *Geographic.* Erim, a classics professor at New York University, has spent more than twenty years digging through the ruins of the city once so beautiful that Emperor Augustus proclaimed, "I have selected this one city from all of Asia as my own." While Erim was working at the site, "a portrait head of a woman from about the time of Christ broke loose from a buttress of Byzantine brick and literally rolled to my feet. A wall built of debris from earlier centuries yielded a massive statue of the goddess Aphrodite, for whom this city was named. And in the odeon, or concert hall," Erim wrote:

> we have exhumed scores of elegant statues and relief fragments.
>
> I have come to know Aphrodisias as a place haunted with memories in marble. Voiceless but eloquent, her faces from the past form a colorful company—Apollo, Artemis, and Dionysus; nymphs, satyrs, centaurs, and bacchantes; pugilists, patricians, magistrates, and helmeted legionaries of Imperial Rome.

Mute though these statues be, Erim eloquently describes their impact:

> While excavating Aphrodisias, I have again and again encountered emotions peculiar to archeology, emotions that elude description. For me, such an experience usually starts with a summons to a particular point; the shovels have revealed yet another fragment of marble, another sculpted head.
>
> Once on the spot, I begin to furrow gently around the head with my bare hands. Silent and expectant, my colleagues watch....
>
> My fingers continue to probe around the head. What have we found? A hero, an emperor, some personage who dominated history? Slowly, as I scoop away the earth, the features reveal themselves. Soon the blind marble eyes stare up at the blue Anatolian sky. And, as I brush the last of the dirt away, we see the face of a fellow man whose memory lives still in this bit of stone, a fellow man whose image we have resurrected from centuries of darkness.

Sometimes, however, in cases of such cataclysmic events as occurred at the popular Roman seaside resort at Herculaneum, archeologists find not just the sculptural representations of the community's citizens, but the citizens themselves.

On August 24, A.D. 79, at one o'clock in the afternoon, according to a sequence worked out by volcanologist Haraldur Sigurdsson, Mount Vesuvius bellowed like a huge, wounded beast and vomited a column of pumice and ash twelve miles into the stratosphere, turning day into night over Pompeii to the south. More than six inches of ash and pumice fell on Pompeii each hour. At 11:30 P.M. the heavy column of volcanic smoke collapsed upon itself and the

For twenty years the National Geographic Society has sponsored Professor Kenan T. Erim's excavation of Aphrodisias, the great ruined Greco-Roman city in southwestern Turkey where, as Erim wrote, "choice sculptures roll out of the sides of ditches...."

Above: *Imperial edicts of Roman emperors were found carved in Greek letters on the walls of the Aphrodisias theater. Emperor Hadrian exempted Aphrodisias from a tax on nails; Octavian ordered neighboring Ephesus to return a looted statue to Aphrodisias—the city he selected as his own after he became emperor.*

Opposite: *The head of a second-century A.D. Aphrodite receives a tender cleansing bath with water-soaked cloths that mime the soft folds of her marble hair.*

first fiery avalanche of superheated gases, pumice, and rock surged down Vesuvius' western flank and killed Herculaneum's population; the slower, ground-hugging pyroclastic flow of ash and debris followed immediately after.

An hour later, at thirty minutes past midnight, there was a second surge and flow. At 5:30, the morning of August 25, a third surge reached the north walls of Pompeii to the south and buried Herculaneum. At 6:30 A.M., a fourth surge suffocated those in Pompeii who had not already fled. A half hour later a fifth surge and flow occurred and at 8:30 A.M., after eleven hours of volcanic fury, a sixth and final surge of ash, rocks, pumice, and superheated gases buried Pompeii's victims. By then Herculaneum lay beneath sixty-six feet of ash.

Just over 1,900 years later skeletons of Herculaneum's fleeing population have come to light. In 1984 then senior writer Rick Gore returned to Herculaneum to see them:

> Out of a timeless, musty dark, an ancient Roman victim of Mount Vesuvius stares into the 20th century, her teeth clenched in agony. Nearby lie charred and tangled remains of scores of others buried in the wet volcanic earth....Macabre new relics of [Vesuvius'] eruption were discovered [in 1982] as Italian workmen began to excavate a series of seawall chambers that lined ancient Herculaneum's beachfront. Since then many other fragments of lost lives have emerged along the beach: a noble lady with her jewels; a Roman soldier carrying sword and tools; lanterns, coins, and even an intact Roman boat. These discoveries do more than reveal the moving last moments of a terrified population. They bring to the light of science a wealth of new details that already are telling us much more about how people lived, as well as died, in the lost cities of Vesuvius.

Gore had first visited Herculaneum in 1982 and been led into the first of the unearthed chambers by Dr. Giuseppe Maggi, then director of the archeological dig:

> As my eyes adapted to the dark, a pitiful cluster of skeletons emerged from the wet volcanic ash at my feet. They seemed to have been huddled together. Maggi is convinced they were a household in flight: seven adults, four children, and a baby lying cradled beneath one of the adults. The most striking skeleton lay with head buried, as if sobbing into a pillow.
>
> "In this chamber nature has composed a masterpiece of pathos," Dr. Maggi told me. "One is deeply moved by the postures. You can imagine each person trying to find courage next to another."
>
> If that chamber was one of pathos, the next was a chamber of horrors. A host of tangled, charred skeletons, including that of a horse, lay chaotically strewn. "I think these people descended the stairs terrified," said Maggi. "In panic they tried to take refuge in this chamber."
>
> As I entered, I could almost sense a collective groan across the ages. I could almost hear the screaming as the fiery avalanche struck. It must have been like being trapped in a furnace.

The appeal of archeology is that it provides the tool with which to break through the silence of centuries; with it one assembles knowledge from jigsaw puzzle bits of debris that have survived the millennia. From Hiram Bingham's 1915 "The Story of Machu Picchu" to Peter Schledermann's 1981 "Eskimo and Viking Finds in the High Arctic" to the most recent stories, Society-supported archeological expeditions have appeared regularly in the Magazine. These reports—part detective story, part adventure, part scholarship, part luck, part search for "the ruins, the lost cities, and the bones"—reflect our need to understand our collective history: ourselves.

A Love Affair with Automobiles, Airplanes, and Autochromes

Preceding spread: *"TEN OF A KIND TAKING THE TWIN PEAKS' GRADE ON HIGH AT SAN FRANCISCO" began the* National Geographic's *caption when this Chandler Motor Car Co. press release photograph made its way into the October 1923 issue on the automobile industry. "A San Francisco distributer decided to show the world what his cars could do on heartbreaking hills," the caption went on to explain. "Ten owners, one a woman, came to the scratch at the foot of the hill and not a gear was shifted after the start. The power of the American-built motor represents an outstanding engineering achievement."*

B y January 1920, the 18th Amendment to the Constitution of the United States was in effect. But Prohibition had no impact on the Magazine's income, since advertisements for liquor, beer, and wine had always been forbidden in its pages.

America in the 1920s was poised between the naïve idealism of youth and a post-war world-weariness. The Senate's vote against our country joining the League of Nations was symptomatic of the nation's eagerness to disengage from Europe's traumas.

"America's present need," presidential candidate Warren G. Harding said in 1920, "is not heroics but healing; not nostrums but normalcy; not revolution but restoration...not surgery but serenity." Harding spoke for the middle-aged; but when F. Scott Fitzgerald wrote, "The uncertainties of 1919 were over. America was going on the greatest, gaudiest spree in history," he was speaking for the Jazz Age young, who didn't want problems; they wanted excitement.

Despite a business slump at the beginning of the decade—in 1921, for the first time since Gilbert H. Grosvenor's employment, membership in the Society actually declined*—what ensued was the most spectacular economic boom this country has ever seen; that boom was followed, however, by an equally spectacular bust.

Between the latter part of October and the middle of November 1929, stocks declined more than $30 billion in paper value (over 40 percent of their total value in previous trading) and on October 29, 1929, "Black Tuesday," stock values plunged $14 billion during that one day alone. Still, while the party lasted, the '20s were great fun.

Bobbed-haired flappers "charlestoned" in short skirts, cloche hats, silk stockings rolled down below their knees; baggy-trousered young men with hip flasks in their pockets strummed ukuleles and serenaded their ladies with tunes like "Barney Google" (with his goo-goo-googley eyes), "Baby Face" ("you've got the cutest little baby face"), "Runnin' Wild" ("lost control/ Runnin' wild, mighty bold"), and ("Your lips tell me no, no, but..") "There's Yes Yes in Your Eyes."

Great sports writers like Grantland Rice, Paul Gallico, Damon Runyon, and Ring Lardner wrote about great sports heroes like Babe Ruth, Lou Gehrig,

* *Circulation fell off 65,867, to 647,341.*

162

Bill Tilden, Johnny Weissmuller, Jack Dempsey, Gene Tunney; they immortalized Knute Rockne's "Four Horsemen" and horse racing's Man O'War. The '20s were not all glitter, however; it was also the decade of the St. Valentine's Day Massacre, the divorce of fifty-one-year-old millionaire Edward ("Daddy") Browning and Frances ("Peaches") Browning, his fifteen-year-old schoolgirl bride, Leopold and Loeb's "thrill murder" of fourteen-year-old Bobby Franks, of Floyd Collins trapped in a Kentucky cavern, and Rudolph Valentino's death.

But more than anything else, the '20s were the era of America's love affair with the automobile. As one writer has observed:

> In the '20s the automobile created the greatest revolution in American life ever caused by a single device....It decentralized cities and created huge, sprawling suburbs, took families off for Sunday outings and decreased church attendance. It gave hard-working Americans an escape to fun and new sights, reasserted their independence and saddled them with debt (by 1925, three of every four cars were bought on the installment plan).
>
> —From *This Fabulous Century,* 1920–1930

And, inevitably, this love affair was lavishly and exuberantly reflected in the pages of the *National Geographic:* In October 1923, the Magazine devoted 79 pages and 76 illustrations to staff writer William Joseph Showalter's "The Automobile Industry: An American Art That Has Revolutionized Methods in Manufacturing and Transformed Transportation."

Not surprisingly, *Geographic* contributors tried to outdo each other by taking their cars where none had gone before—and did so, as in Melvin A. Hall's "By Motor Through the East Coast and Batak Highlands of Sumatra"—with predictably exotic results:

> After an hour of unavailing labor, Joseph and I abandoned the effort to extricate the machine, and as darkness was rapidly falling we...decided to desert the car and attempt to flounder through the mud to the nearest native village. It was a desperate decision, but the only alternative was a night in the car....
>
> About a mile beyond where the car was entombed we came to...half a dozen natives....I was surprised at finding human beings there, and, feeling consequent misgivings over the security of our abandoned car and luggage, I asked the man in charge if he or one of his men would, for a suitable consideration, spend the night in an automobile...to guard it from being molested during my absence. To my astonishment he promptly refused, and, asking the question in turn of his men, met with immediate negatives.
>
> I could not account for their unwillingness. The cushions of the tonneau would surely afford as comfortable quarters as any they were accustomed to; it could not be the storm of which men of the highlands were afraid; and the reward I had offered, though small enough, was probably equivalent to about a week's income.
>
> Then it occurred to me that they were afraid of the automobile itself, and I hastened to assure them that it was not only dry and comfortable, but quite safe; that I had locked it up, and that it could not move until I myself released it.
>
> "Oh, it is not that," said the spokesman, with an air of having slept in automobiles most of his life.
>
> "Well, what is it then?" I was both curious and a trifle annoyed.
>
> "Tigers."

A car pauses for the photograph "MOTORING THROUGH THE FAMOUS WAWONA TUNNEL TREE, MARIPOSA GROVE, CALIFORNIA," from the October 1923 Magazine. The tree survives to this day; the road does not (it has been closed).

"Tigers?"

"Yes, indeed," said Joseph nervously, translating. "He say plenty of tigers here come down sure and eat him up!"

"But not in the automobile," I objected.

"Oh, no; tiger first take him out."

"By Motor Through the East Coast and Batak Highlands of Sumatra" was followed in January 1924 by "The Conquest of the Sahara by Automobile" and its June 1926 Citroën Central African Expedition sequel: "Through the Deserts and Jungles of Africa by Motor: Caterpillar Cars Make 15,000-Mile Trip from Algeria to Madagascar in Nine Months" (caterpillar-tracked ten-horsepower forerunners of the vehicles used for the marvelous Citroën-Haardt Trans-Asiatic Expedition of the next decade).

In February 1925, the Magazine published Felix Shay's "Cairo to Cape Town, Overland: An Adventurous Journey of 135 Days, Made by an American Man and His Wife, Through the Length of the African Continent." Shay's piece contained such wonderful lines as "With evening came relief. A gentle breeze blew from the Nile and we sat on the earth terrace in front of the hotel from dinner until midnight, drinking lemon squashes and whiskeys-and-sodas" and "Each night we heard the sob and boom of the tom-toms lasting into the dawn."

Six months after Shay's journey from Cairo to Cape Town appeared, the Magazine ran Major F. A.C. Forbes-Leith's "From England to India by Automobile: An 8,527-mile Trip Through Ten Countries, from London to Quetta, Requires Five and a Half Months."

By 1929 Geographic readers tiring of automobile adventures could cross Madagascar "by Boat, Auto, Railroad, and Filanzana," or travel from Buenos Aires to Washington by horse—"A Solitary Journey of Two and a Half Years, Through Eleven American Republics, Covers 9,600 Miles of Mountain and Plain, Desert and Jungle."

There was even an "African Queen"-like report by Frank J. Magee, R.N.V.R., on "Transporting a Navy Through the Jungles of Africa in War Time," which Gilbert Grosvenor introduced, saying, "No single achievement during the World War was distinguished by more bizarre features than the successfully executed undertaking of 28 daring men who transported a 'ready-made' navy overland through the wilds of Africa to destroy an enemy flotilla in control of Lake Tanganyika."

But if during the '20s the Geographic was having a love affair with the automobile, it was a mere passing fancy compared to the ardor with which its editors embraced the airplane! During that decade thirty different aviation articles appeared in the Magazine. Brigadier General William (Billy) Mitchell celebrated the future of the airplane and airship in the Magazine's issue of March 1921; Richard Byrd reported his flight with Floyd Bennett across the North Pole in September 1926, the South Pole in August 1930, and across the Atlantic in September 1927. Byrd's transatlantic achievement, accomplished with a four-man crew in his trimotor Fokker America, had been overshadowed, however, by Charles Lindbergh's historic solo crossing in his single-engine Ryan monoplane Spirit of St. Louis just thirty-three days before.

The story of the flight of Sir Ross Smith, his brother Keith, and two crew members from London to Australia appeared in March 1921. "The aëroplane

is the nearest thing to animate life that man has created," Ross Smith declared. An account of Lindbergh's flight in the *Spirit of St. Louis* from Washington, D.C., on a tour of thirteen Latin American countries and back appeared in May 1928.

The July 1924 issue was entirely devoted to aviation and featured "America from the Air," containing aerial photographs of this country (no such views had been printed before), and "The Non-Stop Flight Across America," U. S. Army Air Service Lt. John A. Macready's personal account of the successful completion of the first non-stop coast-to-coast flight across the United States he and Lt. Oakley G. Kelly had made in a large, single-engine, high-winged Fokker *T-2*.

Kelly piloted the plane on takeoff from Mitchel Field on Long Island and Macready landed it at Rockwell Field in San Diego. During the slightly less than 27-hour flight through blinding rainstorms, the darkness of night, and narrow mountain passes, the pilots exchanged places five times—a difficult and hazardous maneuver requiring one man to fly the plane from the near-

"No single achievement during the World War was distinguished by more bizarre features," wrote Gilbert H. Grosvenor in the October 1922 Geographic of this successful effort by twenty-eight men of the British Navy to carry "overland through the wilds of Africa" a small ready-made flotilla.

blind rear controls within the fuselage (there was visibility only to the sides) while the other abandoned the open cockpit:

> On the completion of six hours at the controls in the front cockpit the pilot would signal energetically, by shaking the wheel, for the pilot in the rear to take the controls, and when satisfied that everything was functioning satisfactorily, would open the small door to his rear, pull out the back of the pilot's seat, and drop it on the floor through this hole, together with the parachute and cushion.
>
> Lifting up one side of his hinged seat, he would crawl through this small door and back through the narrow passageway paralleling the gas tank to the rear pilot. By yelling in a very loud voice, the pilots could converse in the rear....
>
> After placing the plane in a safe flying attitude, the change at the wheel was made by the active pilot quickly stepping out and forward and the new pilot sliding in from the rear.
>
> Crawling up over the wires through the tunnel and into the front seat, the pilot on duty took the controls and flew the plane, the other pilot placing the parachute cushion and seat back in position.

Royal Italian Air Force Commander Francesco de Pinedo covered 60,000 miles and six continents by seaplane and, in September 1928, jubilantly wrote, "Sindbad, tied to a roc's foot, flew over no stranger sights than I."

And where airplanes went, so did cameras: for the October 1927 issue, American aviators would "Hurdle the High Andes, Brave Brazilian Jungles, and Follow Smoking Volcanoes to Map New Sky Paths Around South America" in "How Latin America Looks from the Air."

How things looked from the air captured the imagination of Gilbert H. Grosvenor and suddenly every other picture caption seemed to be of one place after another "seen from the air," or "as it looks from the air," or "looking down on," or "viewed from above," or "an airplane view of...." These aerial photographs, of course, were in black and white; but the decade of the 1920s saw the *Geographic*'s picture-minded Editor turn increasingly to color.

As John Oliver La Gorce had stated in a 1915 Society promotional pamphlet designed to attract advertising:

> When, in the judgement of the editor, it is impracticable to create the real atmosphere of a far-away country in black and white, then it is produced in color or photogravure at great expense, so that the readers, young and old, may readily understand the actual conditions without drawing unduly on their imaginations.

Two new employees, Charles Martin and Franklin L. Fisher, had been hired by the Society in 1915 to supervise the Magazine's increasingly sophisticated use of photography. Martin had been put in charge of the Society's new darkrooms and laboratory and Fisher was made chief of the Illustrations Department. By 1920, Gilbert Grosvenor was confident enough in the future of color photography to enlarge the darkrooms and install the first color laboratory in American publishing; but because few magazines in the early 1920s printed color photographs, the vast majority of professional photographers did not find it worth the expense and time it took to learn the process. Therefore the few photographers with any experience in making Lumière Autochromes were eagerly sought out by the *Geographic*. And Editor Grosvenor, aware that the key to his Magazine's progress was the continuing improvement of photo-

Above: *"MEMBERS OF THE NATIONAL GEOGRAPHIC SOCIETY ARRIVING TO RECEIVE TICKETS FOR THE LINDBERGH RECEPTION." Long before the announced hour, members gathered before the Society's headquarters, Hubbard Hall, to obtain tickets. There were requests for 30,000, but only 6,000 members could be accommodated at the Washington Auditorium ceremony, where, on November 14, 1927, Charles Lindbergh was awarded the Hubbard Medal by President Calvin Coolidge. "Disappointment of many was regrettable, but inevitable," the Geographic noted.*

Opposite: *Crowds eagerly mob Lindbergh and his* Spirit of St. Louis *at the Croyden Aërodrome, England, June 6, 1927, more than two weeks after his thrilling New York-to-Paris nonstop solo flight. He had flown to England from Brussels, Belgium.*

graphic quality, was constantly encouraging his small photographic staff to try even harder. As he wrote to Maynard Owen Williams in France in 1926:

> The art of taking photographs in color requires the technique of an engineer, the artistic ability of a great painter, and the news interest of a daily-newspaper photographer, so if you do not strike 100 with every photographic attempt in colors, I hope you will not be discouraged. I am much pleased with the increasing quality shown by all your pictures.

Meanwhile Illustrations Editor Franklin Fisher, who had been combing the United States and Europe for photographers able to provide color plates for the Magazine, was eventually able to assemble a small group of freelancers drawn primarily from the artistic ranks and independently wealthy hobbyists.

The post-war era of color photography commenced in March 1921 with Helen Messinger Murdoch's eight Autochromes of India and Ceylon that accompanied Sir Ross Smith's "From London to Australia by Aëroplane"; her work never appeared in the *Geographic* again.

Two whose work did reappear were Fred Payne Clatworthy, who specialized in photographing the American West, and Franklin Price Knott, who had been a painter of miniatures until he succumbed, in his words, to "the magic of autochromes."

In his March 1928 "Artist Adventures on the Island of Bali" Knott wrote:

> By color photography, millions who read this magazine may glimpse the glories of Nature—God's own great studio. Like an artist's brush, now the camera catches every tint and shade from Arizona desert or Alpine sunset to the gorgeous panoply of Indian rajah courts and the bronze beauty of jungle maids asplash in lotus pools.

Knott spoke for the frustrations of all that period's color photographers with their heavy, cumbersome, forbidding-looking equipment when, referring to his thwarted attempts to capture those shy bronze maids asplash in Bali, he wrote:

> There was one girl in particular on that island, a veritable Venus—straight, slim-limbed, graceful as a deer. Often I got fleeting glimpses of her. Sometimes she would be bathing with other girls in a wayside pool, or walking sedately to the temple with an offering of fruit or flowers. Time and again I sought to take her picture, but caught only the flash of her dazzling smile as she flitted swiftly away.

In order to achieve better coverage of Europe, Franklin Fisher contracted with European photographers Gervais Courtellemont in France, Hans Hildenbrand and Wilhelm Tobien in Germany, Gustav Heurlin in Sweden, and Luigi Pellerano in Italy. As a rule these men would process their Autochromes in the field and send the fragile plates packed in padded crates by steamer to New York. Some of them never saw the National Geographic's Washington offices throughout their careers. These photographers, working independently of each other, produced, as Priit Vesilind noted in his thesis on color photography at the Geographic, perhaps the only complete photographic record of Europe prior to World War Two.

Additional personnel had by now been added to the Society staff. Maynard Owen Williams (1919), Edwin L. Wisherd (1919), Clifton Adams (1922), and W. Robert Moore (1927) had joined Laboratory Chief Charles Martin in Washington and, like their European counterparts, were systematically assembling a color photographic record of pre-war America. Of the 1,818 Autochromes that

Above: *Justification for publishing M. Stéphane Passet's shocking May 1922 Autochrome "A MONGOLIAN WOMAN CONDEMNED TO DIE OF STARVATION" was that it, too, provided "true reflection of the customs of the time."*

Opposite: *Ever since Gilbert H. Grosvenor's 1903 decision to publish the photograph of two half-naked Filipino women working in a field, bare-breasted native women have been a National Geographic staple.*

Franklin Price Knott's Autochromes of Balinese maidens for the March 1928 issue were captioned (opposite, top) "EVEN THE CAMERA COULDN'T COAX A SMILE FROM THIS RESERVED AND DIGNIFIED DAUGHTER OF THE EAST" and (opposite, bottom), with perhaps an odd attempt at ethnographic detachment, "BALI HAS NO BLONDES."

Maynard Owen Williams could not ask the Orthodox Patriarch of Jerusalem to leave his palace to pose in the hot sun in his priceless jewels and rich brocades. And because of the Autochrome film's slow speed, Williams could not photograph inside. So he set up his camera on a sunlit balcony outside the Patriarch's rooms, focused on a stand-in churchman, "and when His Beatitude stood on the same spot," Williams wrote in the December 1927 issue, "I exposed two plates so quickly that he was amazed to learn that the ordeal was over."

appeared in the *National Geographic* between 1921 and 1930, nearly 94 percent were taken by the ten photographers mentioned above.

Martin had come to the *Geographic* following his Army service in the Philippines (Martin's photographs—including the one of the decapitated headhunter, had illustrated Dean C. Worcester's 1913 article on the Philippines) and, although, as Vesilind wrote, Martin "was a hard man to please, ... he initiated the high standards of excellence that have characterized Geographic laboratory operations ever since."

It was Martin who, in collaboration with ichthyologist Dr. W. H. Longley, a professor of biology at Goucher College, made the world's first underwater color photographs. Taken off the Dry Tortugas in the Florida Keys, the photographs represented "many weeks and months of experimentation...[and] necessitated the development of a special technique...because the ordinary autochrome plate would not register the moving life under water." The eight Autochromes and a description of the problems encountered appeared in the January 1927 issue of the *Geographic* as "The First Autochromes From the Ocean Bottom: Marine Life in Its Natural Habitat Along the Florida Keys Is Successfully Photographed in Colors." That piece read, in part:

> It was necessary to hypersensitize all plates used in shallow depths, so that the under-sea exposures might be reduced to a twentieth of a second.
>
> Owing to the dampness, the excessive heat, and the lack of sufficient power at Dry Tortugas to operate an electric fan properly, the sensitizing had to be undertaken each morning at 5 o'clock (the coolest time of day), to prevent the emulsion on the glass plates from melting.

But because of the loss of sunlight at depths of as little as fifteen feet, even the hypersensitized plates proved not to be fast enough, so Martin constructed a flashlight-powder device which he could trigger from beneath the water when a fish swam into view. It worked, but with painful, stress-provoking results:

> A pound of magnesium powder was used for every charge. The ignition of such an amount of dazzling explosive on a dory piloted by two men, who were forced at the same time to follow the shadowy movements of the diver with his camera far below, was more than human nerves could stand. Especially was this true when it sometimes happened that the men in the boat had to wait for two or three hours, every moment anticipating the blinding and deafening detonation. They could never know at what instant the diver would find his quarry in the desired position with respect to his lens....
>
> On one occasion Dr. Longley was seriously burned and incapacitated for six days by a premature explosion of an ounce of powder. Had it been a full charge, the accident would probably have been fatal or resulted in permanent blindness.

Today's magazine reader, familiar with the brilliance of contemporary color photographic reproduction, may not, when looking upon the Autochromes of the early and mid-1920s, understand what all the fuss was about. Often the Autochromes were dull, rigidly posed, limited in their colors; sometimes they were retouched with a heavy hand. But compared to the black-and-white photographs they supplemented, the early Autochromes had an impact in the 1920s similar to the impact created when color television began to supplant black-and-white sets of the 1950s. Still not satisfied with Autochrome's limitations, the Society continued to experiment.

On August 21, 1926, John Oliver La Gorce dispatched the following overnight telegram to Gilbert Grosvenor, then summering in Baddeck, Nova Scotia: "SENT MARTIN TO DRY TORTUGAS LABRATORY. HE OBTAINED EIGHT AUTOCHROMES OF FISHES, FIFTEEN FEET UNDER WATER, FIRST EVER MADE STOP... CAN GET FOUR COLOR PAGES OF THE EIGHT SUBJECTS SUCCESSFULLY. WOULD YOU OBJECT USING TWELVE COLOR PAGES JAMAICA AND THESE FOUR PAGES UNDERSEA FISH, RUNNING AS SIXTEEN PAGE SIGNATURE? IMPRACTICABLE USE FOUR PAGES ALONE STOP REMARKABLE ACHIEVEMENT. SHOULD MAKE BIG HIT. REGARDS. ALL WELL."

Representing "many weeks and months of experimentation," the world's first successful underwater color photographs, including the hog-fish (left, bottom), were a sensation when they appeared in the Magazine's January 1927 issue. But Charles Martin's collaborator, ichthyologist Dr. W. H. Longley, was seriously burned when an ounce of the magnesium flash powder they had been mounting in one-pound quantities on a float (left, top) to provide additional underwater illumination, prematurely exploded.

The Geographic briefly tried the Paget color method. In 1919 Frank F. Jones had been outfitted with Paget plates to record one of the Society's six "Valley of the Ten Thousand Smokes" expeditions into the Mount Katmai region of Alaska. Expedition leader Robert F. Griggs wrote in the September 1921 Magazine of some of the problems photographer Jones encountered:

> ...the obstacles to successful color photography, which are difficult to overcome at best, become greatly intensified in such a region as the Ten Thousand Smokes. The plates are sensitive to the adverse climatic influences, and must be guarded from the hot, damp ground with the most jealous care. The dust clouds which are frequently stirred up by the wind are so all pervasive that it is extremely difficult to keep things clean, and dust is much more serious in color photography than in ordinary black and white work, for, while films can be changed just before exposure, plates must be loaded beforehand.
>
> On a black and white picture it is easy to touch out a spot, but in the Paget color process, which we used, any imperfections on taking-screen, plate, or viewing-screen must remain a permanent blotch on the picture... in color work the number of mechanical factors is greatly increased, and the demands of artistic conception by the operator are greatly increased. It is, therefore, a rare man who can do such work successfully in the rough-and-ready conditions under which we were forced to live...but the results ...reflect the greatest credit on Mr. Jones for his careful patience and his artist's vision.

In 1924 Gilbert Grosvenor's son Melville Bell Grosvenor joined the Society as Franklin Fisher's administrative assistant. Melville had completed his active duty tour in the Navy following his graduation from the United States Naval Academy. "My father believed in color and kept pouring it in, and increasing it," Melville later told Priit Vesilind. "All the staff thought it was too extravagant, overdoing it. But people liked color. They wanted things to look natural, and we tried every method as soon as we could find them—adopted them, tried them out."

In an attempt to improve its color printing and reproduction capabilities, the Society sent members of its growing illustrations staff abroad to study what was being learned in England, France, and Germany, and later to Eastman Kodak in Rochester, New York.

Beginning in September 1927, color was included in every issue of the Magazine—all of it was Lumière Autochrome and all of it often required extraordinary effort to make. Fred Payne Clatworthy, one of Illustrations Editor Franklin Fisher's stable of American Autochrome photographers, specialized in scenic photographs of the West. In the text accompanying his June 1928 piece, "Photographing the Marvels of the West in Colors," Clatworthy wrote, "Few readers of the many who now, as a matter of course, look for the natural-color pictures in the *National Geographic Magazine*, realize the great difficulties experienced in securing the originals." Clatworthy knew of what he spoke: on assignment in Death Valley to photograph desert flowers, he almost lost his life in the sandstorm that suffocated the man traveling behind him.

If the weight of hauling the cameras around, the problems keeping the film plates clean, and the sheer inaccessibility of the sites to which he was being sent weren't problems enough for the *Geographic* photographer, he also had to contend with a notoriously tight-fisted Illustrations Editor. For his first series of sixteen Autochromes used in his April 1923 article, "Western Views in the Land of the Best," Clatworthy had been paid the grand total of $160.

Opposite: "GREEN AS GOSLINGS NOW, BUT PRACTICE MAKES THE GOOSE STEP PERFECT" was the Geographic's coy caption for the photograph at top; beneath was "MEMBERS OF THE NATIONAL SOCIALIST PARTY PARADE AT NUREMBERG." Both accompanied Lincoln Eyre's December 1928 article "Renascent Germany."

Not quite eight years later, on August 5, 1936, Associate Editor John Oliver La Gorce assigned free-lancer Douglas Chandler an article on "the characteristic life and business of the great Berlin metropolis." By then the Nazi party had been in power for four years. "It is realized that present conditions in Germany are decidedly controversial and it is difficult to write about it without impinging upon politics and religion" La Gorce wrote Chandler, "but we leave that for other publications to cover."

The inherent flaw in Grosvenor's editorial guiding principles—to publish only what was of an agreeable nature, and to avoid the critical, the partisan, or the controversial—was never more evident. Chandler's February 1937 article "Changing Berlin," followed La Gorce's dictates, and entirely ignored the brutal, anti-Semitic nature of the Hitler regime.

Over the years, the Geographic's reporting has improved as succeeding editors have replaced the editorial standards of a Victorian gentleman's drawing room with those of the news room.

"That was in '26. Now here I am again in the spring of '29."

"I have felt for several years that in your desire to save The Society money you have been too earnest to purchase your photographic material at the minimum amount that the contributor will accept," Gilbert H. Grosvenor scolded Fisher in an October 27, 1928, memo. Grosvenor's long, alternately congratulatory and chiding letter continued:

> We have discussed this matter repeatedly, and in each case where you have, at my request, increased the honorarium you originally set, whether it was Clatworthy or Courtellemont or Sakamoto, immediate improvement has been obtained in the quality and number of the photographs submitted. By this increased expenditure, the Magazine has derived material far exceeding in value the amount of the advance in cost which I authorized.

Grosvenor reminded Fisher of the original low payment for Clatworthy's Autochromes, pointing out that the price came out to "$10 per page," and that the photographer "was urged to submit another series." Grosvenor's letter continued:

> Several years passed and [Clatworthy] submitted nothing. Finally he called one day and I inquired why he had not sent in any photographs, and he replied that the amount of the honorarium you offered him would not pay his expenses. I therefore made him an offer of $1,000 for a series of 16 full-page negatives satisfactory to us, and within a few months he submitted a remarkable series that we printed in our June 1928 number, entitled "Photographing the West in Colors."
>
> This series of Clatworthy cost us $62.50 per original picture. The sum paid was enough to enable Clatworthy to recoup his actual expenses in making the pictures, but it was not sufficient to enable him to make a financial profit. I consider, however, that it was a fair price, as the great reputation he obtained from this series increases the popular demand for his lectures and the fees he obtains from these lectures.
>
> From The Society's point of view, an expenditure of $62.50 per page, which expense is distributed through an edition of 1,200,000 copies, is almost negligible. So this was an excellent investment for the National Geographic Society.
>
> In the purchase of photographs and other material for the Magazine, we must not lose sight of the fact that advertisers consider space in the Magazine of sufficient value to pay $2,500 for one page, one insertion, and they pay many dollars additional for preparation of their copy. Space so valuable to advertisers is even more valuable to the National Geographic Society.
>
> Common business reasoning, therefore, indicates that the National Geographic Magazine can pay several times $62.50 for a page illustration, as every page of our Magazine is an advertisement of the National Geographic Society's product. An attempt to save a few dollars on the purchase price of pictures is a suicidal policy, as this policy prevents us from obtaining material that we must have, if the National Geographic Society is to exist.

From "Human Emotion Recorded by Photography" (October 1920) to "Through Java in Pursuit of Color" (September 1929) articles celebrating both photography and the photographers appeared in the *Geographic* of that decade. And so, of course, did articles on birds, from "The Crow, Bird Citizen of Every Land: A Feathered Rogue Who Has Many Fascinating Traits and Many Admirable Qualities Despite His Marauding Propensities" (April 1920) to "The Eagle in Action: An Intimate Study of the Eyrie Life of America's National Bird" (May 1929).

There were articles on "The Hairnet Industry in North China" (September 1923); an American woman's trip by houseboat in China—marvelously titled "Ho for the Soochow Ho" (June 1927); and "Stalking the Dragon Lizard on the Island of Komodo" (August 1927); followed by "Mickey the Beaver" (December 1928).

In September 1921 the Magazine published "Our Greatest National Monument: The National Geographic Society Completes Its Explorations in the Valley of Ten Thousand Smokes." Archeologist Neil M. Judd's articles on the riddle of New Mexico's Pueblo Bonito prehistoric ruin ran in 1922; and explorer-botanist Joseph F. Rock's series of adventures in China and Tibet began in 1924.

In 1922, the Society expanded its publishing efforts when it printed *The Valley of Ten Thousand Smokes* in book form as well. In his foreword to the book, Gilbert Grosvenor reemphasized the role of the membership, with his explanation that the expeditions had been:

> ...financed by the 750,000 members of the National Geographic Society, each of whom, millionaire and college professor, captain of industry and clerk, had an equal share in its support. Every member of the organization may thus derive considerable satisfaction that he or she has assisted to bring about such important additions to our knowledge of the young and active planet upon which we live.

Unfortunately, volcanoes were not the only forces active on our young planet in the 1920s and as that decade drew to a close, no article quite so captured the spirit and the innocence of the *Geographic* of this period as did Lincoln Eyre's December 1928 article, "Renascent Germany." Scores of brown-shirted, swastika armband-wearing, "*Sieg Heil!*"-saluting marchers in short pants appeared in a photograph captioned "Members of the National Socialist Party Parade at Nuremberg." And that illustration was followed by one showing little schoolboys in dark uniforms marching in a street; it is captioned "Green as Goslings Now, but Practice Makes the Goose Step Perfect."

The next year Hitler appointed Heinrich Himmler Reichsfuhrer S.S.; Trotsky was expelled from the Soviet Union; Warren Harding's Secretary of the Interior was convicted of accepting a $100,000 bribe in the Teapot Dome scandal; and Herbert C. Hoover took office as the thirty-first President of the United States. Also in 1929, William Faulkner published *The Sound and the Fury*, Hemingway *A Farewell to Arms*, Thomas Wolfe *Look Homeward, Angel*, and Erich Maria Remarque *All Quiet on the Western Front*. Marc Chagall painted *The Milkmaid*; Disney released *Steamboat Willie*, his first musical Mickey Mouse cartoon; and "talkies" killed off silent films. The German *Graf Zeppelin* airship flew 19,500 miles around the world in 20 days, 4 hours, and 14 minutes (and its commander, Dr. Hugo Eckener, received a special medal of honor from the Society for the feat). Kodak introduced 16 mm color film; the U.S. stock exchange collapsed; the world economic crisis began; and the 5th Earl of Rosebery died a happy man at the age of eighty-two after having gained his three lifetime wishes: he had married the richest heiress in England, won the Derby (three times), and become Prime Minister.

As the decade ended, membership in the National Geographic Society reached 1,212,173. The following year, 1930, 34,810 new members would join, bringing the total membership to 1,246,983. It would not be until 1946 that the Society would again meet and then surpass that membership figure.

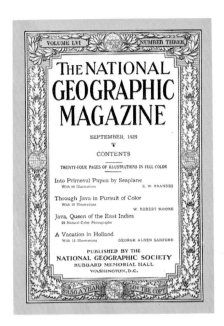

By the end of 1930, membership in the Society had reached 1,246,983. Not until World War Two had ended would membership reach that level again.

"Something lost behind the Ranges. Lost and waiting for you. Go!"

*"Till a voice, as bad as Conscience, rang interminable changes
    On one everlasting Whisper day and night repeated—so:
'Something hidden. Go and find it. Go and look behind the
    Ranges—
'Something lost behind the Ranges. Lost and waiting for you.
    Go!'"*
—From Rudyard Kipling's "The Explorer" (1898)

*"...a task, any task, undertaken in an adventurous spirit
acquires the merit of romance."*
—Joseph Conrad, *A Personal Record* (1912)

Above: *Maynard Owen Williams, shown with an Expedition half-track in the Chinese province of Sinkiang, headed the* National Geographic's *foreign staff for more than thirty years.*

Opposite: *The Expedition's " 'Scarab' nearly came to grief,"* Williams wrote *when a collapsed road left the vehicle teetering forty feet above a gorge in the Himalayas.*

Preceding spread: *Troops outside Farah, Afghanistan, plundered by Genghis Khan, salute the 1931–1932 Citroën-Haardt Trans-Asiatic Expedition as it thunders through en route from the Mediterranean to the Yellow Sea.*

Throughout the history of the National Geographic, men and women have heeded what Kipling called "a voice, as bad as Conscience." Some, like Theodore Roosevelt, followed it into the wilds of Africa; others, like Greely, Peary, and Steger, fought their way through Arctic wastes in an attempt to reach the Pole. Charles and Anne Morrow Lindbergh flew 29,000 miles through the worst possible conditions in a small, single-engine seaplane to search for safe commercial airline routes; and Robin Lee Graham, Tim Severin, and Thor Heyerdahl, seeking something within themselves, put to sea in fragile boats. But during the first third of this century, as adventurers and explorers fanned out across this globe, only the moon seemed more forbidding, mysterious, and "lost behind the Ranges" than China. And although seventy China-related articles appeared in the *Geographic* between 1900 and 1935, none quite so captured the imagination of the Magazine's readers as those written by two very different kinds of men—Maynard Owen Williams and Joseph F. Rock.

Williams, a former missionary in China and Syria, was large, gregarious, unaffected, enthusiastic. He joined the *Geographic* in 1919, contributed the text, photographs, or both to ninety Magazine pieces, and was "Mr. Geographic" to a generation of its readers. Among the first journalists to be present at the opening of Tutankhamun's tomb, and to take natural color photographs in the Arctic, Williams was as much at ease in Bulgaria as in Bali. Chief of the *National Geographic*'s foreign staff for more than thirty years, Williams was once asked how he was able to receive the cooperation of difficult people in out-of-the-way places. He responded, "I get around the world by being a nice little lady. I never carry a gun. I tell everyone I am helpless, and that I cannot speak their language."

In 1930, the amiable Maynard Owen Williams was selected to be the Society's sole representative—and the only American—on a motorized expedition across Asia led by France's Georges-Marie Haardt.

In an era of glorious expeditions, there was none more splendid or undertaken in a more adventurous spirit than the 1931–32 Citroën-Haardt Trans-Asiatic Expedition. During the previous decade, Haardt had traversed the Sahara in half-track vehicles designed by André Citroën. "Then," wrote Williams, with Africa behind them, "Asia's vastness challenged the Frenchmen." And on April 4, 1931, with seven new, specially built Citroën seven-ton, half-track vehicles, Haardt set off from Beirut in an attempt to complete an overland trip from the Mediterranean to the Yellow Sea, following in part the itinerary of Marco Polo's journey more than six-and-a-half centuries before.

The expedition's original plan had been "to cross Asia from Beyrouth to Peiping* in one set of tractor motorcars, avoiding the mountain heart of Asia by cutting north of the Hindu Kush [a mountain range] and driving east through Russian Turkistan." But when the Soviets would not permit the expedition to enter, the expedition was divided in half, one set of seven cars, the China Group, assembled at Tientsin on the Yellow Sea and began rolling west. The other seven, the Pamir Group, to which Williams was assigned, started "going east from the Mediterranean...toward a Central Asian rendezvous."

Williams' account of the exhausting, ten-month-long, dust-ridden, rain-soaked, bandit-bedeviled, 7,370-mile engine-grinding motorized trek over drifting sand dunes and blazing deserts, through icy rivers and across snowy Himalayan mountain passes was serialized in three issues: October 1931, March and November 1932.

From Beirut the Pamir Group's half-tracks proceeded to Damascus and then on to Palmyra, where, Williams noted:

> In the lobby of the Hotel Zenobia a phonograph flippantly played "I Kiss Your Hand, Madame." But, behind the thin door of the salon, desert chiefs, disgorged from shiny motor cars, sat together with cigarette and coffee cup, their real feelings hid behind studied, graceful gestures.

The following morning at Palmyra's lonely French Foreign Legion outpost, Williams wrote, "as our cars swung into line and our hands swept to a mute 'Vive la France!' the Legionnaires stiffened to 'Present Arms.' To the sound of a lone bugle, the Tricolor rose for the first time over this unfinished desert fort."

The Pamir Group crossed the hot, dusty Syrian Desert wasteland between Damascus and Baghdad—where, Williams wrote, "bread dries while one holds it in one's hand. Its rough edges are harsh on cracked and bleeding lips"—and reached the welcome air beside the Euphrates, which, "after the rigors of the desert...had a softness almost feminine." Onward the half-tracks rumbled and clanked through Baghdad, Kermanshah to Siahdehan—where their route joined Marco Polo's trail—then "through Nishapur, city of turquoises and Omar Khayyam, we dashed without stopping." Until, in Afghanistan, a swift river blocked their path:

*Since the 1930s, the standard English spelling of place-names in China has changed. Current usage is as follows: Peiping = Peking = Beijing, Kashgar = Kashi, Tientsin = Tianjin, Sinkiang = Xinjiang; some place-names themselves have also changed: Liangchow = Wuwei, Ningsia = Yinchuan, Suchow = Yibin

Camel supply caravan, supporting the Citroën-Haardt Expedition, skirts Bulunkul Lake's waters in Sinkiang. The Expedition depended upon camels, ponies, and yaks once it crossed the mountains into western China, in addition to its half-track vehicles, to transport its supplies.

Getting a seven-ton radio car across the Helmand was a ticklish task, shared by our mechanics and the rivermen...led by a picturesque graybeard called "Baba Daria,' Father of the River. [He] dubbed [our chief mechanic] "Baba Motor."...
"May my spit cover your faces!" screams Baba Daria.
"What does he say?" asks Baba Motor.
"That the truck is heavier than you said," [replies the translator....]

The expedition pushed on through Afghanistan's Kandahar to Kabul. By late June it had surmounted the Khyber Pass and arrived at Srinagar, then Kashmir's summer capital, 3,445 miles and, counting stops, eighty-one days from Beirut. There, "with the Jhelum on a rampage and the Kashmir Valley a lake, the cars...were stranded..." while "the relief cars from Peiping, which were to meet us in Kashgar, were immobilized...a thousand miles away by an air line which no crow could follow and live."

Expedition leader Georges-Marie Haardt was determined to press on "with two of his seven cars until some definite barrier or lack of time should stop the adventure."

The Pamir Group left Srinagar on July 12 in the two half-tracks, "Golden Scarab" and "Silver Crescent," and reached Bandipur to the north that night.

In the morning...the usual chaos of departure, but with the oddest baggage ever seen. Leather-covered *yakdans* (wooden chests)...small coils of cable on which the porters pounce, only to slink away when they have felt their weight; spare wheels with doughnut tires slung sandwichwise astride half-hidden ponies; axles and gear boxes awkwardly swung between four protesting coolies; cinema tripod, carried upright like a young tree; cameras, sleeping sacks, tool boxes, cases of food, green tents...150 pony loads for our group alone....One porter after another disappears....Servants mount the best ponies and escape before they are detected.

Along the expedition's route waited "grades too steep, hairpin turns too sharp, trails too narrow, underpinning and side walls too infirm." The nine mile "zigzag ascent" from Bandipur Bridge to Tragbal took four and a half hours and was considered "a triumph for the cars."

All around were heavily wooded slopes bathed in morning mist. [Then] an ominous rumble, and...this majestic silence was torn by the roar of a motor. ...a gang of coolies strained the [two] cars around a narrow bend, and on they came, impressive in their slow relentlessness....

Then Williams wrote, "without warning, across the Takla Makan Desert and the Himalayas came the bad news: the China Group definitely stopped in Urumchi; Captain Point a prisoner!"

It was later learned the China Group, under Lt. Cmdr. Victor Point, had crossed the sands of the Gobi and entered Suchow. There they heard reports of an uprising in Sinkiang Province. Despite authorities having closed the "Great Route" to them, Point had continued on and reached the edge of Sinkiang on June 26. A message left by the previous caravan warned of danger; Point, himself, saw "the wreckage of war: horses killed, carts overturned, corpses lining the road and in the ditches, soldiers, women, and children huddled together in utter disorder."

About two weeks later, the China Group was in Turfan when the Governor of Sinkiang Province summoned Point and his party to Urumchi, the provincial capital. Point, unwilling to risk his entire group, left his French members behind and went, with a delegation of Chinese, in one car. Although given a

Opposite: *Maynard Owen Williams was able to persuade a crowd assembled in a Herat bazaar warehouse in Afghanistan to hold perfectly still for the three seconds required for this striking 1931 photograph.*

Asked how he was able to obtain the cooperation of difficult people in out-of-the-way places, Maynard Owen Williams responded, "I get around the world by being a nice little lady. I never carry a gun. I tell everyone I am helpless, and that I cannot speak their language."
"The greatest 'kick' a field man can have is to carry a million and a quarter members up onto a high mountain, show them the world and say, 'It's yours, in a way it could not be without me,' " Maynard Owen Williams wrote Editor Gilbert H. Grosvenor in 1929.

Artist Alexandre Iacovleff accompanied the Citroën-Haardt Expedition and made these drawings of (top) a Vietnamese princess in Hue, (above) a Chinese princess in Urumchi, and (opposite) three princes of Inner Mongolia.

"magnificent welcome" upon arrival, Point was barred when he tried to depart.

"Am I a prisoner?" Point asked. There was no answer. Three days later Point was told he could not leave until the other cars and the Frenchmen of his group arrived from Turfan. But because they would not leave Turfan without Point's express command, he was told to order them to come. Point refused and spent a week under confinement before relenting. Before the cars reached Turfan, a telegram arrived from Nanking "ordering the provincial authorities to stop all activities of the Expedition." The Pamir Group was not permitted to enter. Point's China Group was ordered back to Peking. Tensions grew.

Though forbidden use of his radio, Point was determined to warn Haardt. Soldiers surrounded his camp and "watched our every move." But through Point's radio technician and mechanic "masking the sound of our motors and hiding our antenna in a tree, we were able on July 26, after a month of silence, to renew contact with the Legation of France and the Pamir Group."

Separated by hundreds of miles of desert and rugged mountain passes, there was nothing expedition leader Haardt could do for the China Group but press on toward their distant rendezvous; and so, with their heavy tracked vehicles following paths no machine had ever attempted before, they rumbled on toward the 13,755-foot-high Burzil Pass in the Himalayas, forded icy streams, and plowed through snowfields on their climb until, too exhausted to continue, "mechanics slept in their bags on the snow, ready," Williams wrote, "to start the long wallow to the top at dawn."

> The big day was one of rare beauty, with fine clouds and a blazing sun. The snow was melting rapidly and rivulets of water filled every possible path....The motors, even in this rarefied atmosphere, performed splendidly. So did the gangs of coolies who hauled at long ropes to keep the machines from side slip. Heavy iron teeth were added to the tractor treads...but the front wheels wallowed deep and the whole machine had a tendency to slip sideways toward the river. The fight to cross the pass lasted for ten strenuous hours....
>
> The next day the "Scarab" nearly came to grief. When I caught up, it was perched 40 feet above a mountain torrent, shouldering a precipice and with most of its left tractor poised in midair.

The outside wall had collapsed under the "Scarab's" weight. Once beyond "the Great Himalayas" and the Burzil Pass the expedition abandoned its half-tracks at Mishgar on the northwestern slope of the Karakoram Range of what is now Pakistan. There Haardt's Pamir Group mounted sixty camels and eighty horses for the 420-mile journey to Aksu. They reached Kashgar on September 19, and several days later Haardt's caravan continued on the "Great Route," the ancient Silk Route between China and the Mediterranean that was, Williams wrote, "older than idols or money."

On October 24, Point's China Group at last linked up with Haardt's Pamir Group. As Point later wrote to Williams:

> "It was in the Tokosun Gorge, amid the chaos of high-piled rocks, last obstacles that still separated us, that I saw the black and orange pennant of the 'Golden Scarab.' It was 11 o'clock....That was the car of him toward whom, during seven months, we had not ceased to strive. And there he stood before me, opening wide his arms. My mission was over."

Haardt's men and materials were transferred to Point's four remaining half-tracks for the journey to Peking:

From Urumchi to Peiping [Peking] is 2,300 miles. Two of our caravans had been pillaged. The rebel Ma Chung Ying stood astride the Great Road waiting for us, with tons of our supplies already in his hands. Sand dune and river, desert and rocky defile lay across our path. The cold of the Mongolian plateau was often in our thoughts.

The grueling dash to Peking began: "The tents were seldom pitched," Williams wrote. "Valises and even washbasins were ignored for days at a time. Not only actual cold, but the threat of greater always hung over us. Seldom were we free from fatigue."

They were sheltered by Christian missionaries at short stops before setting off again on the long drives:

> Night after night we felt our way ahead over atrocious trails, with a theoretical rest for men and motors from 2 to 5 in the morning. The driver slumped forward over his wheel. His seat-mate, who could doze during the day, watched the temperature dials for water and oil. Whenever a radiator got cold or a bearing stiff, a motor woke with a roar. A hundred fantastic landscapes tossed up by our ranging headlights are now mingled in what some of us feel was reality and others a bad dream.

It was so cold Williams had to suck on his frozen fountain pen to thaw the ink; but he continued taking notes:

> We left Sinkiang at Mingshui Pass with a cold early-morning wind at 6,600 feet....After a thirty-hour run without sleep, we passed through a flood of refugees fleeing from Ma Chung Ying and arrived at the gates of Suchow....Meanwhile the rebels under General Ma Chung Ying were near—and coming nearer.
> A night attempt to use our wireless brought a colonel down on us. But ...our interpreters...were instructed to lose $29 to him at mah jongg, and the flurry passed.
> The following morning permission to leave came....the city gates at last slowly swung open and our cars filed out. Twenty-four hours later Ma Chung Ying's troops entered Suchow.
> By then we were at Kaotai [175 miles away] with a bandit's head—one sample from 27 fresh ones—dangling by a cord beside the city gate. We... pushed on.

With the rebel chief Ma Chung Ying in close pursuit, the cars of the Citroën-Haardt Trans-Asiatic Expedition rumbled toward Liangchow following the deeply rutted tracks beneath China's Great Wall, begun twenty-two centuries before. On January 5, 1932, as the *Dàhán*, or Great Cold, approached, the combined Pamir and China groups were halfway between Urumchi and Peking, "but the harder portion lay ahead." They turned toward Ningsia. And then:

> North of Ningsia, at midnight, the "Silver Crescent" crashed through thick ice and my camera suddenly went afloat in icy water between my porous leather boots. The headlights were well under water, and still lighted, before I could turn off the current and join my companions by way of a scramble over the roof.

As the car sank lower and lower, Williams and a companion

> bridge[d] the watery gap and drag[ged] to safety the heavy trunk of photographic records. Flashlights threw the chaotic scene into wild relief. [One

Above: *The six-year-old Living Buddha of Guya, photographed in 1927 in a Tibetan monastery in a remote corner of north central China by Joseph F. Rock.*

Opposite: *"THE CHIEF DANCER IMPERSONATES THE KING OF HELL" began the caption for this Joseph F. Rock Autochrome published in the November 1928 issue. "With the scepter of death in his right hand," it continued, "the fearsome Yama, the God of the Dead, appears on the steps of the chanting hall and instills fear into the hearts of his Tibetan audience. Later he is joined in the dance by his retinue of demons, the Bawa."*

Above: *This Rock photograph, published in 1925, arrived at the Society captioned, "The Prince of Choni and yours truly. I am living in the Prince's Lamasery in a quiet place with private courtyard. Below are over 500 Lamas who pray from 4 A.M. to 2 P.M. They are ever busy with trumpets, conch shells, bells and cymbals, which they accompany in a deep basso profundo."*

Left: *Toes of the world's largest sitting Buddha. The 231-foot-high Buddha was carved in the eighth century out of a cliff face overlooking the convergence of three rivers at Leshan, China, in the hope that the Buddha's benign influence would protect boatmen from treacherous currents.*

Botanist-explorer Dr. Joseph F. Rock led two National Geographic Society expeditions (1923–24 and 1927–30) into the remote, unmapped Tibetan borderlands of China and Tibet where no white man had ever been before. Bedeviled by bandits, blasted by blizzards, both a loner and lonely, Rock pushed through desolate defiles and across raging rivers and climbed razor-backed ridges to gather thousands of plants and countless animal and bird specimens. Along the way he sent back reports and photographs that, when published in the National Geographic, gave the world its first look at "the land that time forgot."

Among Rock's submissions was this photograph, which appeared in the November 1928 Magazine with the following caption: "TIBETAN PIPES CALL TO THE DANCE. Monks on the roof of a temple summon the austere Choni lamas to prepare for the ancient pantomime Cham-ngyon-wa. From the 15-foot trumpets are issuing weird, long-drawn notes that can be heard far away in the mountains. These same horns have resounded to the breath of thousands of trumpeters whose very memory is lost in antiquity."

man stood] on the radiator, wielded a piston-like crowbar against the thick ice, the car wallowed lower and lower, finally submerging the entire hood, and after more than an hour of feverish struggle in bitter cold, three other tractors dragged our car, like some submarine monster spewing out water, to the opposite bank. Thirteen hours' delay…and all my [stored] clothing frozen into solid blocks….

In addition to the cold the expedition had to cope with "dust [that] filled our nostrils and mouths. With cold-cracked fingers we scooped out gobs of grit from between our lower teeth and lips. Dust crept through our cylinders as if no air filters were there. Pistons developed unusual wear and oil fouled the spark plugs. While a trifling repair was being made, the water froze about the cylinders."

At Lunghingchang, Chinese officers inspected the group's passports; the army was now becoming more and more evident. In every town uniformed soldiers suspiciously pointed their guns at the half-tracks as they passed.

"Just before nightfall, the 'Silver Crescent' came to a narrow bridge" which the other cars had already crossed. Williams, seated beside the driver, suddenly found himself looking into the barrel of a gun held by a handsome young soldier. "At three or four miles an hour," he wrote, "it took a long time to roll past that swinging rifle barrel." Moments later, as they drove between "chest-high mud walls" lined with troops,

> a salvo of shots rang out. Eleven bullets took effect on "Silver Crescent" and its trailer….
>
> Deploying behind a diagonal bank, we brought our [machine gun] into action.
>
> "Fire high or at a wall," ordered Audouin-Dubreuil [Haardt's second-in-command]….
>
> Four rapid shots. Then a wave of silence. Four more shots. Silence. Four more shots. He had fired twelve cartridges in all when a flag went up over the Chinese headquarters. Men from both sides advanced to a conference, from which Point, bareheaded and laughing, returned.
>
> "This is a terrible country. They had the nerve to call it a 'slight misunderstanding.' "
>
> But after having tea with the officers…both sides agreed to call it that.

Soon after midnight the expedition passed into Inner Mongolia. "The next day a 1919 Dodge, laden with 22 men so deeply coated in dust that they looked like dummies, arrived *en route* to Lunghingchang. Before arriving there, three were killed, the rest stripped of everything."

And then, Maynard Owen Williams wrote:

> At high noon, on February 12, 1932, the Citroën-Haardt Trans-Asiatic Expedition swung into the grounds of the French Legation in Peiping, was welcomed by the élite of many nations, and came into well-earned glory. In 314½ days [the expedition] had blazed a 7,370-mile trail across Asia….

◄ ─────────────────────────────────

Charles and Anne Morrow Lindbergh in the cockpits of the Lockheed Sirius in which, in 1933, they visited twenty-one countries. Their epic 29,000-mile flight took them to four continents and through arctic cold, blizzards, sandstorms, hurricanes, and tropical heat.

From Peking, the expedition was to have continued south through Indo-china, but "the death of our great leader, Georges-Marie Haardt, from pneumonia, at Hong Kong, March 16, 1932, halted the Expedition at a time when French Indo-China had arranged a thousand-mile triumph in his honor."

Rather than being disheartened at the tragic and unforeseen halting of the expedition at this point, Williams wrote:

> …what Haardt and his men might have done doesn't matter. They set themselves a hard and hitherto-unaccomplished task and they did it.

And the *Geographic*'s foreign staff chief summed up his experience, writing,

> …the credit for the present success goes to these gallant Frenchmen whose struggles it was my privilege to observe and to share. To live with these men, to see them in action, to help record their triumph, was the greatest adventure of my life.

In *Twelve Against the Gods* author William Bolitho wrote of Columbus, "…his real glory…is that of all adventurers: to have been the tremendous outsider." Bolitho might have been describing Joseph F. Rock, the Vienna-born scientist-explorer who, between 1922 and 1935, wrote ten articles for the *Geographic*, primarily on the little-known borderlands of China and Tibet. The only son of an Austrian manservant, Rock rejected his family's attempts to direct him into the priesthood, choosing instead to know more of the world about him. Rock's extraordinary facility for languages made it possible for him in the course of his life, apart from his native German, to become fluent in Hungarian, Arabic, English, Italian, Spanish, French, Latin, Greek, Tibetan, and various aboriginal dialects of China, in addition to being able to read and understand Hindi, Japanese, and Sanskrit.

Rock first arrived in America in 1905, moved to Hawaii in 1907, taught himself botany, became a naturalized American citizen in 1913, wrote three authoritative books on the Hawaiian Islands flora, and in 1920 left to spend the next three decades exploring and researching in Asia. There he collected thousands of botanical, ornithological, and zoological specimens; explored, photographed, and mapped parts of Asia that few, if any, Westerners had visited before him. Rock began publishing his findings on the geography, folklore, language, religion, and culture of the people of western China and eastern Tibet in the *Geographic* under such titles as "Seeking the Mountains of Mystery: An Expedition on the China-Tibet Frontier to the Unexplored Amnyi Machen Range, One of Whose Peaks Rivals Everest" (February 1930) and "Through the Great River Trenches of Asia: National Geographic Society Explorer Follows the Yangtze, Mekong, and Salwin Through Mighty Gorges, Some of Whose Canyon Walls Tower to a Height of More Than Two Miles" (August 1926).

Typical of Rock's adventures is the one described in this passage from his September 1925 article, "Experiences of a Lone Geographer: An American Agricultural Explorer Makes His Way Through Brigand-infested Central China en Route to the Amne Machin Range, Tibet":

> I was quartered in the center of the village in a miserable old temple full of coffins….Darkness came on. At midnight the officers of the soldiers

came in and announced that the brigands were outside and that the town could not be held against the impending attack. I never spent such a night in all my life.

I opened my trunks and distributed $600 in silver among my men, wrapped up some extra warm underwear, a towel, condensed milk and some chocolate, besides ammunition for my two .45-caliber revolvers.

Fully clad, I sat waiting for the turn of events. Every minute we expected the firing to commence. The soldiers said that they could protect me, but not my boxes, and that the safest move would be to retreat and try to find a hiding place if the brigands rushed the temple.

The natives of the village began burying their few valuables and great excitement ruled....Outside the hamlet heads of brigands that had been captured some days before were hanging from poles.

I was informed that several hundred bandits were surrounding the village and that capture was inevitable.

However, "at dawn there was not a bandit to be seen," Rock wrote. "They had vanished."

Year after year in China, Joseph Rock was harassed and fired upon by bandits, buffeted and blinded by blizzards, thrilled and humbled by majestic mountains no white man had seen before, exhausted and terrified by encounters with primitive rope crossings over rivers raging far below. And year after year Rock pushed on through desolate defiles and saw-toothed ranges to gather thousands of plants (among them 493 kinds of rhododendron) and mammal and bird specimens.

Joseph F. Rock was capable of great sentiment and equally vast loneliness. In one instance, on a Christmas Eve, he mailed from China some dry gentians, a delphinium, and an edelweiss to Gilbert H. Grosvenor's young niece, and, in another instance, he wrote of General Pereire, a friend and China authority, who "died at Kanze on his way to the great snow mountain of Amne Machin. ...He lies buried within the shadow of the Kanze (Szechuan) Lamasserie. Perhaps I myself will rest in a similar place, alone, as I have lived."

He was a man who played Enrico Caruso records to mountain tribesmen, who carried *David Copperfield* in his luggage to remind him of his unhappy youth, who would not—except under the most dire circumstances—permit his servants to serve anything but an Austrian dinner, cooked over a small charcoal brazier to Rock's specific recipes, on fine china with a linen tablecloth and napkins. He bathed daily in an Abercrombie & Fitch folding bathtub, and wore a white shirt, necktie, and jacket to meet—or be met by—even the most insignificant village or tribal chief. Over and over again in the course of Rock's extended expeditions he was thought to have been "lost," only to reappear. His luck was as mixed as his feelings for the Chinese with whom he lived: in 1941 the plates of a four-volume work on the Nakhi people he was in the process of having printed were destroyed in Shanghai when the Japanese bombed that city; twelve years of research were lost when the ship transporting his papers from Calcutta to the United States was torpedoed by the Japanese in the Arabian Sea. And in 1949, in the Lijiang Valley, the heart of the Nakhi country, in which he had lived and studied for more than twenty-seven years, Rock was proclaimed a public enemy by the Communists and expelled from China forever.

Sea voyages in reproductions of ancient ships have become a popular staple of the Geographic. Tim Severin (opposite) in his December 1977 "The Voyage of 'Brendan'" reported having successfully sailed a leather boat from Ireland to Newfoundland in an attempt to duplicate the legendary sixth-century voyage of the Irish monk, St. Brendan—and by so doing proved at least that such a crossing could have been made.

Before Severin there was Thor Heyerdahl, who, in 1947, built and sailed a balsa-log raft, Kon-Tiki, from South America to Polynesia to show that Western Hemisphere people might have contributed to early Polynesian cultures. In 1969 Heyerdahl built Ra I, a papyrus reed boat "along ancient Egyptian lines" and "sailed her from Africa into the Western Atlantic, where," he laconically reported, "I had to abandon her when the reed bundles came apart."

Ra II (overleaf) was more successful; in 1970 Heyerdahl crossed the Atlantic from Morocco to Barbados in his improved papyrus reed boat, proving that ancient mariners could have done the same.

A Revolution in Color Photography

"They were projecting this loop of film over and over again with its brilliant color, and I asked the man if I could examine the film....I put a loupe on it, a magnifying glass, and I saw there were no flecks in it! *Unlike with an Autochrome or a Finlay the white paper and the white shirts had no residual color in them at all! And I wondered,* How in the devil did they get that color?"

—Luis Marden, on his first look in 1935 at Eastman Kodak's new Kodachrome 16mm film.

A s the Great Depression deepened, magazines fought to avoid going under. *Munsey's* merged with *Argosy, Everybody's* with *Romance,* and even *McClure's*—the magazine Gilbert Hovey Grosvenor had been under such pressure from his Board of Managers to emulate twenty-nine years before—was forced to combine with *Smart Set* in an effort to stay alive. But the *National Geographic,* whose income depended less on advertising revenue than on membership subscription, entered the new decade with the highest circulation in its history.

Still, not even the *Geographic* was immune to outside forces. The membership dipped by 44,362 in 1931, another 168,575 in 1932, and an additional 151,548 in 1933—a loss of 364,485 members in those three years.

Grosvenor was forced to lower salaries of his staff—among them the Society's Foreign Editorial Staff Chief, Maynard Owen Williams, who responded:

April 18, 1933

Dear Dr. Grosvenor:—

I have always felt that when, on occasions of a salary increase, you have said that you announced the increase "with pleasure" that that was no empty phrase. And a sense that such was the case added greatly to the value of the increase.

Since that was true, back there in the halcyon days when we did not appreciate how well off we were, it must have been with a sense of regret that you consented to a necessary reduction of salaries.

In all sincerity, I thanked you for the increases when they came. In all sincerity I thank you for the personal interest and support which, at a time of necessary retrenchment, involves a personal burden and a personal regret to you.

The decrease will call for a change of living, but that should involve no real loss. Sacrifice is a luxury. After so many evidences of *your* interest and support, I cannot but express *my* interest and support in what, because of your personal feelings and sympathy, must be a trying time for you....I hope you don't regard the decreases as in any sense a personal cross. But I hope I'm still enough of a Christian to appreciate a cross, be it mine or yours, and it is a relief to feel that if it is, I'm now having a share in bearing it.

Faithfully yours,

Williams

Preceding spread: *Titled "*STUDENT SMILES FROM FIVE STATES FLASH THROUGH OLD SCROLLWORK GATES AT ASHLEY HALL," *this striking Dufay-color photograph by B. Anthony Stewart of young women in a Charleston, South Carolina, boarding school appeared in the March 1939* National Geographic.

Grosvenor was moved enough to reply that same day:

April 18, 1933

Dear Williams:

I am very much touched by your kind note regarding the salary reductions which it has been necessary to make. Your thoughtful act has eased greatly the aching heart which I have carried as the result of these reductions. Even after the reductions, the salaries of the staff are, I believe, greater than they were in the boom days of 1928.* The National Geographic Society, thanks to the ability and loyalty of the men and women of its organization, has thus far weathered the hurricane of this depression. With trimmed sails we hope to ride it out safely unless some unforeseen cataclysm overwhelms America.

Thank you for your understanding.

Yours faithfully,
Gilbert Grosvenor.

Such flattering communiqués to the "Chief," as Grosvenor was also called by staffers, were not uncommon—especially on the anniversaries of their employment. For sheer adulation, however, no one could surpass those missives sent to Gilbert Grosvenor** by his right-hand man, John Oliver La Gorce, who had named his firstborn son Gilbert Grosvenor La Gorce and had written in 1916:

I wanted to say something to you yesterday but as fat and devil-may-care as I am I couldn't be serious in speaking of it without being a baby— Yesterday was the big double-header—36 on earth but only 11 alive—11 years with you and by your side, absorbing your great broad human viewpoint and your love of your fellows!—You'll never know how you have rolled out many a "petty" kink in my nature and impulse toward others, how often I've stopped to think "Hell no! Grosvenor *wouldn't do that!*"—I owe you a great debt for you have conserved my interests and happiness as your own and I'm ever seeking some way to express it—

This sounds like mush but really it isn't.

The Society's fortunes began to recover in 1934; by 1935, membership had reached more than a million again—in part through the *Geographic's* continuing celebration of the ordinary: "Hunting for Plants in the Canary Islands," "Our Friend the Frog," "Some Odd Pages from the Annals of the Tulip," along with the long-running series on American cities and states. A Society author, however, points out in *Images of the World* that no matter how rose-colored the lenses, reality sometimes crept in. Referring to the preparation of an article that appeared in November 1934, the author wrote:

In planning for coverage of southern California...illustrations editor Franklin L. Fisher realized that there was one place that needed careful treatment: Hollywood.

Writing to the photographer who had the California assignment, Fisher approved the idea of showing how a movie star was made up. But, he added, be sure to "pick someone with whose name scandal has not been connected. Possibly Joan Crawford might be suitable."

The photographer replied that "the only actress out here who hasn't indulged in scandal seems to be Minnie (Mrs. Mickey) Mouse."

---

*In 1928, Maynard Owen Williams' salary was $7,700 per year; it was raised to $9,000 (less an insurance premium of $272) in 1929.

**Grosvenor, as were the other officers of the Society, was also known by the initials with which he signed his memos. Gilbert Hovey Grosvenor, therefore, was "GHG"; John Oliver La Gorce was "JOL," etc.

The Geographic settled for a black and white picture of an actress having an artificial eyelash adjusted. Color photos concentrated on flowers and oranges, while the text tried to live up to its title, "Southern California at Work." But the basic fact of life in America—the Great Depression—could not be avoided. "Men without work walk the streets here now, as everywhere else...."

Despite the world's economic turmoil several significant advances were made in color photographic processes during the early 1930s, and the Society was quick to experiment with them. The Finlay process, a refinement of the Paget method, was developed in England, and eager to create its own color photoengraving division, the Beck Engraving Company in Philadelphia (which made all the black-and-white engraving plates for the Magazine) had invested in the new process, encouraged the Geographic to try it, and even helped pay for Charles Martin's journey to England to learn the techniques.

By 1930, Melville Bell Grosvenor, GHG's son, had been at the Society for six years, writing and editing photograph captions. Although the Finlay image was somewhat grainier than an Autochrome, the emulsion was faster. This enabled Melville Grosvenor, in September 1930, to have published for the first time natural-color photographs from the air.

Because color plates required lengthy exposures and an airplane's vibration and speed would have made a clear aerial color photograph almost impossible, Melville Grosvenor and a crew of technicians flew aboard dirigibles. "Frequently the photographic voyages consisted of shadow-chasing," Melville wrote. "Sometimes, after scurrying over Washington at express-train speed, the ship would arrive at the desired spot just in time to have a wisp of cloud form between the scene and the sun, and thus make the attempt futile."

Melville's photographs of the Statue of Liberty, the Washington Monument, Lincoln Memorial, the Capitol, an ocean liner inward bound to New York harbor were unique only in that they were in color; the technical achievement, not the artistry, continued to be what merited the attention.

Five years later the Finlay process was still being used. B. Anthony Stewart was working in the National Geographic's photographic lab under Charles Martin and was, he told *Geographic* staffer Priit Vesilind, "full of vinegar" and desperately hoping to get a chance to prove himself in the field.

"Old man Martin used to call me the 'Bozo,'" Stewart recalled for Vesilind, "and finally I guess progress enabled me to get a nod from the illustrations chief. So Frank Fisher gave me the astounding assignment of the State of Delaware. And I can remember his saying to Martin, 'Well, hell, he can't do much wrong in Delaware, so go ahead and send him.' It was close enough that 50 cents' worth of gasoline would have brought me back to Washington.

"I had [used] exclusively Finlay...and I found that under ideal conditions ...you could just squeeze through a twenty-fifth of a second with the sun at high noon on a June day, if the motion was coming straight at you. Not across. I was just at the age of trying to make a breakthrough. I wanted to show I could run with the big boys. So I looked up a fox-hunting scene. I arranged—it was all staged by me—that the hounds and riders rode straight at me. Well, the pictures came out and I was an instantaneous success."

Stewart's photograph, "Fox Hunting Near Newark Proves Delaware's Kinship With The Old South," appeared in September 1935 with the note "Such a natural color photograph of moving animals is a rare achievement."

However, if the horses and riders were coming toward Stewart, they were moving very slowly indeed. Although there is the blur of wagging tails, and a couple of dogs are rolling and stretching, there is only a hint of action. That hint was enough, however, to have made Stewart's fox-hunting photograph different from all the stiffly posed, stilted Autochromes that had appeared before.

In 1930, the *Geographic* published 1,328 black-and-white photographs, 366 Autochromes, and 38 Finlays. By 1935, 1,277 black-and-whites appeared, 72 Autochromes and 231 Finlays.

Only slightly faster than the Finlay process, Dufay, introduced in 1937, had the advantage of using film rather than glass as an emulsion base—an advantage the *Geographic*'s world-traveling staff photographer W. Robert Moore was quick to appreciate. In 1931, Moore had taken 150 pounds of bulky, fragile Finlay glass plates to China on assignment; his camera and holders had added another 50 pounds.

No Autochrome, Finlay, or Dufay color prints appeared after 1941, except in later special commemorative issues. The reason for their disappearance was the introduction, in 1936, of Eastman Kodak's revolutionary new 35 mm Kodachrome film. And although Gilbert H. Grosvenor would, in 1963, write, "As each improvement [in color film] became available, the Magazine put it to

"THIS SUDANESE SLAVE GIRL BELONGS TO A RICH ARAB MERCHANT OF MOCHA. Treaties among Christian nations to suppress the slave trade are without effect on human behavior in remote nooks of the Moslem world. When a traveler visits a sheik and admires a slave, his host—true to desert hospitality—may make him a present of his human chattel!"—From Henri de Monfreid's November 1937 "Pearl Fishing in the Red Sea"

Any history of the National Geographic Society is also a history of color photography, for, as Gilbert Hovey Grosvenor wrote in 1963, "In photography—and particularly in the use of color—we have led the way from the first."

The first commercial process in color photography was the Lumière Autochrome, developed in France and marketed in 1907. Although it opened a new world, it required, like the black-and-white photographs of those days, heavy cameras and cumbersome glass plates, and exposures were slow. Franklin Price Knott's Autochrome of a Tunisian carrot peddler ran in the Magazine in September 1916.

The Finlay process photograph, developed in England, appeared in 1930; though grainier than an Autochrome, it offered the increased speed that made it possible for Gilbert H. Grosvenor's son Melville Bell Grosvenor to take the first successful aerial color photographs.

Agfacolor, a German development, allowed exposures of a fifth to a tenth of a second in bright sunlight with lenses set wide open. Although twice as fast as Lumière, Agfacolor was swiftly discarded by the Geographic in favor of the Dufay process, introduced in 1937.

Dufay was only slightly faster than the Finlay process, but it utilized film rather than glass plates—an advantage Geographic photographers in the field, loaded with 150 pounds of cameras, tripods, and glass plates, were quick to appreciate.

Autochromes, Finlays, Agfacolors, and Dufaycolors were all discarded when Kodachrome, introduced by Eastman Kodak in 1936, enabled photographers to capture action in color on 35mm film. In April 1938 the first Kodachromes appeared in the *National Geographic*.

The Autochrome screen (above) *used minute grains of color dyed potato starch. (Autochrome picture,* right)

The Finlay screen (below) *contained 175 lines to the inch. (Finlay picture,* bottom)

The Agfacolor screen (above) was formed by bits of dyed resin. (Agfacolor picture, left)

The Dufaycolor process ruled screen (below) was made up of parallel and crossed lines. (Dufaycolor picture, right)

Unlike every other photographic process of that time, Kodachrome provided a grainless image with a potential for almost limitless enlargement. (Kodachrome picture, left)

Along with seventy-four human skeletons, jewelry, and an offering stand of silver, shell, and gold, this gilded bull's head with lapis lazuli beard adorned a lyre which lay—its musician's stilled fingers spread over its strings—in the "great death pit" of Ur, the site of royal burial and human sacrifice 5,000 years ago in Iraq. M. E. L. Mallowan's account of the excavation led by C. Leonard Woolley, "New Light on Ancient Ur," appeared in the January 1930 National Geographic.

superlative use," there was evidently some foot-dragging in utilizing Kodachrome, since it did not appear in the *Geographic* until two years after its availability.

There is reason to believe that the first to realize Kodachrome's potential was Luis Marden, who had come to the Geographic the year before in search of a job. Marden, then a nineteen-year-old journalist for the *Boston Sunday Herald*, was self-educated, quadri-lingual (English, Spanish, Italian, Portuguese), and a radio broadcaster. At the suggestion of Kip Ross, then a technical adviser for the E. Leitz Company, Marden had just written *Color Photography With the Miniature Camera*, the first instructional book on that topic; and when he learned the Magazine was looking for photographers Marden wrote the Society that he was "experimenting with color."

The Society responded by inviting Marden to Washington. He appeared in Illustrations Editor Franklin L. Fisher's office with a Leica 35 mm camera slung around his neck. "Do you ever use a Leica in your work?" Marden innocently asked the older man.

"Of course we don't," Fisher replied, in effect. "We don't use toys around here. We do serious photography."

To test Marden's prowess with the color process, Fisher said, "Let's see if he can make a color photograph." He made Marden load Finlay plates in the darkroom, gave him an Ica Jewel camera, a car, and driver, and sent the young man out to look for colorful subjects. "I remember it was the blazing African heat of Washington in July," Marden later recalled:

> …everything was burned and sere. Finally I made one exposure on a tripod of a ship being painted red near Haines Point, with weeping willows. I've never made a better one since. Then I came in and I proceeded to develop the film myself and to register it with the viewing screen. It impressed the hell out of them because plenty of… photographers came around to the *Geographic*, pros of many years standing, but they had never shot color.

In those days professional photographers did not bother to learn the color process because there was no market for it; no magazine other than the *Geographic* regularly printed it, and there were only transparencies, not photographic paper, to enable a photographer to print studio portraits.

In 1935, Kodak released its first Kodachrome as a 16 mm motion-picture film. A year later Marden, by then working at the Magazine exclusively as a photographer, saw a demonstration of the film in the window of a downtown Washington camera store. Marden's recollection continues:

> They were projecting this loop of film over and over again with its brilliant color, and I asked the man if I could examine the film. He gave it to me and I put a loupe on it, a magnifying glass, and I saw *there were no flecks in it!* Unlike with an Autochrome or a Finlay the white paper and the white shirts had no residual color in them at all! And I wondered, *How in the devil did they get that color?*
>
> Of course, anyone with a modicum of technical background could have figured out the only way they could do it would be through superimposed layers—one sensitive to red, one to blue, one to green and so on—but the minute I saw that 16mm film I came rushing back to the Geographic, went to the head of the photo lab, Charles Martin, and I said, "I've seen this new film, it's amazing! If it does come out, it's going to open the door for us, it'll be a photographer's liberation, like being let out of prison."

The liberation would be threefold: First, because the Kodachrome's final image was a dye image only, the resulting picture—unlike every other photographic image of that time, including black-and-white—contained no granules of metallic silver. Marden immediately realized that with Kodachrome one had a *grainless* small image with a potential for almost infinite enlargement. Second, the film was faster. Exposures with Lumière Autochrome require bright sunlight; and although the makers of the Finlay plates boasted that with their process it was possible to take "snapshots"—action photographs— "if there was *any* movement you'd get a blur," Marden explained. Third, since both Autochrome and Finlay used glass plates, photographers had to carry a cumbersome load of equipment.

Marden knew that miniature cameras, roll film, dye images would revolutionize *National Geographic* photography—but *only* if he could convince the Illustrations Division's chief to permit him to use them.

Through friends, Marden managed to acquire two rolls of the new 35mm Kodachrome film before it came on the market. The Geographic did not even own a Leica at this time, so one weekend Marden took his, the two rolls of Kodachrome, and, as he said, "exposed them in every possible way: a microphotograph of a postage stamp, street scenes, a close-up of a goldfish, a studio portrait..."

That same weekend, Marden also took a photograph of a woman friend horseback riding in Washington's Rock Creek Park, while on horseback himself. Such a feat would have been nearly impossible with the cumbersome Finlay or Autochrome camera equipment. Two weeks later the developed films came back from Eastman Kodak. "They were color transparencies, the same kind you have today; but, of course, the film came back as a whole strip. Kodak was certainly still a long way from cutting them up and mounting them between little cardboard mats." Marden taped the best of the transparencies onto lantern slide glass plates, blacked out everything around them, then asked Franklin Fisher if he would like to see them.

Fisher came down to the small projection room with his assistant and a couple of others. Marden recalled:

> I was only a lowly lab man in the photographic laboratory and I projected them in our small auditorium. They only filled the center of the screen, but still they were about six feet across. I told them you could take Kodachrome fast, it was portable, and reproduce it any size you wanted.... There was dead silence.
>
> Anyone who knew photography would have understood what I was showing them, but these weren't technical people, these were *illustrations* people. It's hard today to realize how astounding these photographs were! They were sharp, they didn't have extraneous color. "See?" I kept saying. "The white shirt? No flecks of color in it. The flesh tones? No red and green in it."
>
> They didn't say much, just asked a few questions, and that was it. I went back to my kennel, the dark room, and I was in there developing prints and I sensed a presence. It was quite dark under the red lights, and I smelled the tobacco of the head of the photo lab. And Charles Martin—he was a very brusque, at times very graceless old man, and he said, "Say Marden...those pictures you made on that new film....Did you make them on your own time?"
>
> I said, "Yes, Mr. Martin. I made those a few weekends ago. Why?"

Even "The Lion of the Tribe of Judah" willingly posed for the National Geographic *in his coronation attire. W. Robert Moore's "natural color photograph" of Haile Selassie, the newly crowned Emperor of Ethiopia, appeared in the Magazine's June 1931 issue.*

"This has again been a most interesting voyage," Alan J. Villiers (above), wrote Geographic Illustrations Chief Franklin L. Fisher in 1932 of the grain race from Australia to England by twenty big windjammers, in which Villiers had just captained his square-rigger Parma (left) to victory. "We arrived in Falmouth Bay the other day with a passage of 103 days and a few hours, winning...by four hours (not much in 103 days!)...The voyage was marked by the ferocity of the Cape Horn storms. We...blew out our best sails, washed everything movable overboard, gutted the midships house, and drowned the ship's pig...." A great favorite of the Geographic, Villiers produced more than a score of seafaring articles for the Magazine between 1931 and 1976, as well as for the Book Service publication Men, Ships, and the Sea.

"Well, Mr. Fisher just called up and wanted to know. He said those things will never be worth a damn to the National Geographic."

We were in the gloom of the darkroom, and I don't know if I said, "Do you believe that?" or if we just looked at each other and half-smiled. Being a superb technician, Martin, of all people, knew that that was ridiculous.

Still, Martin did send two separate exposures to the Beck Engraving Company from which they made half-tones. To my knowledge, they were the first half-tones made from a Kodachrome anywhere.

Although the test half-tone engravings were successful, over two and a half years would elapse between Kodachrome's arrival on the market and its first appearance in the *National Geographic Magazine*. As Geographic staffer Priit Vesilind wrote in his thesis on color photography at the Society:

It took a successful field assignment by an established photographer to ignite the *Geographic* on 35mm Kodachrome....

In the summer of 1937, a year after Kodachrome had been available, Bud Wisherd, the assistant chief of the photo lab, requisitioned a dozen more rolls of the film. He mailed five to W. Robert Moore on assignment in Austria, and Marden took the remaining rolls on an expedition in Mexico. But Marden's Mexican Kodachromes were not published until September 1940....Moore's appeared in April 1938....

B. Anthony Stewart may have described Marden's position aptly: "To be new and an originator—that, of all things, is resented. Wait a little before you originate, because you're going to be stepping on people's toes."

Moore, given the privilege of being the *Geographic*'s Kodachrome pioneer, later described [how]...within a few days [he] had shot all of his five rolls of Kodachrome, capturing, among other things, a swirling band of Austrian dancers. "The men swung their partners round and round to the swiftly changing tempo of guitar, mandolin, and wheezy accordion," he later described. "I took pot shots, standing up, and lying down on my stomach in the courtyard. Rough boots kicked up dust, dresses and petticoats swished, the dancers had fun; so did I.

"Routine now, yes, but the dance was one of the first color subjects I had ever shot without having my camera securely anchored to a tripod."

The realization that magazine photography would never be the same hit home at the *Geographic* as Moore's 35mm photographs ran through the processing and editing mill.

"Everyone just went wild over them," remembered B. Anthony Stewart. "The dancing girls, the iridescence of color. It was just something that color photographers had never dreamed of."

"Without doubt," Edwin Wisherd said at his retirement in 1971,[*] "that's the most important thing that has happened in photography during my time here—the development of the 35mm camera and Kodachrome's fast emulsion color film."

In 1938, 3 Lumière Autochromes, 222 Finlays, 69 Dufays, and 18 Agfacolors were published in the Magazine, compared to 62 Kodachromes; the following year, the Magazine published 8 Autochromes, 47 Finlays, 93 Dufays, 11 Agfacolors, and 317 Kodachromes. A revolution in color photography was occurring.

As Melville Bell Grosvenor, by then an Assistant Editor of the *Geographic*, recalled: "Frank Fisher was very old-fashioned in his ideas, but he saw the point instantly. And my father did. And we just threw out our other pictures

"...it was evident that we had discovered one of the richest and most important fossil fields in all the world," *Roy Chapman Andrews* (above) *wrote in* "Explorations in the Gobi Desert," *for the June 1933* Geographic. "The dinosaur eggs alone made it famous, but they were by no means the most important of the thousands of specimens we brought from the Gobi. We were surprised at the universal popular interest that the dinosaur eggs aroused."

Opposite: *Roy Chapman Andrews, 1931 Hubbard Medal winner for geographic discoveries in Central Asia, inches out over a Gobi Desert aerie to capture an eaglet.*

---

[*]*Edwin L. (Bud) Wisherd joined the Society Photographic Laboratory in 1919 and, upon Charles Martin's retirement in 1941, replaced Martin as chief.*

from the field—just scrapped them and replaced them as fast as we could with Kodachromes. I'll never forget that—that was really a thrill."

There were other thrills to be found at the *Geographic* in the 1930s, for it was the decade of the great expeditions, the "Glory Years."

The decade had led off with M.E.L. Mallowan's January 1930 "New Light on Ancient Ur" account of one of archeology's great moments—the discovery in Iraq of the "death pits" of Ur, where 5,000 years earlier human sacrifice and royal burials had occurred:

> One of these graves consisted of a shaft dug to a depth of 30 feet below surface level. At the bottom was a single chamber surmounted by a dome... erected over a wooden ceiling which had eventually crashed onto the brick floor, covering up the five occupants. Four were servants and the fifth was a woman, most probably of royal blood...dressed in the brilliant court costume of the period. She wore a headdress of gold ribbons radiating in seven strips from the center of the head, a wealth of gold poplar leaves strung with carnelian and lapis lazuli beads, and around the neck gold chains and carnelian beads....
>
> The treasures lavished on the dead were remarkable for splendor and number. As we cleared the shaft to the level of the floor, it appeared almost as if we were treading on a carpet of gold....The [68] women, lying in ordered rows, were decked out after the fashion of the principal occupant of the domed chamber. Hair ribbons of silver or gold were almost invariable and many of the gold ribbons bore marks of exquisitely fine network, the veiling now entirely decayed that had once shrouded the head.
>
> So great was the weight and quantity of jewelry on the heads that the women must have worn wigs that were both large and substantial....
>
> The six men, perhaps the funeral bodyguard, were ranged in a row against the front wall of the shaft. The women lay in rows across it, and in three corners were buried the principal treasures.

The June 1930 issue contained "The First Airship Flight Around the World," Dr. Hugo Eckener's account of his 19,500-mile, three-week circumnavigation of the globe as commander of the giant German *Graf Zeppelin,* a journey for which he was awarded the Society's Special Gold Medal. In August the Magazine ran Admiral Richard Evelyn Byrd's account of the first conquest of the South Pole by air.

That same year, from the deck of the ocean liner in which he was crossing the Atlantic, Gilbert H. Grosvenor spied the Finnish windjammer *Grace Harwar.* An avid blue-water sailor himself, and one to whom sailing ships "gave a glorious hint of the romance and mystery of the sea," Grosvenor was able to photograph the full-rigged ship as it passed. When he later learned that Australian writer-photographer Alan Villiers had sailed on it when it had carried grain from Australia to England, Grosvenor asked him to contribute sailing pieces to the Magazine. The following year Alan Villiers wrote "Rounding the Horn in a Windjammer," the first of many Villiers articles on sailing to appear in the *Geographic.* It appeared in the February 1931 issue and contained this extract from the diary of Ronald Walker, a young reporter friend of Villiers who had served with him as one of the crew:

> "Great seas come up to meet the ship, thrusting at her, shouldering one another to get at her, like footballers in a mad footer 'scrum.' Up and up they heave, gathering for the blow. You turn to watch them. The wind howls in your face and the sea spits at you spitefully, driving its spray above and

Above: *"SHOCK ABSORBERS SERVE DUAL PURPOSE: Professor [Auguste] Piccard, (right), and his companion of the first flight sat on two cushions and kept certain delicate instruments in two baskets. These baskets and cushions were designed also for head protection against a bump in case of a sudden landing." —From "Ballooning in the Stratosphere," March 1933 issue.*

Opposite: *First woman to fly alone across the Atlantic, Amelia Earhart acknowledges cheers of a crowd upon touchdown at Londonderry, Northern Ireland, on May 21, 1932. The following month, for her accomplishment Earhart received a Special Gold Medal—the first ever awarded to a woman by the Society.*

This photograph of a male and female ivory-billed woodpecker—the first ever of a nesting pair of these rare birds—was taken by ornithology professor Arthur A. Allen for his June 1937 Geographic *article "Hunting With a Microphone The Voices of Vanishing Birds." In this landmark ecologically-aware article in the Magazine, Dr. Allen warned of civilization's inroads on birds such as the trumpeter swan.*

around. A great sea, a liquid mountain of menace, hangs poised above the ship. Up, up, it leaps, shouldering its smaller children aside, the splendid crest whitening where it breaks, lending a touch of color like the plume of a warrior's helmet.

"Down, down, sinks the ship, shuddering already at the impending blow. A hundred lesser blows she has avoided; this mighty one she cannot beat. She writhes like a living thing, in fear and trembling. She heels over heavily; she hovers frighteningly....

"The stars shoot suddenly past the spars—not so bad with them out— careening madly across the sky. The ship receives the blow full, staggering at the impact. A tremor runs through the laboring hull....

"But the shattered sea crest has met its match. The warrior's plume has dropped; the ship rises again, tumbling hundreds of tons of roaring, fighting water from her gushing wash ports. The sea sweeps her furiously end to end, murderously intent upon human prey. Balked of that, it shifts whatever is movable and snarls and hisses at the hatch breakwaters, maddeningly intent at breaking them down....

"But the ship wins. Under her load of hundreds of tons of seething water, she rolls on, recovering her poise, steadying herself to meet the next onslaught, and the next, and the next after that...."

Villiers sadly reports that six days after young Walker made that entry in his diary, he was killed after loosing the three-masted sailing ship's fouled fore upper topgallantsail. While the men on deck heaved to lift the rain-sodden, swollen canvas, the halyards carried away and the yard fell on Walker, below.

In June 1931, William Beebe's "A Round Trip to Davy Jones's Locker" told of his 1930 descent in a heavy steel bathysphere to the depth of 1,426 feet. (Four years later, in 1934, Beebe reached a depth of 3,028 feet—a record that would stand for fifteen years.)

That same month appeared an article announcing the start of the 1931–32 Citroën-Haardt Trans-Asiatic Expedition. With its motor caravan of half-track vehicles, the expedition blazed a 7,370-mile trail across Asia, an overland exploration from the Mediterranean to the Yellow Sea.

Roy Chapman Andrews told of discovering "one of the richest and most important fossil fields in all the world" in his June 1933 piece, "Explorations in the Gobi Desert." Andrews discovered not just dinosaur eggs he estimated to be ninety-five million years old in Mongolia, but also the skulls of tiny rat-size mammals that coexisted with the dinosaurs, and, among others, the thirty-million-year-old bones and "a huge tooth, nearly as large as an apple" from "the giant *Baluchitherium*...the largest land mammal that ever lived upon the earth...an aberrant browsing rhinoceros [that]...stood about 17 feet high at the shoulders and was 24 feet long, and weighed many tons."

For his findings Roy Chapman Andrews in 1931 was presented the Society's Hubbard Medal by its president Gilbert H. Grosvenor, who stated such discoveries "have pushed back the horizons of life upon the earth and filled in gaps in the great ancestral tree of all that breathes."

Alan Villiers appeared again in January 1933 with "The Cape Horn Grain-ship Race," his stirring account of a big square-rigger ship race "Through Raging Gales and Irksome Calms 16,000 Miles, from Australia to England."

Two months later the Magazine published Dr. Auguste Piccard's "Ballooning in the Stratosphere," Piccard's report on his ascent with a companion to 53,152 feet in a spherical aluminum cabin similar in appearance to the bathysphere in which Beebe had descended in the sea. The purpose of his ascent to an altitude of over ten miles, was to study cosmic rays.

In 1934, Anne Morrow Lindbergh became the second woman to be awarded a Society medal. Two years earlier, Amelia Earhart had been presented the Society's Special Gold Medal for the first solo transatlantic flight by a woman. Anne Morrow Lindbergh was awarded the Hubbard Medal for the notable flights she had made in 1931 and 1933 as co-pilot and radio operator on the Charles A. Lindbergh aerial surveys. During their 1933 flight they faced the worst possible flying hazards—blizzards, sandstorms, hurricanes, arctic cold, and tropical heat—while flying their pontoon-equipped Lockheed Sirius to four continents and visiting twenty-one countries. Their 29,000-mile adventure was an incredible feat of flying for that era. And her own account of that trip, laconically titled "Flying Around the North Atlantic," appeared in the Magazine in September 1934.

The following month's issue contained Captain Albert W. Stevens' story of the National Geographic Society–U.S. Army Air Corps Stratosphere Expedition in *Explorer*, the largest free balloon until then ever constructed.

While dazzling heights were being reached in balloons and dazzling distances in aircraft, there was still room in the *Geographic* for old-fashioned adventures such as Henri de Monfreid's November 1937 article "Pearl Fishing in the Red Sea":

> In ten minutes we were outside the island, the mainsail bellying. Night had come and I set the course by the compass....We had to cross the central part of the Red Sea....It was 2 a.m. and all eyes were fixed on the horizon....
>
> With the aid of a night glass I guided us closer and Ali pointed out to me an isolated black point in the sea to the south of the island. It was a zarug [a small Red Sea sailing craft used by smugglers] moored to... a reef.
>
> Our rifles were quickly loaded with five cartridges each. I had only 50 all told. Then the ax, a jumper bar, and a big iron hammer were put in battle order as if we were going to board....the die had been cast; I could no longer retreat.
>
> We were half a cable's length from the sleeping zarug when the old Sudanese recognized it. I put the helm up, and as the sail fell we drifted to within a few yards of our quarry....I called the usual "Hooooo" as if we were an innocent boat coming there by accident, though this was a very unusual hour to drop anchor.
>
> Meanwhile I had lighted the fuse with my cigarette. The little light from the jet of flame passed unobserved and the black fuse smoked on ominously in the dark.
>
> I plunged the long boat hook into the sea as if about to anchor, but held the bomb under the vital parts of the boat. I carefully counted the seconds up to ten. Then I cried to the nakhoda, who was ready:
>
> "Call your men."
>
> All together we shouted to the men in the zarug to jump into the sea if they could.
>
> The Zaraniks awoke; the breech of a rifle clicked; I continued to count—18, 19....
>
> Then a greenish flame spurted from the center of their boat as the dynamite exploded.
>
> I had held the explosive under the mast, where there are usually no sleepers. The mast crashed down, and a few seconds later stones rained on all sides—the pebble ballast had been blown sky-high....
>
> The zarug sank in a few seconds and men floundered in the water. There were wild shouts. The Sudanese swam heavily by their arms, as they

"Almost within the memory of men still living," Arthur A. Allen (above) wrote, "four species of North American birds have become extinct. In our museums will be found the dried skins or mounted specimens of the great auk, the Labrador duck, the passenger pigeon, and the heath hen. The Carolina parakeet seems about to follow them...." Determined to preserve the voices of birds still living, Allen spied on, among others, "the ivory-billed woodpecker...now perhaps the rarest North American bird," and managed to record both its antics and its voice.—From "Hunting With a Microphone the Voices of Vanishing Birds," June 1937 issue

Wotta Life! Wotta Life! : : By Gaar Williams

WELL,—I DON'T THINK I WAS SO DUMB AFTER ALL.!

THE FLAG POLE SITTER

Geographic expedition leaders Piccard, Byrd, and Beebe are lampooned in this 1934 drawing by Gaar Williams. By the 1930s the National Geographic *was a rich target for newspaper and magazine cartoons: three appeared between March 1930 and December 1931 in* The New Yorker *alone. In 1932, when one of the* Geographic's *editors began preparing an album of humorous references to the Magazine, Gilbert Hovey Grosvenor complained that some of the "good ones" had been left out— among them a* New Yorker *cartoon of a person telephoning the* Geographic *to say that he "had a photograph of the other side of the mountain illustrated in the current* Geographic."

were chained two and two by the legs. The Zaraniks fled toward the island in small boats. The panic was complete.

I lighted a big torch of gasoline, which illuminated the scene....The Zaraniks were paddling off in three boats, bailing out the water meanwhile. I fired in their direction while my Somalis gave chase....

Day was breaking and soon we could see through the transparent water the hull of the sunken zarug....

One of the rescued Sudanese, a Hercules with rather thin legs, took one end of a rope and dived. I saw him climb over the wreck, feeling each object, his arms stretched out, the soles of his feet pointing to the surface.

With a supple movement of his back he turned up again and came straight to the surface, blowing streams of bubbles through his nose as he rose. He had made fast the rope and told us to haul on it.

A box came up....It was the coffer of the Zaranik nakhoda, ornamented with inlaid copper figures. I opened it by forcing the padlock.

The Sudanese swore the pearls were inside....

Inexplicably, an illustration accompanying de Monfreid's account was of a strikingly beautiful partially nude woman with the caption "This Sudanese Slave Girl Belongs to a Rich Arab Merchant of Mocha," which then went on to explain, "Treaties among Christian nations to suppress the slave trade are without effect on human behavior in remote nooks of the Moslem world. When a traveler visits a sheik and admires a slave, his host—true to desert hospitality—may make him a present of his human chattel!"

The Sudanese slave girl appeared in 1937, the year Franco's rebel forces took Malaga, the year Guernica was destroyed by German aircraft and Spain's Republican government moved to Barcelona. It was the year the Japanese seized Peking, Shanghai, and Nanking; and that Chiang Kai-shek joined forces with Mao Tse-tung and Chou En-lai. England's King Edward VIII had abdicated the year before and now, as the Duke of Windsor, married Wallis Simpson. In 1937 President Franklin Delano Roosevelt signed the second of three Neutrality Acts; Mussolini visited Libya and Berlin. Picasso painted *Guernica*; Ernest Hemingway wrote *To Have and Have Not*; and John P. Marquand's *The Late George Apley* won the Pulitzer Prize.

Disney premiered *Snow White and the Seven Dwarfs*, and Greta Garbo starred in *Camille*. Amelia Earhart disappeared on her flight over the Pacific; the German zeppelin *Hindenburg* exploded as it approached its Lakehurst, New Jersey, mooring mast; Joe Louis gained his heavyweight title by knocking out James J. Braddock; the Golden Gate Bridge opened; and the New York Yankees beat the New York Giants in the World Series, 4–1.

A serious economic recession continued in the U.S.; the popular songs were "Nice Work If You Can Get It" and "The Lady Is a Tramp."

If the *Geographic* managed to avoid showing people struggling through the Depression, it could not dismiss the emerging political forces in the world.

The year before, in September 1936, Mrs. Kenneth Roberts reported in "Sojourning in the Italy of Today" that when the modern Italian army's victories over the Ethiopians "were achieved in the face of bitter opposition from the League of Nations—which to Italy means England, the first power in the world—Mussolini, in Italian eyes, became almost a god."

Ruth Q. McBride's "Turbulent Spain" told how the summer offices of the American Embassy located in San Sebastián's Hotel Continental were cut off from the outside world: "They could not even reach Ambassador Bowers at his summer place at near-by Fuenterrabía. They had no means of defense. Leftist soldiers, merely peasants armed with rifles, many of them women, swept past the doors—killing."

Omnipresent in the background of photographs chosen to illustrate Douglas Chandler's February 1937 "Changing Berlin" are the banners of the Nazi government. One photograph's caption was "May Day Masses Jam the Berlin Lustgarten to Hear Adolf Hitler Speak" and continued, "Decorated with the swastika sign and guarded by troops, the speaker's stand appears in the far background, on the broad steps of the Old Museum....The Maypole, decorated with bunting and swastikas, represents a revival of an old folk custom formerly observed chiefly in the rural districts."

That same issue contained Gretchen Schwinn's "We Escape from Madrid," which opened,

> Stewed cat cost mother and me a dollar a plate in Madrid....The dark, tendon-laced meat, little like the rabbit for which it masqueraded, was our first, except horseflesh, in six weeks. Audible to us as we ate, rebel artillery, some 20 miles away, blasted slowly toward Spain's capital, where the first bombs had fallen two months before....
>
> Outside, we wore red, white, and blue arm bands with the symbol "U.S." in large letters, the Embassy stamp, and an identification number, "just in case."
>
> We dressed in our oldest, plainest clothes, never wearing hats. One rainy day a menacing crowd forced mother to throw away a hat she had recklessly worn. Our arm bands weren't enough. Only aristocrats wore hats. We had seen a woman stoned by girls because of her hat. For a young man, a mustache would have been as dangerous.

John Patric's March 1937 "Imperial Rome Reborn" celebrated Italy's new glories, but the photographs were chilling: one of gasmasked uniformed children was captioned "Weird Visitors From Another World? No; Schoolboys Preparing For War." Another, of children marching, read, "Chins High, Shoulders Squared, Boy Black Shirts Emulate Il Duce's Posture."

But through the decade that began with the Depression and ended in the midst of world war, *Geographic* readers could still take comfort in birds— "Birds of the High Seas," "Birds of the Northern Seas," and "Birds That Cruise the Coast and Inland Waters."

There were "Canaries and Other Cage-Bird Friends," "Crows, Magpies and Jays," "The Eagle, King of Birds, and His Kin," "Far-Flying Wild Fowl and Their Foes," "Game Birds of Prairie, Forest, and Tundra," "Humming Birds, Swifts and Goatsuckers," "Parrots, Kingfishers, and Flycatchers," "The Shore Birds, Cranes, and Rails."

There were "Thrushes, Thrashers and Swallows," "Winged Denizens of Woodland, Stream and Marsh," there were "Sparrows, Towhees, and Longspurs: These Happy Little Singers Make Merry in Field, Forest, and Desert Throughout North America," and "The Tanagers and Finches: Their Flashes of Color and Lilting Songs Gladden the Hearts of American Bird Lovers East and West."

And comfort they found; readers flocked back to the *Geographic.* Membership in the Society returned to a million in 1935 and, with the exception of 1938, continued to climb until 1944. But if the *Geographic* was to be remembered for anything during the first part of the coming decade, it would be for the ubiquity of its maps. The National Geographic Society's Cartographic Section, born in one world war, came into its own in the next.

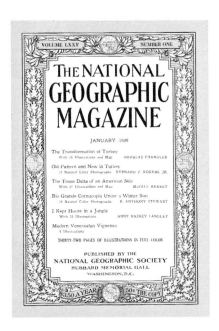

By the end of 1939, membership in the National Geographic Society had risen to 1,093,578.

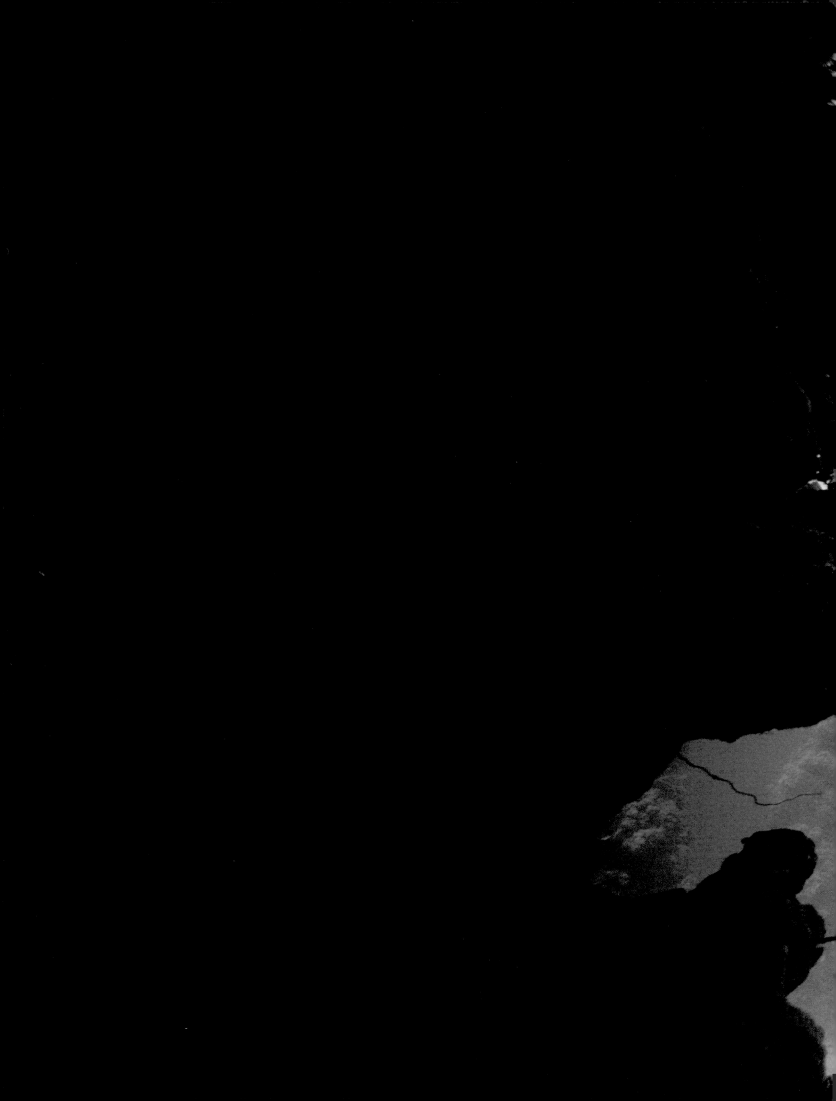

"These Marvelous Nether Regions"

"...the only other place comparable to these marvelous nether regions, must surely be naked space itself, out far beyond the atmosphere, between the stars,...where sunlight has no grip upon the dust and rubbish of planetary air, where the blackness of space, the shining planets, comets, suns, and stars must really be closely akin to the world of life as it appears to the eyes of an awed human being, in the open ocean, one half mile down."

—William Beebe, *Half Mile Down*

William Beebe, writing of that moment on June 6, 1930, when he and Otis Barton first peered out of the portholes of their bathysphere into the murky depths at six hundred feet, noted, "Ever since the beginnings of human history, when first the Phoenicians dared to sail the open sea, thousands upon thousands of human beings had reached the depth at which we were now suspended, and had passed on to lower levels. But all of these were dead, drowned victims of war, tempest, or other Acts of God."

It was not surprising that such a thought had crossed Beebe's mind, for during the last three hundred feet of their bathysphere's descent, he had had good reason to suspect that he and Barton might join them.

Shortly after they had been lowered into the water Beebe heard Barton cry out that their two-ton steel sphere had sprung a leak. Beebe turned his flashlight on the interior of the sphere's 400-pound door and saw a "slow trickle of water beneath it. About a pint had gathered in the bottom. I wiped away the meandering stream and still it came. I knew that the door would fit more tightly the deeper we got, but there remained a shadow of worry as to how much the relaxed pressure of the ascent would allow the door to expand."

The bathysphere was suspended in the waters off Bermuda at the end of a seven-eighths-inch, non-twisting steel cable attached to a seven-ton, steam-powered winch mounted on a barge. The 4-foot-9-inch-diameter steel sphere contained enough oxygen to last the two men eight hours, a rack of carbon dioxide absorber, a searchlight, and a telephone—none of which would do them any good should the door not hold.

After a brief pause at 600 feet, the two men descended again until, as Beebe wrote, " 'Eight hundred feet' came down the [telephone] wire and I called a halt. Half a dozen times in my life I have had hunches so vivid that I could not ignore them, and this was one of the times. Eight hundred feet spelled bottom and I could not escape it."

Five days later, Beebe and Barton were lowered over the side in their bathysphere again. Wrote naturalist Beebe:

Preceding spread: *The onion-shaped portholed dome of Captain Jacques-Yves Cousteau's hangar for his two-man hydro-jet diving saucer, DS-2, was part of Conshelf Two, a submerged colony of four prefabricated steel structures with fish pens and several antishark cages located thirty-six feet beneath the surface of the Red Sea on a coral reef twenty-five miles northeast of Port Sudan, Sudan. Nearby was the Starfish House, shelter for five French oceanauts who lived and worked there for a month in June 1963 without coming to the surface. Conshelf Two was part of Cousteau's long-range research program to test man's ability to build and maintain an underwater village. For more than a decade the National Geographic Society supported Captain Cousteau's underseas studies.*

Opposite: *Dr. William Beebe, oceanographic naturalist, squeezes past the protruding steel bolts that seal the 400-pound door of the bathysphere in which he and Otis Barton descended 3,028 feet in the waters off Bermuda in 1934. That dive, sponsored by the New York Zoological Society and the National Geographic Society, set a depth record that remained unbroken for fifteen years.*

223

Jellyfish, large and small, drifted past....Fifteen-inch bonitos darted past in trios and once a small, stodgy triggerfish strayed from his water-logged sargassum shelter to peer in at me....At four hundred feet there came into view the first real deep-sea fish...lanternfish and bronze eels....Pale shrimps drifted by, their transparency almost removing them from vision. ...At five hundred feet...for the first time I saw strange, ghostly dark forms hovering in the distance,—forms which never came nearer, but reappeared at deeper, darker depths....At six hundred feet...I saw my first shrimps with minute but very distinct portholes of lights. Again a great cloud of a body moved in the distance—this time pale, much lighter than the water....

At 1,426 feet on the June 1930 dive, Beebe wrote of "the very deepest point we reached":

> ...There came to me at that instant a tremendous wave of emotion, a real appreciation of what was momentarily almost superhuman, cosmic, of the whole situation; our barge slowly rolling high overhead in the blazing sunlight, like the merest chip in the midst of the ocean, the long cobweb of cable leading down through the spectrum to our lonely sphere, where, sealed tight, two conscious human beings sat and peered into the abysmal darkness as we dangled in mid-water, isolated as a lost planet in outermost space.

Four years later, in 1934, Beebe and Barton returned again to the depths off Bermuda. At 2,100 feet Beebe saw "...ghostly things in every direction." At 2,800: "Here's a telescope-eyed fish." At 3,028: "Long, lacelike things again." His narrative continued:

> A cross swell arose and on deck the crew paid out a bit of cable to ease the strain. We were swinging at 3,028 feet. There were only about a dozen turns of cable left on the reel, and a full half of the drum showed its naked, wooden core. Would we come up?...Through the telephone we learned at this moment we were under pressure of 1,360 pounds to the square inch. Each window held back more than 19 tons of water, while a total of 7,016 tons was piled up in all directions on the bathysphere itself.

Beebe and Barton's descent to a depth of 3,028 feet was a record that would remain unbroken for fifteen years.

Today, more than fifty years later, still only a pitifully small amount of the vast underwater world has been seen. "This world ocean, this vast culture broth and spring of life," James Dugan wrote in the National Geographic's Special Publication *World Beneath the Sea,*

> blankets seven-tenths of the globe and shrouds awesomely spectacular topography. But soundings and electronic magic, rather than personal inspection, have revealed most of the little we know about the grandeur of the depths....
>
> These sketches [made by the stylus of recording sonar] form mere shadows of reality. Most of the underwater canyons, trenches, valleys, mountains, and plains remain a territory not yet invaded by human explorers. And what a territory!
>
> Earth's longest mountain range, the Mid-Oceanic Ridge, meanders between the continental land masses for 35,000 miles through all the oceans. Rising 6,000 to 12,000 feet above the bottom, the pinnacles of the ridge occasionally break through the surface to form islands, such as the Azores

Above: "Silver Hatchetfish Drifting Through the Abysmal Darkness" was the June 1931 caption for this painting by Else Bostelmann of the tiny, pop-eyed fish never seen alive until Beebe spied one through the window of the bathysphere during his June 1930 record dive.

Earlier Beebe had written of the moment he and Barton reached 1,426 feet, the deepest point of that 1930 dive: "There came to me at that instant a...real appreciation of...the whole situation; our barge slowly rolling high overhead in the blazing sunlight,...the long cobweb of cable leading down through the spectrum to our lonely sphere, where, sealed tight, two conscious human beings sat and peered into the abysmal darkness.... isolated as a lost planet in outermost space."—From "A Round Trip to Davy Jones's Locker," June 1931 issue

Opposite: Else Bostelmann's watercolor, part of a series in the December 1932 Geographic, was captioned: "Hawaiian Reefs Present a Sunburst of Color: In the shallows...of the Pacific, great monoliths of coral lift their heads,...seafans enhance the beauty seen through their waving, purple veils; and graceful fish show clashing yet harmonious patterns and colors."

and Ascension. In expanse the ridge almost matches the areas of all the continents.*

Although 29,028-foot Mount Everest is the roof of the land above the water, earth's greatest known height has its roots in the Pacific off Peru. There the Andes ascend 25,000 feet before surfacing, then climb another 23,000 feet—a total rise of more than nine miles.

The deep trenches claim my admiration most of all, for they have no rivals on dry land....

Little was known about the deep ocean trenches when Beebe made his descents. The depth of his bathysphere's descent was limited to the length of cable by which it was suspended, its mobility limited to the location of the barge to which the cable was attached; its maneuverability was that of an underwater yo-yo. The next development after the bathysphere was the bathyscaph designed by Auguste Piccard beginning in 1937.

Piccard attached what looks like a bathysphere to the bottom of a large, gas-filled float containing iron pellets for ballast. Like underwater balloons, bathyscaphs could descend by releasing gas, and ascend by dropping pellets. Freed of cables, the bathyscaph was able to explore the ocean's depths with only a telephone link to the outside. On January 23, 1960, in the bathyscaph *Trieste,* Auguste Piccard's son Jacques and a young American Navy lieutenant, Don Walsh, descended into the Challenger Deep in the Mariana Trench on man's deepest dive—35,800 feet. "Like a free balloon on a windless day, indifferent to the almost 200,000 tons of water pressing on the cabin from all sides ...slowly, surely, in the name of science and humanity, the *Trieste* took possession of the abyss, the last extreme on our earth that remained to be conquered."

Bathyscaphs and bathyspheres were still nothing more than "marine elevators," and modern Navy submarines were not yet designed for reaching great depths.

In June 1963, Jacques-Yves Cousteau's divers Albert Falco and Claude Wesly spent seven days under thirty-three feet of water near Marseilles in an undersea dwelling called Conshelf One, performing what Cousteau referred to as a "logistical experiment more than a physiological one."

It was Jacques-Yves Cousteau who, in 1942, had with engineer Emile Gagnan invented the Aqua-Lung. Reported Cousteau in a 1952 *Geographic*:

The best way to observe fish is to become a fish. And the best way to become a fish—or a reasonable facsimile thereof—is to don an underwater breathing device called the Aqua-Lung. The Aqua-Lung frees a man to glide, unhurried and unharmed, fathoms deep beneath the sea. It permits him to skim face down through the water, roll over or loll on his side, propelled along by flippered feet....In shallow water or in deep, he feels its weight upon him no more than do the fish that flicker shyly past him.

*Dr. J.R. Heirtzler, U.S. Chief Scientist of Project FAMOUS [French-American Mid-Ocean Undersea Study] describes the Mid-Oceanic system as "the largest mountain range on this planet—a system greater than the Rockies, the Andes, and the Himalayas combined."

▶

The most extensive series of pictures of the undersea world were made by Luis Marden and published in the February 1956 National Geographic.

This photograph of divers from Cousteau's ship Calypso *was captioned: "Aqua-Men Cross the Equator by Swimming Under It...."*

Six-foot sand tiger sharks probe a cave near Japan's Bonin Islands in this photograph made by David Doubilet to accompany Eugenie Clark's August 1981 National Geographic *article* "SHARKS: *Magnificent and Misunderstood."*

But there was no misunderstanding the sharks' intentions when, during that assignment, Doubilet was lowered in a cage off Dangerous Reef, South Australia, and for the first time in his life was alone with a great white shark:

"I see the shark's head easing out of the murk, a pig-faced beast with an underslung grin," Doubilet wrote in the Society's 1981 Book Service publication Images of the World. *"Light dapples his back. Slowly he turns away, scattering the few small fish that hang near my cage. Near its top and on all four sides, the cage has a camera port, a horizontal space 24 inches wide between the bars. I lean out as the shark comes close. The wide-angle 15mm lens takes in all 16 feet of the great white. He makes his turn and passes by the cage again.*

"Then I see the eye, silver-dollar size, black, bottomless. It is not the eye of an animal but the porthole of a machine. The shark bangs the cage with his tail. I whirl, my air tank clanging against the metal bars. The shark is gone. Where? Suddenly, he comes again, this time from above, raking his teeth along the yellow floats atop the cage.

"The shark sinks to my eye level. The teeth drop down. He is opening his jaws. I am looking down the animal's throat.

"Now the shark wants a closer look at me—perhaps a taste. He pushes his head into the cage, coming in at the camera port, but...he cannot open his jaws. The shark turns sideways. The pointy nose pushes further in. The dome of my camera housing is one inch from the shark's snout. For the first time, I take my eye from the camera. I'm transfixed. All I can see is the flat black eye.

"Then, inexplicably, the shark is gone, a myth, black-magical animal vanished in a darkening sea."

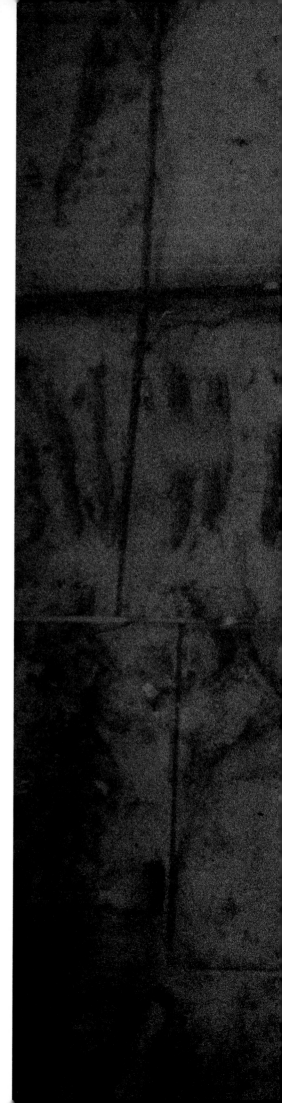

With the development and perfection of the Aqua-Lung a new, and often rapturous, era of undersea exploration began. Cousteau later wrote of his first tests: "I reached the bottom in a state of transport....To halt and hang attached to nothing, no lines or air pipes to the surface, was a dream. At night I had often had visions of flying by extending my arms as wings. Now I flew without wings."

Cousteau has not dreamed he was flying since that dive in the Aqua-Lung; but pioneer, visionary, promoter, and poet that he is, he continued to dream of other things. His next step after the Aqua-Lung was to develop a device that would combine the freedom of the Aqua-Lung diver with the protection of a pressure-resistant enclosure for deep dives. "But a regular submarine or a deep-diving bathyscaph," Cousteau pointed out, "would be too big and clumsy for intimate reef exploration. We needed a radically new submarine, something small, agile."

Cousteau's answer was the rudderless diving saucer, which by 1960 was capable of jetting around reefs and along cliff faces to 1,000 feet. Among this strange craft's early passengers was Harold Edgerton*—"Papa Flash" to the men of Cousteau's scientific research ship *Calypso*. Edgerton logged these impressions from his maiden voyage: "Like an airplane we descended to where the reef dropped off into deeper water....We were getting good, clean oxygen. Being in the saucer was no different from being in an automobile, except that we had more room and lolled comfortably on our foam mattresses like Romans at a banquet....Falco [the pilot] spotted a squadron of squid, swimming against the bottom in perfect formation....A host of fishes of many colors circled us...."

Propelled by two battery-powered hydrojet nozzles, the saucer could maneuver like an underwater airplane. James Dugan was given a ride by Falco, who, free of having to ferry scientists back and forth, put the diving saucer through its paces: swooping, climbing, hovering, banking, zooming, swerving like an acrobatic pilot.

On April 10, 1963, the U.S.S. *Thresher*, pride of America's nuclear submarine fleet, apparently dove out of control during her sea trials—her hull collapsed and the submarine sank with her 129-man crew in 8,490 feet of water 260 miles off New England's rocky coast. The difficulty encountered in even finding the lost submarine underscored the scientific gap that existed between underwater technology and man's ability to operate freely in that domain. Rescue or recovery of *Thresher* and her crew was simply beyond our means. The bathyscaph *Trieste* was the Navy's only submersible capable of reaching the depth at which the sunken nuclear submarine lay (and only two

*The name of Dr. Harold E. Edgerton, Massachusetts Institute of Technology engineer, inventor of a strobe-light system and innumerable sophisticated underwater camera devices, appears over and over again in the annals of undersea exploration. Edgerton inventions have ridden on submersibles, been lowered on cables, and been carried by a generation of divers, all to provide photographs of objects, creatures, and natural occurrences at such depths and in such darkness that they could not otherwise have been seen.

▶

A diver swims above an iron grid constructed by underwater archeologist George F. Bass and his party to facilitate the precise plotting of the finds from an ancient shipwreck in the Aegean, reported in the June 1978 article "Glass Treasure From the Aegean."

other bathyscaphs, both French, had ever operated at that depth). Reaching the depth at which *Thresher* lay was one thing; finding it was another. Four months passed before the *Trieste* could locate *Thresher*'s wreckage. And once there, the bathyscaph was able only to take photographs of the sunken submarine, recover a few pieces, and do no more.

The need to develop submersibles capable of searching large areas was emphasized three years after *Thresher*'s loss when, in January 1966, a U.S. Air Force K-147 refueling a B-52 bomber in flight over Palomares, Spain, suddenly exploded into flames. Fire engulfed the B-52 and it fell out of the sky. On board that bomber, in addition to its crew, were four H-bombs. Three fell on land; the fourth disappeared into the sea off the coast of Spain and sank 2,550 feet to the ocean floor. Though the bomb was unarmed, it was essential that it be recovered. "In the end," James Dugan wrote, "the submarinos *Alvin* and *Aluminaut* provided the key to the suspenseful search."

Alvin, commissioned by the Navy in 1964, was the first and most famous of the new submersibles. The 22-foot-long craft could carry a crew of three, was equipped with an arm-claw, and could travel at a full speed of one-and-a-half knots. Since improved, *Alvin* has been lengthened to 25 feet, been equipped with a second arm-claw, and been given a new pressure hull that enables the tiny submersible to operate at a depth of 13,120 feet. *Alvin* has now made more than 1,700 dives. In 1977, while the *Alvin* scientists were exploring the Galápagos Rift as part of their continuing study of plate tectonics, they were astonished to discover giant clams, bizarre white, thread-like worms, dandelion-shaped creatures never before seen, and long tube worms that defied description.

" 'The crabs were unbelievable!' " said noted underwater photographer Al Giddings. " 'We watched them tiptoe up to the giant tube worms and snip off bites before the worms could retreat into their shelters! Most beautiful was the Rose Garden. There were hundreds of red-plumed tube worms, some nearly as big around as your wrist.' "

"The animals Al described belong to a complex ecosystem," wrote marine botanist Sylvia A. Earle, "all of them heretofore unknown and the most conspicuous—giant tube dwellers—so different that they are being classed as a separate subphylum of animals." Earle, a contributor of numerous articles to the Magazine, continues in the Special Publication *Exploring the Deep Frontier*:

> A significant discovery is that large amounts of hemoglobin—the oxygen-carrying substance in human blood—courses through the bodies of the Galapagos tube worms and accounts for the red color.
>
> The most exciting biological discovery concerning the newfound ecosystem is how it seems to sustain itself. On land, plants derive energy from the sun through photosynthesis. But here, close to the vents, the basic food chain begins instead with sulfur bacteria that obtain their energy from a chemical reaction. Such systems, moreover, may not be so rare in the deep sea, if—as can now be assumed—more vents exist. Later dives in *Alvin* in

◀ ─────────────────────────────

"A weapon of war? A symbol of authority? On the seabed off Ulu Burun near the Turkish town of Kaş, a diver in 1984 ponders a mace head of stone from a vessel that sank about 1375 B.C."—From the 1985 Book Service publication *Adventures in Archaeology*

Above: *Two-foot-long* Metridium
senile, *sea anemones in the Strait of
Georgia, an inland sea separating
mainland Canada from Vancouver
Island. These most primitive of multicellular animals, photographed by
David Doubilet, thrive in areas of
swift current.*

Right: *This brilliantly colored jellyfish,* Olindias formosa, *was encountered by Anne Levine Doubilet at the
entrance of a cave in Japan's Izu
Oceanic Park at a depth of 130 feet.
The Japanese nickname for it,* hana
gasakurage, *means cherry blossom.*

Above: *This nearly six-foot-tall tree-like coral, an alcyonarian, grows in Izu at a depth of 165 feet. The foliage and trunk of the "tree" are actually a colony of millions of soft-coral animals. And what appear to be leaves are clusters of polyps, opened to feed on current-borne plankton.*

1979 near the tip of Baja California resulted in the discovery of another system, including many groups of organisms similar to those in the Galapagos Rift.

The Baja dives caused even greater excitement when geologists observed chimneys of rock spewing forth black clouds of minerals: iron, copper, zinc, and silver sulfides. "It may be that minerals we find on dry land today were formed by a process such as we witnessed," [scientist-explorer] Robert Ballard commented. "The deep-sea discoveries made during the coming decade should help solve more of the mysteries of earth's origin—and may provide valuable new sources of minerals as well."

The foot-long clams and bright red worms, encased in tubes as long as eight feet, that exist at depths of 9,200 feet around the Pacific vents seem not to exist at the 3,280-foot-deeper vents in the Atlantic. These vents, formed like those in the Pacific as the plates that make up the sea floor slowly spread apart, are home to free-swimming animals such as blind shrimp, an eel-like fish, and a strange six-sided animal that researchers believe is related to an ancient organism known only from Alpine Eocene-era deposits. And from these vents, mineral-rich water spews at temperatures of 662° F—hot enough to melt protective plastic coverings on *Alvin*'s claws. More recently, researchers in *Alvin* discovered off the coast of Oregon colonies of tube worms that feed not off sulfides but methane.

To the layman, *Alvin*'s most dramatic voyage of discovery was not to observe strange new forms of life, but, in 1986, to bear witness to the grim specter of a legendary death. A year earlier, on September 1, 1985, in the North Atlantic, *Argo*, a search vehicle containing video cameras, side-scan sonar, a computerized timing system, and a host of other electronic gear, was being towed behind the U.S. Navy research vessel *Knorr*. Suddenly the first ghostly images of a once-majestic ship were displayed on the *Knorr*'s video screens. "I cannot believe my eyes," Robert Ballard reported in the December 1985 *Geographic*:

> From the abyss two and a half miles beneath the sea the bow of a great vessel emerges in ghostly detail. I have never seen the ship—nor has anyone for 73 years—yet I know nearly every feature of her. She is S.S. *Titanic*, the luxury liner lost after collision with an iceberg in 1912 at a cost of 1,522 lives.
>
> The sea has preserved her well...the lines of the deck's teak planking are visible beneath a thin coating of "snow" formed by remains of marine organisms. Other features stand out in the strobe lights of our towed undersea vehicle. Twin anchor chains run from windlasses...beneath a tangle of cables to hawsepipes near the bow. A ventilator shaft lies open between the chains, and capstan heads stand on either side....

On different camera passes across the field of debris, bottles of wine, bed springs, chamber pots, bits of twisted railing, boiler coal, a metal serving platter all showed up on film. But it wasn't until the return trip, a year later, that *Alvin*, and its pilots, Ralph Hollis and Dudley Foster and Robert Ballard, de-

---

In 1983, diver Doug Osborne, encased in "a submarine you can wear," called WASP, prepared to lift the coral-encrusted wheel of the Breadalbane, *a British ship lost in the Arctic 130 years before.*

236

scended 12,500 feet to view the *Titanic* for themselves. "My first direct view of *Titanic* lasted less than two minutes, but the stark sight of her immense black hull towering above the ocean floor will remain forever ingrained in my memory," Ballard wrote in 1986, then added:

> In a way I am sad we found her. After 33 hours of exploring her dismembered hulk, we know her fate, and it is not a pretty sight. Though still impressive in her dimensions, she is no longer the graceful lady that sank a mere five days into her maiden voyage, in 1912....Her beauty has faded, her massive steel plates are dissolving in rivers of rust, and her ornate woodwork has been devoured by an army of countless wood-boring organisms whose hollow calcium tubes now litter her barren shape. After years of gluttony the creatures starved and dropped dead at the table. I have no sympathy for them; they robbed *Titanic* of her last touch of elegance.
>
> *Titanic*'s band has long since ceased to play. She is gone, home-ported at last. She will surely never be raised.

As stunning and romantic as was the adventure of seeing the *Titanic* with his own eyes, Ballard recognizes the limitations man's presence imposes:

> The day is fast approaching when that job can be done faster and better without man's physical presence in the sea....Until recently there has been no way of duplicating...man's sophisticated eyes and brain and articulating hands to solve complex problems or perform difficult tasks. Whatever the cost, and the risks, of transporting man into the deep, it has been worth it.

For the previous thirteen years Ballard had spent an average of four months out of the year at sea and logged countless hours crouched beneath the surface in one submersible or another. Many of his dives were made in partnership with the National Geographic Society—exploring the Mid-Atlantic Ridge, descending 20,000 feet into the Cayman Trough, studying the unique forms of life surrounding the warm springs in the Galápagos Rift, and discovering on the East Pacific Rise "black smokers"—hydrothermal vents that spew fluids hot enough to melt lead. "Certainly the dives were exciting," Ballard said, noting that they were "historic achievements, every one of them. But," he added,

> in 13 years I had managed to explore a mere 40 miles of undersea mountain range. There are more than 40,000 miles of such ranges throughout the world's oceans. Did I really want to spend the rest of my life in the hope of exploring another 80 or 100 miles at best?
>
> It seemed to me we had a choice. We could continue indefinitely with manned submersibles, which are limited in the time they can spend below by both their passengers' endurance and their expendable power supply.... [or] we could begin thinking of remotely operated deep-sea vehicles, sophisticated robots that could give man...a "telepresence" in the sea, an extension of his unique senses and capabilities to extreme depths without physically transporting him a foot below the surface.
>
> Through such robots man could remain under the sea for weeks instead of mere hours at a time, extending his reach immeasurably into earth's last great uncharted frontier. Equally important, via live television these machines could bring the wonders of the deep to countless millions rather than to the lucky few who are able to ride in submersibles.

When that happens, those pictures, too, are certain to appear in the *Geographic*.

Above: *One of the most widely used research submersibles ever built, the manned underwater vehicle* Alvin, *laden with instruments and remote-control sensors, has dived more than 1,700 times since being commissioned in 1964.*

Opposite: *The* Alvin's *crewmembers' unexpected 1977 discovery on the sea floor of bright red worms encased in white tubes—some as long as eight feet—clustered near a warm-water vent in the Galápagos Rift prompted scientific study of these oases of life in 1979. But to the layman the most dramatic voyage of discovery was* Alvin's *descent in 1986 to a depth of 12,500 feet to view the sunken remains of the once majestic White Star ocean liner* Titanic, *which sank with a loss of 1,522 lives after its collision with an iceberg on its maiden voyage in 1912.*

Overleaf: *Aboard* Alvin, *shown in this painting by Pierre Mion hovering over* Titanic's *crumpled bow, was* Robert D. Ballard, *who wrote in the December 1986* National Geographic, *"The stark sight of her immense black hull towering above the ocean floor will remain forever ingrained in my memory."*

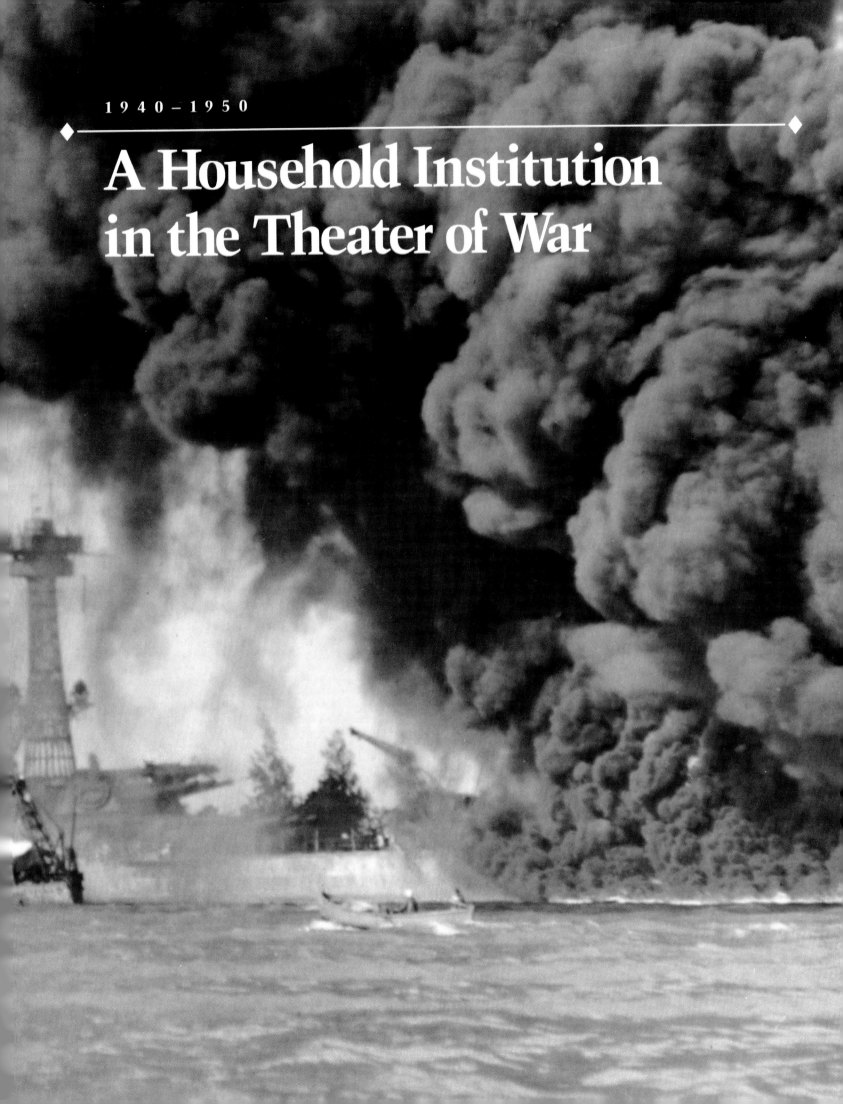

A Household Institution in the Theater of War

"...Under your direction The Society's Magazine has become a household institution in the homes of America and throughout the nations of the world—in short, wherever there is a postal system and wherever geographic knowledge is esteemed."

—From a letter written to Gilbert H. Grosvenor by President Harry S Truman, read aloud at Constitution Hall to the 4,000 guests attending Grosvenor's fiftieth anniversary as Editor celebration, May 19, 1949

On July 26, 1942, the following editorial, "Footnote to Geography," appeared in *The New York Times:*

For years Americans have been learning geography by the most pleasant method imaginable. They have learned about their own country at first hand and from automobile road maps. They have learned about the far corners of the world from newspaper accounts of famous flights and from the National Geographic Society, which itself seemed always to be sending out a new expedition to find and report on some unknown corner of the planet.

Since the war began we have been learning still more geography, less pleasantly. Laconic military communiqués have told, often in veiled language, what was happening on remote battle fronts. Correspondents have gone with the armies to places remote even in the encyclopedias. And the newspapers have striven mightily to keep up with them for the daily readers.

Inevitably, the newspapers have had to lean on many sources to make intelligible the brief and sometimes cryptic dispatches and communiqués. And chief among those sources is the old reliable National Geographic Society. Not only has it given generously of its information and opened its magnificent geographic library; it has published splendid war maps and it issues daily bulletins to assist the newspapers and news services.

It is highly reassuring, when such place names as Staryi Oskol and Zivotin crop up in the communiqués, to be able to ask somebody what and where they are—and get the right answers. The N. G. S. hasn't failed us yet.

The first map supplement to have appeared in the *National Geographic* was the "Theatre of Military Operations in Luzon," distributed in the June 1899 Magazine, which marked Gilbert H. Grosvenor's first appearance on the masthead as Assistant Editor. The plates for that map had been borrowed from the government. But as the Society's membership and income grew, the young Editor was able to cease borrowing map plates or ordering maps from commercial cartographers and, eventually, as Grosvenor later recalled, "I was able to organize a highly competent research and cartographic staff to design and produce distinctive maps and to contribute original techniques and projections to the science of cartography."

Preceding spread: *Mortally wounded U.S. Navy battleships* West Virginia *and* Tennessee *lie shrouded in smoke after the December 7, 1941, Japanese carrier-plane surprise attack on the U.S. fleet anchored at Pearl Harbor, Hawaii.*

244

In 1916, Albert H. Bumstead, an accomplished mathematician, was appointed the Society's first chief cartographer. And it was under his tutelage that a corps of Geographic mapmakers began turning out the Society's remarkable maps—a production that, perhaps, reached its pinnacle of worldwide distinction during World War Two.

On January 15, 1945, while U.S. forces were recovering from the Battle of the Bulge, *The New York Times*, again in an editorial, saluted the Geographic, this time calling its mapmaking "probably the most ambitious cartographical undertaking on record," adding:

> The maps are to be found at the front, in the air, in our embassies and consulates, in business and newspaper offices, in schools. As a whole they constitute the most comprehensive atlas and gazeteer [sic] ever compiled—no mean achievement at a time like this....The Society deserves thanks for having undertaken voluntarily an important task in education. Its maps not only enable us to follow the war's progress but convince us, as never before, that China, Australia, and Europe are our next-door neighbors.

The *Times* editorial might have mentioned that the Society's maps had found their way into two other very high offices, as well. "One morning, just two weeks after Pearl Harbor, an aide to President Franklin D. Roosevelt called upon me at The Society's headquarters and asked for a map showing a town near Singapore then under Japanese attack," Gilbert H. Grosvenor later wrote.

The aide confided that he had come at the request of the President, who liked to see on a map the location of places in the war news. That morning, however, Mr. Roosevelt had not been able to find the headline-making little town on charts available in the White House. However, we had a map...and [I] gave it to the aide.

That afternoon, I ventured to send President Roosevelt a cabinet containing National Geographic Society maps mounted on rollers. These rollers were so conveniently arranged that the President, while sitting at his desk, could pull down the map of any area in the world that he wished to study....

Within an hour of its arrival at the White House, President Roosevelt had the cabinet mounted directly behind his study chair so that it was within easy reach....The President's collection of National Geographic maps attracted the attention of more than one distinguished visitor. Prime Minister Winston Churchill was so keenly interested that President Roosevelt telephoned me and asked if The Society would make a duplicate cabinet for him to give Mr. Churchill.

Inscribed "WSC from FDR, Christmas 1943," the case with its maps covering the world was taken by President Roosevelt in his airplane to Cairo and given Mr. Churchill as a personal Christmas gift from the Chief Executive.

National Geographic Society maps, tacked up in kitchens, in dens, in youngsters' bedrooms all over this nation, enabled an entire generation of Americans to chart the daily progress of World War Two. Often, when battlegrounds shifted, that month's issue of the *Geographic*, with uncanny prescience, supplied the appropriate map. When the Germans seized Austria in March 1938, the Society issued a map of "Europe and the Mediterranean" in April. In October 1939, one month after the Nazis invaded Poland and annexed Danzig, and Britain and France declared war on Germany, members received a map of "Central Europe and the Mediterranean, as of September 1, 1939."

"Europe and the Near East" was issued in May 1940 as some 900 vessels of all types and sizes evacuated nearly 350,000 British and French troops from the beaches at Dunkirk. In March 1941 the Society issued a map of the "Indian Ocean Including Australia, New Zealand and Malaysia"; in May, the German battleship *Bismarck* was sunk. In June, the Germans invaded Russia; two months later, Churchill and Roosevelt met and signed the Atlantic Charter. The map of the "Atlantic Ocean," issued in September, covered the seas in which the German U-boat "wolf packs" were operating at peak, and, in time for Pearl Harbor three months later, members received a new "Map of The World."

The "Theater of War in the Pacific Ocean" reached members with the February 1942 issue. That following September, Fleet Admiral Chester W. Nimitz, flying to Henderson Field on Guadalcanal in a B-17, became lost in a driving rainstorm over the Coral Sea. "Studying the big 'Pacific Ocean' map's Solomon Islands inset," Gilbert H. Grosvenor later reported, "the Wartime Commander in Chief of the Pacific Fleet and his ten American and New Zealand aides identified the shore line of San Cristóbal Island in the southeastern Solomons below them. They were thus enabled to put the plane back on its course."

The Pacific Ocean map was followed by a map of "North America" in May 1942; the map of the "Theater of War in Europe, Africa, and Western Asia" in July; "South America" in October; and, in December 1942, "Asia and Adjacent Areas."

Fourteenth Army

U.S. military propaganda intended to mislead the enemy duped the Society as well. Included in its 1944 Insignia and Decorations of the U.S. Armed Forces *booklet were twenty-one shoulder sleeve insignia for nonexistent U.S. Army units. Among the fraudulent patches depicted were those for the XXXI Corps, XXXIII Corps, Sixth and Ninth Airborne Divisions; and the Eleventh, Fourteenth, Seventeenth, and Twenty-second Divisions and the Eighteenth and Twenty-first Airborne Divisions—all part of a fake Fourteenth Army's equally fictitious two corps and nineteen divisions.*

XVI Corps

XVIII Corps

XIX Corps

XX Corps

XXI Corps

XXII Corps

XXIII Corps

XXXI Corps

XXXIII Corps

Antilles Department

Hawaiian Department

Panama Canal Department

Philippine Department

First Division

Second Division

Third Division

Fourth Division

Fifth Division

Sixth Division

AIRBORNE

Sixth Airborne Division

Seventh Division

Eighth Division

Ninth Division

AIRBORNE

Ninth Airborne Division

ARMY SHOULDER SLEEVE INSIGNIA
(Pages 89-96)

Tenth Light
Division

Eleventh Division

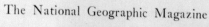

Eleventh Airborne
Division

Thirteenth Airborne
Division

Fourteenth Division

Seventeenth Division

Seventeenth Airborne
Division

Eighteenth Airborne
Division

Twenty-first Airborne
Division

Twenty-second Division

Twenty-fourth
Division

Twenty-sixth
Division

Twenty-seventh
Division

Twenty-eighth
Division

Twenty-ninth
Division

Thirtieth
Division

Thirty-first
Division

Thirty-second
Division

Thirty-third
Division

Thirty-fourth
Division

Thirty-fifth
Division

Thirty-sixth
Division

Thirty-seventh
Division

Thirty-eighth
Division

Fortieth Division

Forty-first Division

Forty-second Division

Forty-third Division

Forty-fourth Division

ARMY SHOULDER SLEEVE INSIGNIA
(Pages 89-96)

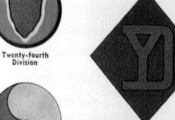

"GI Joe Shares 'Mess Kit Luck' with Two Young Admirers. The shoulder sleeve insignia identifies him as a member of Lt. Gen. Mark W. Clark's Fifth Army in Italy. Lt. Gen. Lucian K. Truscott, Jr., succeeded General Clark as commander of the Fifth Army," was the caption that ran with this photograph in the Magazine.

By then Lt. Col. Jimmy Doolittle's sixteen carrier-launched B-25s had bombed Tokyo; the Bataan Death March had taken place; Russians and Germans were fighting in the ruins of Stalingrad; and the murder of millions of Jews and political prisoners by the Nazis had begun.

In 1943, the Allied forces in North Africa were placed under General Dwight D. Eisenhower's command; the Japanese were driven off Guadalcanal; U.S. forces landed in New Guinea; and the Allies invaded Sicily, then mainland Italy. Society maps of "Africa" were issued in February, the "Northern and Southern Hemispheres" in April, "Europe and the Near East" in June, and the "Pacific Ocean and the Bay of Bengal" in September, the month Italy surrendered. An updated "World Map" came out that December.

In April 1944, at the suggestion of the United States Army, the Society issued a map of "Japan and Adjacent Regions of Asia and the Pacific Ocean." The center of the map was the heart of Tokyo.

"So highly did the Army Air Forces regard this map of Japan," Gilbert Grosvenor recalled, "that they immediately borrowed the original drawings from The Society and made 5,000 enlargements on sheets 50 by 65 inches for use in over-all planning in the air offensive...."

On June 6, 1944, D-Day, the Allies invaded France; 1,000 transports and gliders dropped paratroopers on Normandy early that morning while 1,000 R.A.F. and 1,400 U.S. bombers provided air support and attacked installations. At 6:30 A.M. assault troops from British, U.S., French, and Polish divisions waded ashore along the sixty-mile-long Carentan-Bayeux-Caen beachline and fought their way inland.

In Italy, Monte Cassino and Rome were occupied by the Allies.

And the following month a National Geographic Society "Map of Germany and Its Approaches" was issued to members. The Army Corps of Engineers made enlargements of this map for use by ground and air forces in Europe and posted copies at road intersections along the Allied forces' routes of advance into Germany.

Lt. Gen. Eugene Reybold, then Chief of Engineers, U.S. Army, wrote Gilbert H. Grosvenor: "Our dependence upon the National Geographic Society for the type of map it issues has, in fact, led us to consider The Society an integral part of our military mapping establishment."

Nine times during a single year the *National Geographic Magazine*'s plates for its big ten-color wall-map supplements had to be put back on the presses to supply the needs of the Army, Navy, and other government agencies.

In August 1944, Allied troops liberated Paris.

In October, as the membership received its map of "Southeast Asia and Pacific Islands From the Indies and the Philippines to the Solomons," General MacArthur returned to the Philippines. That same month the Battle of Leyte Gulf took place—the biggest naval action ever waged. American and Japanese ships fought in three separate engagements: in the Surigao Strait, off Samar and off Cape Engaño. And Japanese naval power suffered a devastating blow.

By February 1945, Manila was recaptured; Corregidor was reoccupied by the first of March. That month's issue had a new map of "The Philippines."

Hitler committed suicide on April 30, 1945; German units began surrendering on May 4, and unconditional surrender was signed three days later, on May 7, 1945. V-E Day was proclaimed the following day.

On August 6, 1945, the United States dropped an atomic bomb on Hiroshima, Japan, population about 350,000. On August 9, a second atomic bomb

Left: *"No Time to Prink in the Mirrorlike Tail Assembly of a Liberator"* stated the unblushing caption for this photograph, and it went on to point out: *"About a third of the country's aircraft workers are women. Many jobs they do as well as men; some they do better. Riveting, a kind of needle point in metals, is one of women's standout operations."*— From the August 1944 issue

Below left: *Wartime gas rationing was absent from this photograph's cheery caption, "Soldiers, WACS, and Government Girls Sight-seeing by Bicycle Make Washington Look Like Amsterdam-on-Potomac."—From the September 1943 issue*

Below: *This photograph of Jane Russell wearing a Dorothy Lamour–type sarong and posed in front of several National Geographic Society maps never appeared in the Magazine.*

was dropped, on Nagasaki, population about 250,000. More than 150,000 Japanese were killed. On August 14, V-J Day, Japan surrendered. The formal surrender ceremony took place on board the U.S.S. *Missouri*, anchored in Tokyo Bay, on September 2, 1945.

In the course of the war the armed services "drew heavily on The Society's collection of 53,000 maps from all over the world," Gilbert Grosvenor later wrote, "many of them large-scale charts of strategic cities, harbors, railroads, air routes, and caravan trails; the vast file of unpublished manuscripts containing geographical data unavailable in print; and our library of some 20,000 travel books. Scores of Government agencies consulted The Society's research departments daily for elusive geographical information needed in war planning."

Even before the Japanese bombed Pearl Harbor the Army and Navy intelligence services were given access to the Society's entire collection of photographs. Although originally taken for "peaceful educational purposes," these photographs provided invaluable documentation of harbors, beaches, industrial sites, bridge crossings, airfields, waterfronts, railroad marshaling yards and other potential targets. Suddenly all those photographs taken over the previous decade showing European sites "as seen from the air" and "an airplane view of" were invaluable when compared with reconnaissance photographs because camouflage and new construction could be readily distinguished. By the war's end over 35,000 prints of National Geographic photographs, selected by the government, had been delivered to the appropriate intelligence agencies.

At its July 2, 1949, convention in Chicago, the Air Force Association presented the National Geographic Society with a Citation of Honor:

> World War II introduced global conflict for the first time in history, and with it the demand for cartographic information to guide American airmen and airplanes to the far corners of the earth and back again.
>
> That the military services could not meet this demand was part of national unpreparedness. That a non-military agency, the National Geographic Society, could help fill the breach was at once a tribute to the significance of this organization and a testimonial to the civilian contribution that made victory possible.
>
> The maps and charts of The Society guided men of the Air Force over the waters of the Atlantic and the ice caps of the Far North; over the islands of the Caribbean and the jungles of South America; helped build the air routes of Africa; went with the men of the Air Force over the Himalayas from India to China; took airmen up the long, hard island route of the Pacific from near defeat in Australia to victory in Japan.
>
> For its ability to meet the emergency requirements of its country, for its invaluable contributions to the Air Force in accomplishing a global mission, the National Geographic Society is awarded this Citation of Honor for outstanding public service.

Despite the fact that scores of editorial and clerical workers had entered the armed services, the *Geographic* continued to print the sort of articles it had run during World War One—timely, informative, chatty, often relentlessly cheerful pieces like: "Lisbon—Gateway to Warring Europe" ("Lisbon is filled with representatives of foreign powers....Some are spies, but the majority are engaged in ordinary commercial work") and "Lend-Lease is a Two-way Benefit" (with a photograph of smiling English school children, spoons raised over bowls of vegetables, or waiting in line for breakfast, plates in hand, captioned "Who Wouldn't Share His Vegetables to Keep These Little Allies Healthy and

Above: *Utah Beach just after D day, June 6, 1944. Landing craft unload tanks, trucks, and supplies on the beach in Normandy, France, as part of the largest invasion armada in history. The following month the Society issued its "Map of Germany and Its Approaches" and the Army Corps of Engineers posted enlarged copies of it at road intersections along the Allied Forces' routes of advance into Germany.*

Left: *"The armed services...drew heavily on The Society's collection of 53,000 maps from all over the world," Grosvenor wrote in his 1957 history of the Society. Chief of Staff General Dwight D. Eisenhower, shown here in his post-war Pentagon office, consults a Society map of Germany like that which traveled with him during his 1945 offensive.*

Smiling!"). And, too, there were the sort of upbeat military service-oriented articles seen before: "QM, the Fighting Storekeeper," "Your Dog Joins Up," "When GI Joes Took London," and "Okinawa, Threshold to Japan," with photographs by a young Lt. David Douglas Duncan, U.S.M.C., who would later gain fame as a photojournalist for his *Time* and *Life* magazine combat photographs of Marines in Korea and Vietnam. (The Okinawa photographs, however, were post-battle.)

In the meantime, the Society's News Service was sending daily news bulletins to hundreds of newspapers with information on the battlefields (a typical bulletin might have discussed the terrain, the flora and fauna of the battle scene and provided a guide to the pronunciation of the more difficult area place names); and the familiar *School Bulletin* was furnishing much the same service for its tens of thousands of teacher and student recipients. Though short of manpower, the Society's "chin up in the face of adversity" was expected by the membership. The Society, in turn, expected no less of its family of members, and in July 1943 the *Geographic* "did something that no other publication would dream of," *Newsweek* reported. "To an experimental block of 100,000 readers Dr. Grosvenor calmly suggested that 1944 dues be paid now so as to lighten year-end billings. Four days later, 2,000 favorable replies had come in. The naturalist-editor wore a pleased smile."

What sort of people were the 1943 Society's 1,199,738 loyal members? In a 1941 Society booklet, under the heading "How National Geographic Family Heads Earn Their Incomes," 529 different categories were listed, among them: dentists and dental surgeons (11,715), executives (13,710), clerks (32,589), barbers (1,557), undertakers (3,000), bartenders (39), politicians (228), senators (159), congressmen (123), college matrons (261), college presidents (456), masseurs (156), poets (15), tropical-fish raisers (3), clergymen (27,843), farmers (36,816), bankers (15,084), philanthropists (9), physicians and surgeons (53,514), road builders (207), lighthouse keepers (126), housewives (39,543), and royalty (114).

And it is perhaps because of the Editor's deliberate effort to make the membership feel that they were all part of one big, happy family that in addition to the thousands of affectionate letters the Magazine received, it also received its share of rebukes: "Don't take us through any more Himalayas or over any more glaciers."

Others were equally blunt:

"Captions poor."

"Text poor."

"Don't care for North and South Pole articles."

"Don't care for prehistoric bones, carvings, or figures."

"Did not care for the Cats."

"Less tiresome discourse."

"Europe, Asia, and Africa do not interest us."

"Authors are not all they should be."

And one that must have cut the Chief to the quick: "Birds—less."

All letters, of praise or rebuke, were answered, although some, such as the reply to the boy who wrote simply, "I am studying the world...send me everything you have," must have given pause.

On February 15, 1942, the Washington *Star*'s Sunday section reported on the preventive measures taken by the Society against air raids in a piece headlined GEOGRAPHIC SOCIETY RAID DRILL EMPTIES BUILDING IN 3 MINUTES and with

" 'Uncle Sam Is a Good Egg,' British
School Children Agree. The black-
board tells the story as hungry
youngsters take bacon and eggs to
frying pans in September, 1941. Then
an egg was an event, and it was still
in its shell. Not so today; dehydration
removes shell and moisture, too."
— From the June 1943 issue

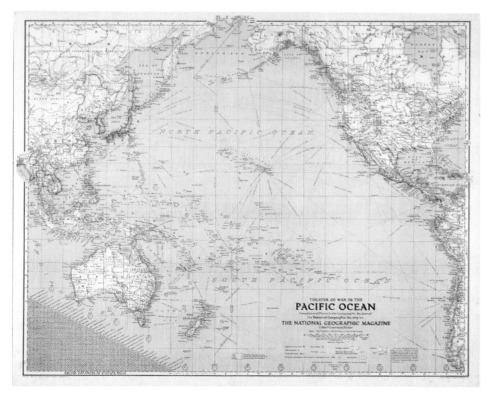

Above: *General Douglas MacArthur receives the Japanese surrender on September 2, 1945, aboard the battleship U.S.S.* Missouri *anchored in Tokyo Bay. Signing for Japan is Major General Yohijiro Umezo.*

Fleet Admiral Chester W. Nimitz, standing farthest left with other senior military and naval officers of the Allied powers, would later write Gilbert H. Grosvenor that the Society had "lent an unexpected but most welcome hand" when the Commander in Chief of the Pacific Fleet was aboard a B-17 flying from Espíritu Santo and became lost in a pouring rainstorm. Nimitz and his companions, with the help of the Society's "Theater of War in the Pacific Ocean" map (right), were able to recognize islands beneath them and landed safely at Guadalcanal's Henderson Field.

the subhead: *226 Employes* [sic] *Fill Evacuation Room; Sand Bags, Pumps Kept in Readiness.* The article went on to say:

> Probably no building in Washington is better prepared to protect its occupants from death or injury in an air raid than the Geographic Society headquarters. The drills going on are part of a well-planned program which the society believes will provide ample protection to its employes if air raids should start.
>
> Charles Althoff, an employe in the shipping department, merely pushed a button and three long blasts blared from seven automobile horns placed at strategic locations. Before the series of three blasts could be repeated, the first employes were filing in the safest room in the building—the cafeteria in the basement.
>
> While the alarm was being sounded again, Dr. Gilbert Grosvenor... reached the evacuation room....
>
> Few things have been missed in making the evacuation room safe from almost anything but a demolition bomb. The society's defense chieftains claim the cafeteria is practically bomb-proof and concussion-proof. Windows on the side next to an alley have been bricked on the outside. Doors leading into the cafeteria have been covered with monk's cloth, and a 16-inch-thick brick wall, 1½ feet higher than the basement windows, has been built along the entire front of the building to prevent bomb fragments from crashing in the windows.
>
> Dozens of big bags full of sand are ready for instant use in barricading three doorways leading from the cafeteria to the kitchen....
>
> A special room has been constructed in another section of the basement for the defense work. Here are located the mechanism which will set the automobile warning horns in operation, as well as all sorts of supplies such as shovels, pick-axes, hoes, radio equipment, bottles of distilled water, asbestos shields and stirrup pumps for fighting fires, flashlights, hundreds of feet of hose and all the other necessary equipment....
>
> Each secretary, stenographer, file clerk and typist has a small canvas bag tacked on the side of her desk. When an alarm sounds, they quickly jam all their important documents into the bag and drop it into large galvanized cans placed in each room. The cans then are picked up by the warden staff and taken to the basement.

The article—illustrated by photographs of a woman employee looking at a stack of sandbags placed in front of the Society headquarters' kitchen door, of Dr. Grosvenor watching a male employee pretending to telephone Society plane spotters on the headquarters' roof, and of the building warden pointing at the sixteen-inch-thick brick wall—then went on to report how the Society's "priceless collection of photographs...are in fireproof vaults on an upper floor," and that the Explorers' Room exhibits were considered safe on the first floor.

However, when air raid drills weren't being held, life in the Society's Washington headquarters continued seemingly undisturbed.

In the fall of 1943 *The New Yorker* magazine published "Geography Unshackled," a three-part profile of Gilbert H. Grosvenor written by Geoffrey T. Hellman. In it, Hellman described the *Geographic*'s Editor as "a kindly, mild-mannered, purposeful, poker-faced, peripatetic man of sixty-seven, endowed with the sprightly air of an inquiring grasshopper, with a clear, pink complexion, and with the mixture of business sagacity, intellectual curiosity, regard for tradition, and tolerance of temperate innovation that is sometimes found in the president of a fairly wealthy college," and told how, "around four or five in the afternoon, just before he leaves for the day, the editor of the *Geographic*,

"A BIKINI FAMILY SAYS FAREWELL AT THE GRAVE OF A LOVED ONE" was the title of a picture illustrating Carl Markwith's July 1946 National Geographic *article "Farewell to Bikini." The Bikinians were being moved from their atoll so that their homeland could be used for atomic tests. "Civilization and the Atomic Age had come to Bikini," Markwith wrote, "and they had been in the way."*

who does not touch alcohol or tobacco or accept advertisements for them in his periodical, pours some powdered Ovaltine onto a shoehorn, deposits it in a Lily cup filled with...[warm milk], stirs it with a penholder, and drinks it down with a pleased expression."

Nowhere, perhaps, was life at the Society during this period better captured than in this description of Hellman's:

> The twenty or so editors of the *Geographic*, all of them notably polite individuals who are forever bowing to one another through the sculptured bronze doors of their elevators, move in a leisurely welter of thick rugs, bookshelves stocked with British and American *Who's Who's*, encyclopedias, bound volumes of the *Geographic*, and geographical books published by the Society; interoffice memorandum slips with "Memorandum from Dr. Grosvenor," "Memorandum from Dr. La Gorce," or the like printed at the top; incoming correspondence stamped, in purple ink, "Commendation," "Criticism," or "Suggestions;" and photostatic copies of appreciative letters from generals, admirals, coördinators, and ex-Presidents to whom the Society has presented maps or turned over geographic research, much of it bearing on the war. The editors of the *Geographic* freshen up in rooms where the towels are hung under a hierarchy of name cards; in one room the cards read, from left to right, "Dr. Grosvenor, Dr. La Gorce, Mr. Fisher, Mr. M. B. Grosvenor, Mr. Simpich, Mr. Hildebrand, Mr. Bumstead, Dr. Williams, Mr. Borah, Mr. Riddiford, Mr. Canova, Mr. Vosburgh, Mr. L. J. Canova, Guests, Guests, Mr. Nicholas." For business errands around town, such as trips to the magazine's circulation annex or to its printers, the firm of Judd & Detweiler, the editors have a fleet of five chauffeur-driven cars. They lunch in three private dining rooms, drinking buttermilk, exchanging puns, and ordering excellent à-la-carte meals from menus on which prices are discreetly omitted. One of these dining rooms is reserved for Dr. Grosvenor, and he sometimes eats there alone, like the captain of a ship. This room adjoins a bathroom containing two marble shower baths, which in sticky weather are occasionally patronized not only by Dr. Grosvenor but other of the more important staff members. Next to the private dining rooms is a cafeteria where lesser employees lunch well and inexpensively but without access to as wide a selection of food as is afforded their betters. Here the sexes are segregated, a partition separating the men from the women. "We prefer not to mix 'em up," an officer of the Society once explained. "We feel the men are freer in their own room. If they want to swap a risqué anecdote or two, they can do it." Every now and then, private-dining-room delicacies are displayed, by mistake, in the cafeteria. On one of these occasions, a stenographer, sliding her tray along, pointed to a dish which had captured her fancy. "My goodness, no," said the counter maid. "That's the officers' liver."

Hellman, noting also in his *New Yorker* profile Gilbert Grosvenor's interest in birds, pointed out:

> This war, like the last one, has made the editor of the *Geographic* go easy on birds in his magazine, recent issues of which have featured articles on United States food production, the Alaskan highway, the Coast Guard, Army dogs, convoys, aircraft carriers, women in uniform, and military and naval insignia, but he knows that wars are fleeting affairs compared to birds and he does not propose to wait for the end of the conflict before again doing justice, editorially, to his favorite topic. "The Chief has been begging us to run a series of color photographs on the wrens of Australia for the past three years," a *Geographic* editor recently told an acquaintance. "Everyone conspires to keep him from using these goddam birds. We keep putting him off, but he'll sneak them in any month now."

The first recipient of the Grosvenor Medal was Gilbert H. Grosvenor, to whom it was awarded on May 19, 1949, for "outstanding service to geography as editor of the National Geographic, 1899–1949."

This Selz caricature accompanied Geoffrey T. Hellman's "Geography Unshackled," a three-part profile of Gilbert H. Grosvenor that appeared in The New Yorker *in the fall of 1943. Hellman described Grosvenor as "a kindly, mild-mannered, purposeful, poker-faced, peripatetic man of sixty-seven, endowed with the sprightly air of an inquiring grasshopper."*

He did. But not until the October 1945 Magazine article "The Fairy Wrens of Australia."

The post-war cartographic staff was kept busy, since the upheavals of World War Two had resulted in the realignment of nations, the redrawing of national boundaries, and the renaming of countries and cities. The remapping of almost the entire face of the globe was required.

The Atomic Age, ushered in by Hiroshima and Nagasaki, was marked in October 1945 by "Your New World of Tomorrow" and in July 1946 by Carl Markwith's "Farewell to Bikini," a piece about moving the native inhabitants of Bikini Atoll in the Ralik Chain of the Marshall Islands to another island so their home could be used for atomic tests. "As good-byes were being called back and forth, I found myself wishing that I could say, as I had each time before, *Kim naj drol ilju*—'We shall return tomorrow.' I refrained," Markwith wrote, "because there would be no returning for me—nor perhaps for them. Civilization and the Atomic Age had come to Bikini, and they had been in the way."

In January 1947, the Magazine published Melville Bell Grosvenor's "Cuba—American Sugar Bowl," illustrated by forty-two of his color prints, a record for the number of color photographs used in a single story.

"The article, full of hope," the current *National Geographic* Editor Wilbur E. Garrett wrote in 1984, "gave little hint of the forces and problems even then incubating in Cuba that would soon lead this seemingly semicolonial island nation to become a focus of concern and frustration for United States Latin American policy for decades and even to this day. Yet Cuba was only symptomatic of the postwar storm of change that would toss nations, colonies, and continents like rag dolls in the hands of angry children. Sweeping change was overtaking the map of Africa. Iron, bamboo, and no-name curtains cut off hundreds of millions of people from one another."

No such changes swept over the *Geographic*; instead it was more of the same: pieces like "Carefree People of the Cameroons," "Weighing the Aga Khan in Diamonds," "'Flying Squirrels,' Nature's Gliders," "Yemen—Southern Arabia's Mountain Wonderland," "First American Ascent of Mount St. Elias," "Rhode Island, Modern City-State," and, in April 1949, an entire issue devoted to "The British Way."

One month later, on May 18, 1949, at the Chevy Chase Club, just outside Washington, in Chevy Chase, Maryland, the Society's trustees honored Gilbert Hovey Grosvenor's fiftieth anniversary as Editor of the *National Geographic Magazine* with an unofficial dinner attended by those of the staff whose names appeared on the Magazine's masthead, by department heads, and by a few personal friends. The following evening, at the anniversary celebration held in Washington's Constitution Hall, 4,000 guests witnessed the first presentation of a newly designed gold disk with a bas-relief profile of Gilbert H. Grosvenor on one side and signs of the zodiac and a globe on the other. Only an outsider to the Society might think it odd that the first recipient of the new Grosvenor Medal should be Gilbert Hovey Grosvenor himself. Within the Society, however, it raised no more questions than would a monarchy based on the divine right of kings.

"Always accepting praise modestly," Albert W. Atwood wrote in "Gilbert H. Grosvenor's Golden Jubilee" for the Magazine's August 1949 issue, "this generous, gracious, and gentle Editor invariably gives unstinted credit to others. He said at the anniversary celebration in Constitution Hall:

I would not have you exaggerate my part. I realize more keenly than anyone else possibly can that the success of the National Geographic Society, which you generously ascribe so largely to my humble efforts, was brought about by the wise counsel and unswerving support always given me by the distinguished gentlemen of the Board of Trustees and by the faithful and brilliant services of the many remarkably able men and the wonderfully skillful women composing The Society staff...."

Unknown in advance to Dr. Grosvenor, twelve notable leaders of National Geographic Society expeditions had been gathered on the platform to see him receive the medal. Among these expedition leaders were:

Rear Admiral Richard E. Byrd (the first man to fly over the North and South Poles); former U.S. Senator Hiram Bingham (discoverer of the lost Inca city of Machu Picchu); Dr. Dillon Ripley (leader of the Nepal expedition completed only weeks before and future Secretary of the Smithsonian Institution); Dr. Robert F. Griggs (Valley of Ten Thousand Smokes); Maj. Gen. Orvil A. Anderson (*Explorer* balloon stratosphere flights); Dr. Matthew W. Stirling (archeological discoveries in Panama and Mexico; he was present with his wife, Marion); Dr. Maurice Ewing (explorer of the Mid-Atlantic Ridge); and Dr. Arthur A. Allen (who had solved the 163-year-old mystery of the bristle-thighed curlew's nesting place).

A motion picture of the highlights of each of the expeditions, with brief commentaries by their leaders, was shown, and then Trustee and Treasurer of the Society Dr. Robert V. Fleming read a letter of congratulations from President Harry S Truman, which said in part, "Under your direction The Society's magazine has become a household institution in the homes of America and throughout the nations of the world—in short, wherever there is a postal system and wherever geographic knowledge is esteemed."

National Geographic Society Treasurer Fleming then introduced Dr. Charles F. Kettering, former Director of Research and Vice President of General Motors, and a Trustee of the Society, who made the formal presentation of the Grosvenor Medal, which had been created by the Board of Trustees especially for the anniversary celebration.

"Presenting this medal is the most appreciated of all the many pleasures life has brought me," Dr. Kettering said. "If there ever is 'one world,' a copy of the *National Geographic* will be on the center of the table!"

Dr. Kettering presented the medal to Dr. Grosvenor, who, Atwood later reported in the *Geographic*, "held it up for the vast audience to see, in a friendly, unstudied gesture as if to share it with his friends, which brought the crowd to its feet for prolonged cheering.

" 'Every morning when I look into my mirror,' he said, 'I am going to say to my mirror, "You lie." Then I shall take this beautiful medal, look at the idealized Grosvenor face on it which Mrs. [Laura Gardin] Fraser has modeled with fingers of genius, and chuckle to myself! My descendants happily will not know the difference between fact and fiction.'"

A basket of fifty golden roses stood on the speaker's platform. They had been presented to Grosvenor by his National Geographic colleagues. The inscription, "To Our Beloved Chief," Grosvenor told the audience, meant as much to him as "this glorious medal."

The *Geographic*'s circulation at the end of 1949 was 1,831,588; its circulation would continue to climb during the five more years that Gilbert H. Grosvenor would serve as Editor. Its reputation, however, had begun to decline.

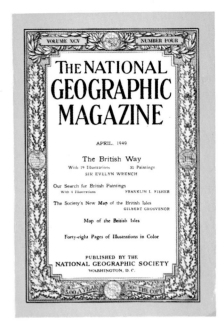

At the end of 1949, the Magazine's circulation was 1,831,588.

"Young Men Put Into the Earth"

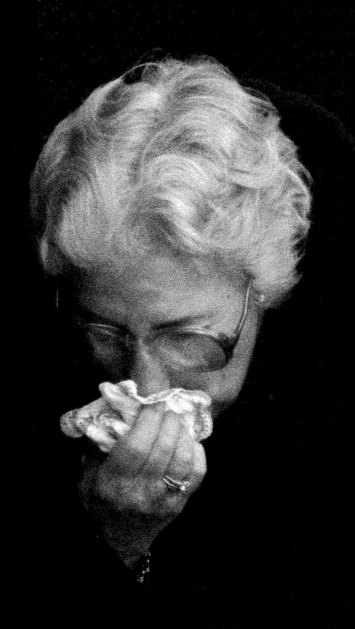

"The names have a power, a life, all their own. Even on the coldest days, sunlight makes them warm to the touch. Young men put into the earth, rising out of the earth. You can feel their blood flowing again."

—On the Vietnam Veterans Memorial, in Joel L. Swerdlow's "To Heal a Nation," May 1985

The *National Geographic*'s coverage of wars and conflicts during Gilbert Hovey Grosvenor's editorship was determined, for the most part, by how loosely the staff was able to interpret his fifth and sixth editorial guiding principles:

- Nothing of a partisan or controversial character is printed.
- Only what is of a kindly nature is printed about any country or people, everything unpleasant or unduly critical being avoided.

Geographic writers' efforts to produce responsible journalism within the restrictions imposed by Grosvenor's principles were what led, over the years, to accusations that the *Geographic* viewed the world through rose-colored glasses. And the example repeatedly given to support that charge was Douglas Chandler's 1937 article, "Changing Berlin," which in hindsight one now recognizes as appallingly innocent:

> To develop boys and girls in body and mind, and thus insure a sturdy race to defend Germany in the future, is a policy of the present government. …As a substitute for Scout training, German youngsters now join an institution known as the Hitler Youth organization. Its emblem is the swastika, and its wide activities and political training are enormously popular with all classes.

Curiously enough, Grosvenor's policy had its defenders even in those journalistic camps one would assume most opposed. For instance, on June 21, 1940, *The Detroit News* contained the following editorial:

> How the National Geographic Magazine carries on, month after month, with only an occasional word to suggest its awareness of the war is, on its face, something of a miracle, as publishing goes these days. To write anything now for public consumption which is in no way associated with a war that rocks all mankind is an effort. To assemble any quantity of nonwar material that can lure us even momentarily from our dark preoccupation is no mean feat. To be able to get together a periodical on geography, of all things, and leave it out, borders on the supernatural.
>
> It is a policy of the Geographic and we should be grateful to it for trying, and for succeeding so well. And, of course, it is right. There is not much war can do to change physical geography. The Alps will be the Alps when the Caesars return to dust. Old Faithful will spout; the great sea tortoises will waddle ashore on Galopagos [sic]; the Mediterranean will be as

Above: *France, World War One. "An American Ambulance Rolling Through A Ruined Town In France."* — From the May 1917 issue

Opposite: *France, World War One.* The Rampart of Verdun, *drawing by the eminent French artist Lucien Jonas (1880–1947). "Those who held the pass at Thermopylae…Horatius and his comrades at the bridge….First among the immortals of history now stand the defenders of Verdun, who said: 'They shall not pass.'"*—From the April 1918 issue

Preceding spread: *"The names have a power, a life, all their own,"* wrote Joel L. Swerdlow of the Vietnam Memorial. *"Even on the coldest days, sunlight makes them warm to the touch. Young men put into the earth…."*—From "To Heal a Nation," in the May 1985 issue

265

Above: *London, World War Two. Prime Minister Winston Churchill examines the bomb-shattered ruins of the legislative chamber of Great Britain's House of Commons in 1941, one year after standing before its members to say, "I have nothing to offer but blood, toil, tears and sweat."—From the August 1965 issue*

Opposite: *"December 29, 1940: London's burning. The pall of smoke from 1,500 fires enshrouds her homes, her churches, her dead. But, its golden cross burnished by the flames, St. Paul's dome rides above the searing storm. Like London, St. Paul's was scarred, hit by a bomb two months earlier. Like London, the cathedral would endure."*

blue, and the Sahara the same restless sea of sand. The land will not change, and neither—over much of the earth—will the people who go with it. They will draw their life and sustenance from the nature around them. The races that spring from it will be molded, for good or ill, by necessities imposed on them by nature.

Coverage of World War One had never been anything but supportive of the Allied powers. The *Geographic* took measure of Germany's aspirations and wrote, "Paltry indeed seem the dominions of all the tyrants of the past, who attempted to 'wade through the slaughter' to the throne of world empire, compared with the vaulting ambition of the Hohenzollerns for Prussianizing the earth...."

Bloody World War One battles were hinted at, kept in the distance. "Plain Tales from the Trenches" takes place not along the Somme or in Verdun, but "As Told Over the Tea Table in Blighty—a Soldiers' 'Home' in Paris":

All the long tables are ready for tea. The cloths are blue and white and so are the dishes. The milk pitchers are full to running over, the jam bowls too, and the large plates of fresh, sweet-smelling bread and butter are just where they ought to be. And there's cake—the good kind, full of raisins and currants and nuts. Why, there's even plenty of *sugar!* So, as I tie on my absurd little apron I say to myself that it doesn't look like a war-time tea at all.

But it is, in the fullest sense of the word....

The first three to come to my table are "Kangaroos"—tall and straight, freshly shaven, uniforms brushed and pressed, boots of a dazzling brilliance....a fourth man wearing the same divisional colors on his sleeve joins our group to the gay shout of "Hello, Digger," which is only another name for "mate," you know.

Then: "Where's Barty? Didn't he come along with you?"

The newcomer shakes his head, and when he is asked "Why not?" answers simply, "Dead." To a further question of "When?" he replies, "Monday." And Barty's only requiem from three husky throats is, "He was a good bloke."

The author, Carol K. Corey, temporarily leaves her table in the tea-room. When she returns she finds:

a hot-headed chap storming indignantly: "There you go again, talking about the war. There ought to be a law"—

"That's so," interpolates his neighbor. "What else do we know after three years of it? You pick a nice, new, interesting subject and tell us about it. Why not give us a little lecture on *mud?* That's always interesting to the ladies. They say 'Poor dear' so sweetly that you forget to tell them the one good thing about it, which is that it keeps you warm. I never had a cold till I got here and cleaned it off. And look what a 'beaut' I've got now. I tell you, Missus, the mud's never hurt me. Neither has the war. Why, I used to have asthma somethin' fierce; but now it's all 'partee.' If we get home with all our arms and legs and eyes—or just enough to get on with—this here war's goin' to be a good thing for a lot of us.

"Of course, I ain't sayin' it's pleasant; far from it. There's the route marches and the everlastin' salutin' and the bully beef and the bumps on the ground at night. But there's compensations. Take my case. I had *three* sisters all learning the piano at once, and all of 'em dubs at it. Yeah, it could be worse. Pass the cake, Kid."

"...It could be worse. Pass the cake, Kid" ran in the March 1918 issue. And should the Society membership want some cake for themselves, the April 1918 issue of the *Geographic* asked its members to do without. "...there is a new

266

opportunity for helping by personal deprivation," the article read, "...opportunity of *pledging himself and his household to eat neither wheat bread, wheat cereals, nor pastry made of wheat flour until the new wheat crop is harvested.*"

Gilbert Hovey Grosvenor remained in place as Editor through World War Two—as did his editorial principles. The January 1944 article, "At Ease in the South Seas," was typical of the conventional homefront morale-boosting articles the *Geographic* was running. Written by Frederick Simpich, Jr., son of longtime *Geographic* Assistant Editor Frederick Simpich, the article told of soldiers' efforts in the South Pacific to keep amused between battles. His piece contained photographs of shirtless GIs sitting around a crate—"Even on Guadalcanal Card Games Have Kibitzers"; their pets—"On Coral Sands, 'Scuttlebutt' the Pup Wears Camouflage on All but Ears"; a sailor dozing in a barnyard bunk—"'Turn Out and Turn To!' Says This Army Horse Nuzzling a Sleeping Sailor"; and a helmeted soldier posed in his foxhole which has the sign "Beneath These Portals Pass the Fastest Men in the World!" is captioned "Laughs Aimed at Themselves Keep Marines' Morale High in Oft-bombed Guadalcanal."

War was seen as little more than an inconvenience.

"We had an extremely heavy raid one night just before I left London in January, 1941," Harvey Klemmer wrote in "Everyday Life in Wartime England" for the April 1941 *Geographic*:

> Bombs came shrieking down at the rate of one a minute. A number of fires were started, and a good share of the City—London's financial district—was wiped out.
>
> The crash of bombs and the glow of the fires gave us the feeling of living through some sort of medieval nightmare. Few got any sleep that night.
>
> It was almost with dread that I opened my curtains in the morning. But there was no reason for dread, then. The sun was shining brightly. Traffic moved in Berkeley Square as usual. I noticed that an old street sweeper, with whom I had become acquainted, was on the job. The attendants in apartment houses stood on the sidewalk, resplendent in their various uniforms. Models and seamstresses tripped into the gown shop up the street. Large posters in a travel agency window advertised cruises to Australia.
>
> In my own building the valets went about preparing breakfast and laying out clothes. When I went through the lobby, I noticed that one of the porters was very carefully shining the brass about the main entrance.
>
> The difference between that morning and the experiences of the night before is symbolic, to me, of the two kinds of life that now exist in England.

There were even opportunities during the war to have great fun. When Annapolis graduate Melville Bell Grosvenor was invited to take a cruise on a U.S. Navy escort carrier, he grabbed at the chance. Strapped into the rear-seat ball turret of a TBF Avenger torpedo plane, MBG was catapulted off the deck. After a mock-torpedo run on the ship, he wrote,

> Quickly the four planes joined the landing circle, wheeling gracefully around the carrier.

Pacific warfare, World War Two. Aboard the carrier U.S.S. Yorktown, near Truk in April 1944, artist William F. Draper records the York- *town's fight for life against attacking Japanese dive bombers and torpedo planes.*

When our turn came, I felt a distinct braking effect as the landing flaps and wheels were lowered. Our plane seemed to slow down almost to a walk!

I craned my head around to see our approach. The once tiny flight deck loomed larger and larger.

Finally out of the corner of my eye I saw the Signal Officer give the cut and we dropped softly to the deck. The plane stopped abruptly, like an automobile quickly braked. If I had not leaned back against the armor plate, my head and shoulders would have snapped forward.

Our TBF was taken down the elevator smartly and I climbed out in the hangar. Friends crowded around to ask how I liked it. "Greatest thrill of my life," I said....

"We sail soon on another mission," the Captain remarked. "I wish you could make the long cruise with us."

"There's nothing I would like to do better," I replied. "Here's wishing you luck and a full bag of subs."

"We'll do our best!"

Few casualties were shown in the *Geographic*'s war. Death was sanitized, seen from thousands of feet. One of the most vivid pieces to appear during World War Two was 1st Lt. Benjamin C. McCartney's account of his visits to Florence, Italy. In that city first as a pre-war student, he returned as a bombardier in the nose of a B-26 over Florence's railroad marshaling yards:

> The cars and tracks were very near now under the absolute black cross hairs—all freight cars, I noticed. Then, still watching the judgment of the cross hairs riding evenly, slowly on one freight car, I felt the ship jump, and pressed the microphone button. "Bombs away! Bombs away! Bomb-bay doors going closed...."
>
> Far ahead and all through the air the bombs were falling in languid, reluctant strings; everywhere under the bellies of the planes the bomb-bay doors slowly closed. Beneath me now the bombs from my own ship fell away slowly, fat and yellowish....
>
> All the bombs were flinging along beneath us, twisting slightly, keeping up with us, and then slowly straightening out, dropping away and sliding fast, far down until I lost them....Ahead in the yard, halfway up, there was already a brown broil of smoke from the first bombs of the planes in the other squadron.
>
> I waited while we came over the bottom of the yard and then suddenly, just inside the yard, there was the instant rip and black spurt of my own bombs down among the railroad cars.
>
> I pressed the microphone button violently. "We hit it! We hit it! We got it dead center!"

The day after his manuscript was received, the *Geographic* noted that the War Department had announced Lieutenant McCartney's death from wounds received when his B-26 was struck by flak on a subsequent raid over an Italian rail bridge. Death had suddenly intruded.

On June 24, 1948, the Soviets closed off all land and water access to Berlin from the West. The Iron Curtain had descended. The Western powers responded with the Berlin Airlift. Between June 26, 1948, and September 30,

◄ ─────────────────────────────

The Berlin Wall, begun Saturday night, August 12, 1961, by East German troops and People's Police as a barrier to exclude "revenge-seeking politicians and agents of West German militarism," instead imprisoned East Berliners and kept millions from seeking freedom in the West.

1949, American and British aircraft made more than two hundred thousand flights into that beleaguered city with supply-carrying planes sometimes landing as often as one every forty-five seconds. More than two million tons of goods were brought in during that period. Post-war Berlin tensions were typified by pieces like "What I Saw Across the Rhine" (January 1947), "Airlift to Berlin" (May 1949), and "Berlin, Island in a Soviet Sea" (November 1951). Such articles were, however, seemingly written with détente in mind.

Gilbert H. Grosvenor and his cartographic department scored a timely triumph when the Magazine's "Roaming Korea South of the Iron Curtain"— containing a map of the Korean peninsula—appeared in June 1950, the month the North Korean Communist forces launched their attack across the Demilitarized Zone that had separated the two countries since the end of World War Two. Korean War reporting, however, remained of the old school: "Our Navy in the Far East," by Arthur W. Radford, and "The GI and the Kids of Korea," by Robert H. Mosier.

The 1956 Hungarian Revolution was covered by "Freedom Flight from Hungary" with Robert F. Sisson's moving photographs of Hungarian families processing through Austrian refugee centers.

And then on August 12, 1961, East German troops and People's Police— giving the excuse that it would keep "revenge-seeking politicians and agents of West German militarism" out of their territory—began construction of that monument to Communism's failure, the Berlin Wall. It was a Saturday night, a Berliner friend of *Geographic* freelancer Howard Sochurek recalled in 1970:

> A million people over there in East Berlin went to bed without any idea of what was going to happen. But I will bet that many of them were planning to get out of Communist Germany. More than three million people had fled the Soviet zone up to that time, and the rate was increasing to about 2,000 a day. You simply took a 20-pfennig train ride to the refugee camp at Marienfelde in West Berlin, since freedom of movement had been guaranteed by the protocols governing the city.
>
> At 2 a.m. on Sunday, tanks and trucks rolled in with East German troops and People's Police. Train service to the West stopped, stations were sealed, and building of the Wall began. A million people woke up in jail....

Not long after, Sochurek was in jail, too. His account of his arrest and that of Jürgen Toft, his East German interpreter, in East Berlin, was chillingly familiar:

> ...at about five o'clock one evening, while Jürgen and I were standing on the Elsenbrücke—a bridge across the Spree River—we were both arrested.
>
> I had been photographing the sunset on the river and the ducks paddling furiously to avoid a large coal barge that was passing by. I had just packed my cameras into their cases and was walking off the bridge when a member of the *Volkspolizei*—People's Police—stepped from a half-concealed guard post under the bridge and called, "*Halt.*"
>
> The uniformed "Vopo" approached and said, "Give me your credentials!"
>
> Jürgen handed over the identity card which every citizen of the GDR must carry, and a letter from the Reisebüro explaining my purpose in East Berlin. I presented my United States passport, complete with the East German visa and the police registration stamp recorded on the day of my arrival.

"You have violated the people's law," the "Vopo" said. "It is forbidden to photograph this port. It is forbidden to photograph the border. It is forbidden to photograph the railroad yard."

Across the street, faces began to appear in the windows of an old, six-story apartment building. They stared down at us—impassive and impersonal. Two sergeants in a dark-green Volga police car drove up. Our "Vopo" had obviously summoned them before he had even halted us.

"This document from the Reisebüro," he now said coldly, "is forged." Jürgen tensed visibly. His face paled. He turned to me and remarked in English, "This is getting serious."

One of the sergeants took our credentials and carried them to a call box across the street. We waited there for an hour until a call was returned.

"You will come with us."

Jürgen and I were instructed to sit in the back of the police car, and we were then driven away.

Sochurek and interpreter Toft were taken to the State Security Service offices in a former SS Headquarters building, up three flights of stairs beneath huge portraits of the German Democratic Republic's then Chairman of the State Council Walter Ulbricht and "Progress Through Work Builds Socialism" banners to a hallway where they were detained for several hours. Finally they were released with the apology that since their arrest had taken place outside of normal business hours, it had taken longer to contact the appropriate authorities.

Sochurek was not unknown to the Magazine's readers; he had already written some of the *Geographic*'s strongest stories on Vietnam—among them "American Special Forces in Action in Vietnam," a piece about the first stages of a revolt against the Republic of Vietnam by 3,000 heavily armed *montagnard* tribesmen in five Special Forces camps. Caught up with American Special Forces Capt. Vernon Gillespie in a war within a war, Sochurek wrote in 1965:

> Frankly, our situation, as I pieced it together, unnerved me. Here we are, I thought, surrounded by Viet Cong who nightly harass the camp with mortar and sniper fire. That isn't enough. The camp itself is indefensible against a large-scale attack. It could be overrun by a full battalion any time the Communists chose to do so—as they had done in July at another camp exactly like Gillespie's. Even that isn't enough. The camp is full of armed montagnards who resent the Government of South Vietnam and the Vietnamese officers on the spot.
>
> In a capsule, I thought, our situation here mirrored the complexity of the problem that the United States faces in Viet Nam.

One does not think of the *Geographic*, with its slow, leisurely, cautious, researched manner of publishing as a journal in which controversial war coverage might emerge. However, the first published photographs of armed American servicemen fighting in Vietnam appeared in Dickey Chapelle's "Helicopter War in South Viet Nam," a November 1962 piece which appeared the year President John F. Kennedy announced that U.S. advisers would fire if fired upon—the same year America's attention was turned not to Vietnam, but to the Cuban missile crisis.

And yet, even a year before Dickey Chapelle's article, W. E. Garrett and Peter T. White had gone to Vietnam to report on how "South Viet Nam Fights the Red Tide." This piece, published three years *before* Congress passed the Gulf of Tonkin Resolution authorizing presidential action in Vietnam and

Opposite: *Saudi Arabia, 1980.* *"Noontime summons princes and subjects to prayer," began the caption for this photograph of guards laying down their weapons and kneeling with petitioners alongside Prince Salman ibn Abdulaziz, Governor of Riyadh, and member of the Al Saud family, which has ruled the kingdom of Saudi Arabia since its founding in 1932.—From "Saudi Arabia: The Kingdom and Its Power," in the September 1980 issue*

four years before the large-scale commitment of U.S. troops, accurately and eerily warned of what was in store. While Garrett's photographs were, for the most part, of the conventional travelogue variety—Tet celebrants, temple interiors, seaport life, cyclo drivers, beautiful girls in flowing gowns—White's text was conspicuously grim. He wrote of the growing Viet Cong strength: "Quietly and relentlessly—with the world hardly aware of it yet—the rich country in the south was slipping ever deeper into a calculatedly cruel civil war....From dusk to dawn, the Viet Cong ruled nearly half of South Viet Nam."

But it was in the closing paragraphs of his article, with its hints of America's growing involvement and the hopelessness of such an action, that White's prescient appraisal appeared. "You've undoubtedly met the Viet Cong, you know," a South Vietnamese friend tells author White,

> "An elevator boy, or a shopkeeper, or a driver for MAAG. They are everywhere. They collect 'taxes' even in Saigon, by persuasion, or by threats. They give a receipt. If a man takes it to the police station, the Viet Cong kills him. Let the police catch him with it, and he might go to a re-education camp."
>
> I asked, "What will happen to Viet Nam?"
>
> "I hope for a miracle to save us," he said. "Otherwise the Viet Cong will get stronger. Will the Americans go home? Maybe they'll let their own soldiers fight. But how could they do better in the swamps and jungles than the French?"
>
> And what about my six girls, and the eager cadets? [White was asking of the six young women he had met at the Vietnamese-American Association and the cadets he'd spoken with at South Vietnam's Military Academy.]
>
> "Some will die, some will escape. But most people will have to stay and try to get along. One day you may face those boys across a conference table. Or across a battlefield. As our old primers say: Man is born good, but life makes him bad."

It is estimated that between August 4, 1964, and January 27, 1973, approximately 8,744,000 Americans saw service in Vietnam. Nevertheless, on April 30, 1975, South Vietnam surrendered to the Communists.

In the meantime, however, there had been and would continue to be other wars—wars between Israel and the Arab nations, India and Pakistan, Iran and Iraq, the Soviet Union and Afghanistan; there was continuing violence in Northern Ireland, Central America, and the Far East, where starved, terrified, bedraggled refugees fled Cambodia and its genocidal war in which an estimated one to three million people had been killed. This latter horror was covered in such pieces as Garrett's May 1980 "Thailand: Refuge From Terror" and White's "Kampuchea Wakens From a Nightmare," which appeared in May 1982.

Over the years *Geographic* writers and photographers returned again and again to the Middle East: Joseph Judge's "Israel—The Seventh Day," appeared in December 1972 and quoted a Jerusalem innkeeper who said, "I'm not a deeply religious man. When I was young in Vienna, every time there was a holiday my father would tell me about Jews being killed on this day, or being driven out, or being massacred. Every holiday was the celebration of a tragedy. I said to myself, 'This religion is a dangerous business!' " Several days later, Judge is speaking with a young leader of the Druze sect, who tells him, "At some point a man must say to himself that he cannot hate forever."

Opposite: *Iran, 1984. "The ayatollah speaks....And men listen, go to the front, and earn martyrdom—or return, many with horrible wounds from mines or shrapnel. Rehabilitation centers are well stocked with braces and artificial limbs. This center was a mansion of one of the shah's generals, whose paintings were torn from frames and replaced, here with Khomeini and a poem by a Shiite mystic urging dedication to Allah." —From "Iran Under the Ayatollah," in the July 1985 issue*

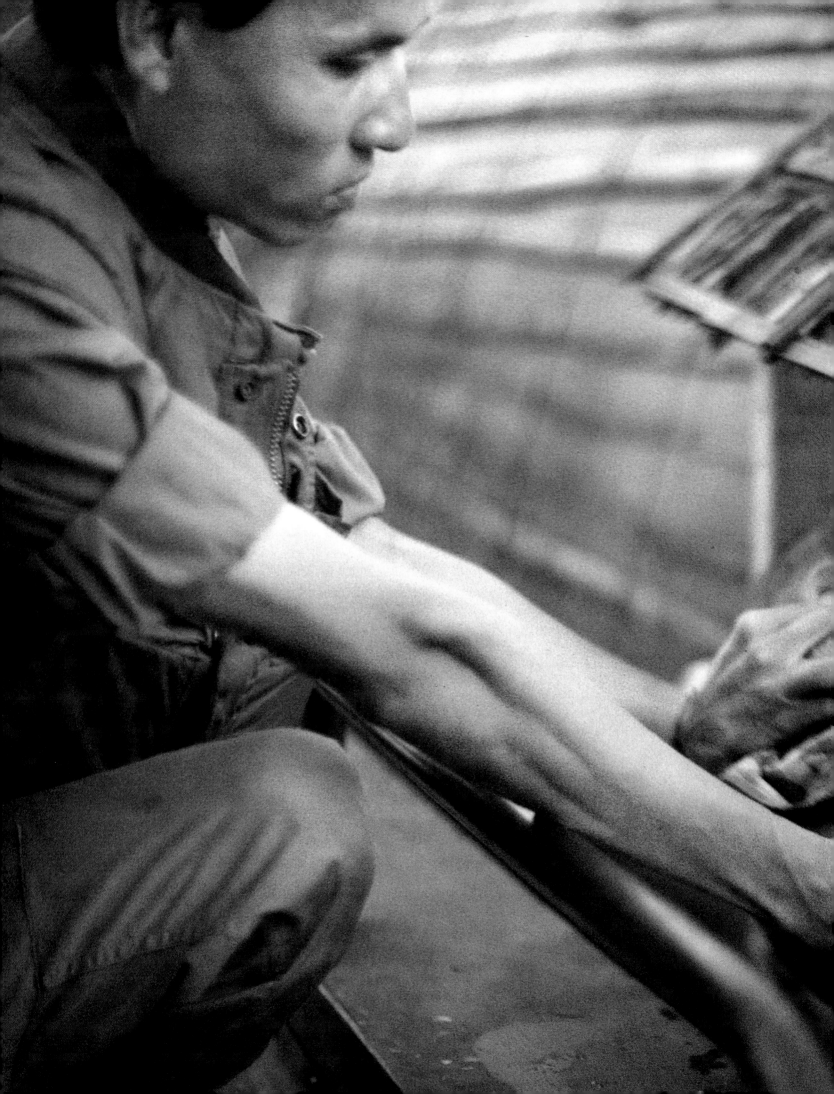

Left: *South Vietnam, 1968. A Vietnamese mother, aboard a sampan intercepted by a U.S. Navy patrol boat (searching for Viet Cong and contraband in the delta before the 7 A.M. curfew prohibiting river traffic ended), pleads through an interpreter for help for her sick baby. The sailors helped as best they could from their vessel's first-aid kit, then continued on.
—From "The Mekong, River of Terror and Hope," in the December 1968 issue*

Some of the first published photographs of armed American servicemen fighting in Vietnam appeared in Dickey Chapelle's Geographic *article "Helicopter War in South Viet Nam" in November 1962— the year America's attention was turned not to Vietnam but to the Cuban Missile Crisis. Chapelle was killed covering the war in Vietnam in 1965.*

Following spread: *Veterans Day Weekend 1984. A veteran, standing above the Vietnam Veterans Memorial in Washington, D.C., blows taps for fallen comrades. The surface of the stark polished black granite wall is inscribed as a lasting tribute with the names of the men and women who lost their lives as a result of America's longest war.—From "Vietnam Veterans Memorial: America Remembers," in the May 1985 issue*

1959

IN HONOR OF THE MEN AND WOMEN OF THE ARMED FORCES OF THE UNITED STATES WHO SERVED IN THE VIETNAM WAR. THE NAMES OF THOSE WHO GAVE THEIR LIVES AND OF THOSE WHO REMAIN MISSING ARE INSCRIBED IN THE ORDER THEY WERE TAKEN FROM US.

TRIMBLE · ORVIN C JONES Jr · CHARLES D LANDIS · ALAN P MATEJA · ROBERT A STERLING
· TOMMY JOE BECKER · GREGORY W HERMANN · DAVID D STOVER · RONALD H WIGGINS
E K BARSOM III · ALBERTO ORTIZ Jr · PAUL J PESCE · THOMAS S POWELL · ARNOLD J RAHM
MASON J BURNHAM · WILBUR H CHILDRESS · JOHN J STEGELAND III · CHARLES TURNER
ORLOVICH · MICHAEL D BAKER · GEORGE W CARTER · MICHAEL D CLEAVES · WADE L ELLEN
ICKER · JOHNNY M JONES · KENNETH J YONAN · LEROY B AIKEN · ROBERT W BROWNLEE Jr
Y Jr · CALVIN C COOKE Jr · RICHARD E DUNN · DONALD R HOSKINS · RICHARD L RUSSELL
EISMAN · THOMAS K DUFFY · THOMAS F SHAW · ROBERT L SOWERS · CLAUD P STROTHER
LLCUFFER · PETER R MILLER · FRANKLIN EAST · WILLIAM A HAINES Jr · PAUL V MARTINDALE
JOHN J MOTT · LARRY D EPSTINE · MELVIN D SEAGRAVES · ROY J DAY · JOSEPH M BERKSON
VN · TERENCE F COURTNEY · WILLIAM C JESSE · CHARLES V MORGAN · JOHN J PETRILLA Jr
DAVID R SLAGLE · THOMAS C WIDERQUIST · JOSEPH C HOPPER · JOSEPH W McDONALD
ANDER McIVER · FREDDIE L SLATER · DON L UNGER · LESTER BRACEY · DAVID B WILLIAMS
WRIGHT · GLEN S IVEY · MARVIN B C WILES · JOHN W CONSOLVO Jr · JOHN M LEAVER Jr
ROBINSON · EDMUND B TAYLOR Jr · JOHN T CONRY · MIKE J AGUILAR · OSCAR AGUILAR
· WILLIAM A BOATRIGHT · STEVEN E BLOWERSOCK · EDWARD D BURNETT · CLINT E CARR
NNING · ALVIN R ELENBURG · DAVID CRUZ FLORES · DIETER K FREITAG · JAMES D GROVES
HARRELL · JEFFREY L HARRIS · DALE L HAYES · WILLIAM F HENAGHAN · FRANK T HENSON
WELL · FREDDIE JACKSON · KENNETH ROSENBERG · THOMAS A LAHNER · ROBERT A LODGE
O A LYDIC · GARY R MONTELEONE · LARRY S MILSTIN · TERRY D NEISS · DEAN A PHILLIPS
CHARD RIDGEWAY · EFRAIN RIVERA-AGOSTO · JAMES C JENSEN · JOHN TENERIO SABLAN

DALE R BUIS · CHESTER N OVNARD · MAURICE W FLOURNOY · ALFONS A BANKOWSKI · FREDERICK T GARSIDE ·
RALPH W MAGEE · GLENN MATTESON · LESLIE V SAMPSON · EDGAR W WEITKAMP Jr · OSCAR B WESTON Jr ·
THEODORE G FELAND · GERALD M BIBER · JOHN M BISCHOFF · ODIS D ARNOLD · WALTER H MOON ·
BRUCE R JONES · FLOYD STUDER · JAMES T DAVIS · HERMAN K DURRWACHTER Jr · FRED M STEUER ·
THEODORE J BERLETT · MILO B COGHILL · FERGUS C GROVES II · ROBERT D LARSON · JOSEPH M FAHEY Jr ·
FLOYD M FRAZIER · STANLEY G HARTSON · EDWARD K KISSAM Jr · JACK D LE TOURNEAU · GLEN F MERRIHEW ·
LEWIS M WALLING Jr · ROBERT L WESTFALL · CHARLES A PULLIAM · AL SUMINGGUIT PADAYHAG ·
IVAN P WHITLOCK · MILTON D BRITTON · BARNEY KAATZ · JAMES GABRIEL Jr · WAYNE E MARCHAND ·
BILLIE L BEARD · RONALD E LEWIS · HEWETT F E COE · GEORGE E COLLIER · ROBERT L GARDNER ·
WALTER R McCARTHY Jr · WILLIAM F TRAIN III · ROBERT L SIMPSON · DON J YORK · JOSEPH A GOLDBERG ·
HAROLD L GUTHRIE · JAMES E LANE · ANTHONY J TENCZA · WILLIAM R BUNKER III · RICHARD L K ELLIS ·
THOMAS E ANDERSON · GERALD C GRIFFIN · RICHARD E HAMILTON · GERALD O NORTON · JERALD W PENDELL ·
MICHAEL J TUNNEY · MIGUEL A VALENTIN Jr · HERBERT W BOOTH Jr · TERRY D CORDELL · RICHARD L FOXX ·
JOHNNIE GENE LEE · GARRY C McFETRIDGE · ROBERT D BENNETT · WILLIAM B TULLY · RICHARD D BENZEL ·
CHARLES E HOLLOWAY · JACK M LISLE · DONALD L BRAMAN · WILLIAM L DEAL · KENNETH N GOOD ·
CLAYTON A FANNIN · CHARLES M FITTS · LAWRENCE C HAMMOND · BOYCE E LAWSON · JAMES D McANDREW ·
JAMES A STONE · DONALD R STORM · RAYMOND G WEBE · JOHN DUARTE · LEON LARIMER ·

PAUL F McNALLY ·
HAROLD K STEVE
NELSON E VAN G
ROBERT L CURLEE
CRAIG L HAGEN ·
ZOLTAN A KOVAC
MARVIN G SHIELE
SAMUEL J GANCI
DONALD L BAKER
MICHAEL L WILDE
JACKIE W SANFO
LEROY A BOURGE
ROBERT L ARMON
FRANK P WATSON
ROBERTO SAMAN
ALBERT S KNIGHT
EUGENE D FRANK
DOUGLAS H D'O
PETER MONGILA
ROBERT A BUTZ
GEORGE P ZUPA
EARL G DOWNE

Nicaragua, 1985. "One at a time is how they usually die in a guerrilla war..." began the caption for this December 1985 Geographic photograph ("NICARAGUA: Nation in Conflict"). It continued: "Last July photographer Jim Nachtwey went on patrol with one of several special Sandinista battalions charged with engaging contras in the north. Not far from San José de Bocay his 200-man group surprised about 300 rebels, called in artillery, and attacked....As the bullets fly overhead, a soldier lies dying (right), the sole Sandinista fatality." Nachtwey, a veteran photographer who has covered combat from Northern Ireland to El Salvador, says of his work: "It never gets easier. It only gets harder."

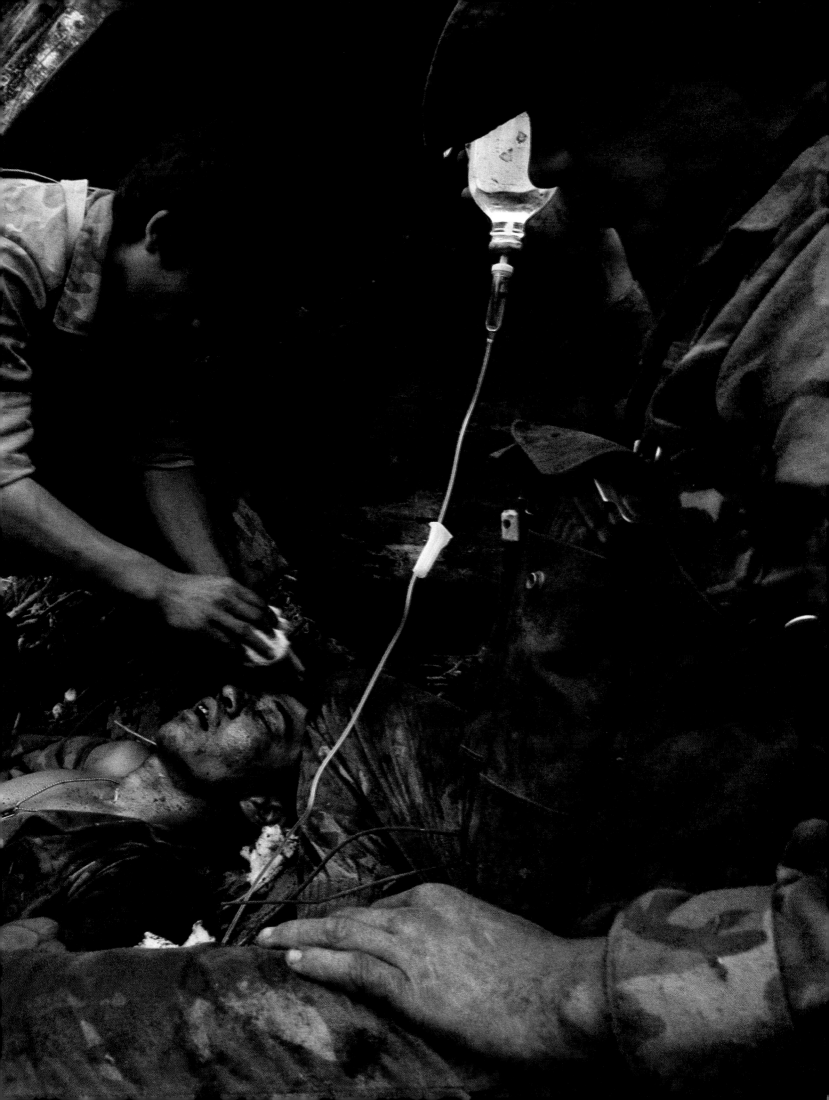

Above: *Afghanistan, 1985. "'The helicopters hovered,' says Jalad Khan, 'and they picked out their targets.' Salvo after rocket salvo later, the Soviet helicopters had destroyed Khan's village of Doubandi, about 30 miles from the border with Pakistan. Frontier villages have served as refuges for the freedom fighters and, as a consequence, have been increasingly subject to aerial attacks."—From "Along Afghanistan's War-torn Frontier," in the June 1985 issue*

Opposite: *Afghanistan, 1985. The face of war is visible in the haunted eyes of this young girl who fled her homeland to seek shelter in a refugee camp in Pakistan.—From "Along Afghanistan's War-torn Frontier," in the June 1985 issue*

Middle Eastern history is filled with such lessons ignored.

John J. Putman's October 1975 piece, "The Arab World, Inc.," asked, "Who are those oil-rich Arabs, and what are they doing with all that money?" The answer was given in Robert Azzi's September 1980 article, "Saudi Arabia: The Kingdom and Its Power," subtitled "An awesome inrush of wealth brings the Saudi kingdom—founded on Islamic principles and stern desert ways—to a whirlwind clash of tradition and change." Azzi's piece opened with a Saudi government official's poignant self-appraisal: "In your terms I'm a great success," he told the author:

> "a symbol of all the potential of Saudi Arabia—barefoot Bedouin boy from barren desert goes to U.S.A. to find happiness and a Ph.D."
>
> You're wrong.
>
> "Sometimes I think I'm the greatest failure in the world. I, Faisal... Saffooq al-Bashir, of the Al Sabaa tribe, was raised to be the next leader of my tribe. They expected it of me, and I failed them. Do you know that it took me 11 days to find my family when I first returned from the U.S.A.? Eleven days in a small hotel trying to find a few thousand Bedouin somewhere in the desert.
>
> "What am I now? A technocrat—deputy minister of planning. Trying to find a way to spend a five-year-plan budget of 236 billion dollars. Dollars that will only speed up the disappearance of the nomads. I act against all the forces that created me."

Not all conflicts are dependent upon weapons.

Lebanon's problems continue. William S. Ellis, the shy *Geographic* staff writer whose 1977 piece, "South Africa's Lonely Ordeal," created such a tempest, was in Beirut in 1969 to write "Lebanon, Little Bible Land in the Crossfire of History."

"George Mobley was the photographer," Ellis recalled, "We were at a funeral for a Palestinian [Arab commando killed earlier in a skirmish near the Israeli border] and suddenly thousands of angry Palestinians around me turned on George Mobley and yelled, '*Get the American! Get the American!*' Mobley had jumped up on a rooftop to escape and he looked down at me and yelled, 'Do something! *Do something!*'

"I did," Ellis said, smiling. "I yelled, 'Get the American!' "

Because photographers attract the wrong kind of attention Ellis tended to stay away from them whenever possible. But thirteen years after his "Little Bible Land in the Crossfire of History" story, Ellis was back in Beirut again where, as Ellis said, "Something happened that changed my mind. Steve McCurry—a freelance photographer here—and I were doing Beirut together and we were going south out of the city when Steve said he had to make a phone call. So we stopped the car and went into a hotel to use the telephone. And where *we would have been* had we *not* stopped to make the call, a huge car bomb went off. We would have been right there at that time. So ever since then, I have felt a little differently about photographers."

Looking at the photographs in this chapter, one cannot help but feel a little differently, too, about the photographers from the *National Geographic*. There are no "Smile, and point at the mountain" photographs here, no "red shirts," no painted natives, no cute little animals, stunning sunsets. There is only the face of war, whether it be the expression of a grieving mother at the Vietnam Memorial or in a bamboo hut, or of a haunted young Afghan refugee.

The Red Shirt School
of Photography

"...they'd go on these expeditions and everybody would be in khaki because that's the color of field uniforms. And they'd come back with the dullest bunch of pictures you ever saw! You couldn't use them editorially because they had no color. So, we decided to have people wear colorful shirts...."
—Melville Bell Grosvenor on the genesis of "The Red Shirt School of Photography"

Preceding spread: *"Movies and motorcars, two American institutions which were born about the same time and grew up over the same years, finally met at the drive-in theater.... Charlton Heston [in* The Ten Commandments *(1956)] spreads Mosaical arms against the evening sky of Salt Lake City, Utah."*—From the 1975 National Geographic Society Book Service publication We Americans

The 1950s was the decade of the atomic bomb shelter, of Soviet expansionism, the stockpiling of nuclear weapons, and the fear of Communist subversion. In August 1953, nine months after the United States exploded its first H-bomb and obliterated Enewetak in the South Pacific's Marshall Islands, the Soviet Union detonated its first H-bomb in Siberia. By then President Harry S Truman had been succeeded by Dwight D. Eisenhower. It was the time of Senator Estes Kefauver's Special Committee to Investigate Crime hearings; and, on television, a mesmerized nation watched mobster Frank Costello's writhing hands. It was the time, too, of the televised Army–McCarthy hearings, with Boston lawyer Joseph Welch pitted against Joseph McCarthy, the Communist witch-hunting Republican Senator from Wisconsin. What one publication referred to as "a four-year binge of hysteria and character assassination" came to an end when McCarthy attempted to besmirch the reputation of a young man in Welch's law firm and Welch responded, "Until this moment, Senator, I think I never really gauged your cruelty or your recklessness....Let us not assassinate this lad further, Senator. You have done enough. Have you no sense of decency, sir, at long last? Have you left no sense of decency?"

It was the decade of Elvis Presley, of Rock and Roll, of Peggy Lee singing "Fever," Rosemary Clooney "Come On-a My House," and Dick Clark's American Bandstand; it was the decade of Davy Crockett hats, UFOs, the hoola-hoop, of paint-by-number kits of the *Last Supper*, and of seeing how many bodies could be stuffed into phone booths and Volkswagens. It was the decade in which city-dwelling parents of the booming post-war baby population flocked out to the suburbs (1,396,000 brand-new houses were built in 1950, alone; and what was once a Long Island potato farm became, in 1951, a community of 17,447 houses called Levittown). It was the decade of Broadway hit musicals like *West Side Story*, *My Fair Lady*, and *The Music Man*; books like *The Caine Mutiny*, *From Here to Eternity*, *Peyton Place*, and *The Catcher in the Rye*. It was the time when Hollywood, fighting back against the devastating inroads made by television, proclaimed "Movies are better than ever" and proved it with films like *High Noon*, *The African Queen*, and *The Man With the Golden Arm*. But it was the "foreign flicks" that swept the college campuses—movies such as *The 400 Blows, La Strada, The Seventh Seal, The Lavender Hill Mob, Rashomon,* and *M. Hulot's Holiday, Hiroshima Mon Amour,* and Brigitte Bardot in *And God Created Woman.*

By 1950, Gilbert Hovey Grosvenor had been President of the National Geographic Society for thirty years and first Assistant Editor, then Editor of its Magazine for fifty-one; he was now seventy-five years old. GHG's right-hand man, John Oliver La Gorce, for many years Associate Editor of the Magazine and Vice President of the Society, was seventy-one. Franklin L. Fisher, chief of the Illustrations Division, was sixty-five. J. R. Hildebrand, the Magazine's third-ranking editor, a veteran of the Society for thirty-one years, was sixty-two. Maynard Owen Williams, chief of the foreign editorial staff, was sixty-one.

Lyrics like "Awopbopaloobop-alopbamboom! Tutti-frutti! All rootie!" were not heard in the hallways of the Society. Bermuda shorts were not seen in the ladies' section of the cafeteria. And the sort of ambitious young Turks one might expect to find in organizations the size of the Geographic had yet to be hired.

The Magazine did, however, still reflect some of the forces at work in the world outside the Society's heavy brass doors. "Roaming Korea South of the Iron Curtain," a thirty-two page article by Enzo de Chetelat, with thirty-four illustrations and a map, was published in the June 1950 issue:

> Imagine the United States divided into North and South by a border from east to west at about the latitude of San Francisco, with a Communist curtain over the North and guerrillas raiding the South. Then you will have some idea of the difficulties faced by Korea.

That issue had just been delivered to the Society's membership when, on June 25, the North Koreans launched a massive invasion of the South and by the next month had occupied most of the Korean peninsula. For the second time in five years America was at war.

The October 1950 issue ran "Seeing the Earth from 80 Miles Up," a piece on the testing of captured German V-2 rockets at the White Sands (New Mexico) Proving Ground that was totally innocent of the convulsions the nation's scientific community would go through when the Soviet Union launched its 184-pound *Sputnik* seven years later. The issue also contained "Strife-torn Indochina," one of twenty articles that would appear during the next three decades about Vietnam. The Korean War was touched on again in Robert H. Mosier's "The GI and the Kids of Korea," a May 1953 article subtitled "America's Fighting Men Share Their Food, Clothing, and Shelter with Children of a War-torn Land." It was a piece largely about the heartwarming relationship between a Marine Corps sergeant and his plucky houseboy Kim whose

> father had been killed and his home bombed out, and he had wandered up toward the front in search of work or food or both. He'd been pushed around a good deal, but he wore a grin that looked as if it were stuck on to stay.

"Nevada Learns to Live with the Atom," one of the 1950s' most chilling articles, reflected the nation's major concern: the threat of nuclear war. Despite the resignation implied in its title, and the optimism of its subtitle, "While Blasts Teach Civilians and Soldiers Survival in Atomic War, the Sagebrush State Takes the Spectacular Tests in Stride," the text provided a vivid

Above: *The impact of television as a new social force was recognized during the tempestuous nationally broadcast Army–McCarthy hearings beginning in April 1954, when millions came to share U.S. Army counsel Joseph Welch's revulsion from the smear tactics of Wisconsin's Republican Senator Joseph R. McCarthy—here covering the microphones as his aide Roy Cohn whispers.*

Right: *On March 9, 1954, the critical report of CBS news's pioneering television journalist Edward R. Murrow on McCarthy's anti-Communist tactics helped bring about McCarthy's downfall. The Wisconsin senator later called Murrow "the cleverest of the jackal pack."*

eyewitness account of an atomic test of a device similar in force to the bombs dropped on Hiroshima and Nagasaki nearly eight years before.

The article's author, Samuel W. Matthews, had come to the Society two years earlier and had been working in News Service when *Geographic* science editor F. Barrows Colton, to whom the piece had been initially assigned, fell ill. Matthews was given the assignment and went out to the desert as one of twenty media representatives picked to accompany 850 soldiers and about 600 officers and observers into narrow, sandbag-lipped, two-foot-wide, five-foot-deep slit trenches reinforced with tar paper, chicken wire, and timber, "twice as close to the forthcoming blast," he wrote, "as men ever had deliberately gone before." The "Trembling Twenty" newsmen had also been dubbed "Men of Extinction."

"Good morning, gentlemen," the Army spokesman's voice had blared through loudspeakers across the Nevada desert as the first pale green band of dawn had silhouetted the rugged mountains to the east. "Welcome to Yucca Flat, valley where the tall mushrooms grow...."

At H minus two minutes they were told, "Kneel down in your trench. Look down. Brace yourself against the forward wall," Matthews wrote. And then,

> Two miles out across the flat, a bright white light shone from the top of a 300-foot tower. At that point, the 22d atomic explosion within the United States was a hundred-odd seconds away.
>
> "I don't mind admitting it," the dark shape next to me said abruptly into the gloom. "I'm scared...."
>
> The siren howled just behind us. We knelt in the dust, heads down, muscles tense....
>
> "H-minus 20 seconds." I took a deep breath. The "count down" of seconds began.
>
> "Zero minus ten...nine...eight...seven...six...five...four...three...two...one—"
>
> Half-night in the trench turned suddenly into blinding, pure-white noon. It was impossible not to blink. Grains of sand leaped into detail and vanished in the brightness, as if a giant searchlight had been switched on just at the back of my head.
>
> Through closed eyelids the world turned orange and, a split second later, crimson. There was no sound; the flash had come and gone in utter silence. I opened my eyes and saw the sergeant huddled in a ball, his face lifted in startled wonder.
>
> Earth rocked beneath us. A giant hand seemed to jar the trench from side to side, heaved it, shook it; then all was steady again.
>
> Still no sound. How long?
>
> Like a clap of thunder directly overhead, the shock wave came. Instantaneous, solid as a physical blow, the sound beat down. The trench seemed to jump convulsively.
>
> A gale of dust and sand and brush roared across the opening above and was gone.
>
> "You may stand and look," I heard beyond the roaring in my ears.
>
> Thick brown dust covered the desert like churning fog. Beyond and above, the atomic fireball rose in the sky, a giant sphere of orange and black, tongues of fire amid billowing soot.
>
> A column of gray tinged with tan lifted behind it. Then the fire dimmed, and, somewhere within the massive cloud, paint was spilled in pastel shades.

Panelists Dorothy Kilgallen, Steve Allen, Arlene Francis, and Bennett Cerf exchanged quips with moderator John Daly on TV's popular quiz show "What's My Line."—All photographs on these two pages are from We Americans

The spreading mushroom became bright pink. Purple shaded to lavender, orange turned dusty rose, and the hues folded and overlapped in great
rolls and waves of cloud.

Higher and higher the cloud boiled against the bright blue-green of
dawn. On the very summit ice crystals formed, cascading over the rim like
pure-white surf in the sky.

The towering column silhouetted soldiers climbing from the trenches
into the murk of dust. Odors of scorching filled the air, acrid and sharp.
Along the ground it was impossible to see more than 100 feet.

Only above, huge and clear, towered the trademark cloud of atomic explosion, leaning with the wind toward the east, shearing from its stem—a
dirty brownish cloud now topped by pink and gold where the sun hit it
40,000 feet above us.

Within 24 hours this cloud, carrying its invisible spitting radioactive
particles, was to be tracked across Utah, Colorado, Kansas, Missouri, and
into southern Ohio. Airborne and ground teams would follow it with delicate instruments, tracking "fall-out"—the descent of the "hot" particles
back to earth.

Although Matthews mentioned the nervousness of the newsmen and
some of the troops, the basic mood of the piece was *Geographic*-style upbeat:
"The Army is proving that human beings, properly dug in and protected, can
survive atomic blasts at quite close range," he wrote. "Yet such demonstra-
tions by the Department of Defense and the Federal Civil Defense Administration are secondary to the main purpose of the Atomic Energy Commission
in Nevada. Its goal is to provide better weapons to ensure America's security
against attack."

Another reason for the test—a reason made evident more by the photographs than by the text—was to evaluate the impact of an atomic explosion on
civilian structures. Department store-window mannequins were placed fully
clothed in living rooms and kitchens of two wood-frame test houses sited at
different distances from the blast. Plaster babies were tucked into nursery
cribs; dummy drivers were seated behind the steering wheels of cars. The
house nearest to ground zero, Matthews later wrote, had been "crumbled into
matchwood, crushed into the desert, what remained was only wreckage,
charred where it faced the blast."

Six months later the *Geographic* published "Man's New Servant, the
Friendly Atom," a tribute to all the atom's anticipated peacetime benefits,
among them atomic power plants and atomic submarines.

During the early 1950s the Magazine was still running articles like "Our
Home-town Planet, Earth," "Work-hard, Play-hard Michigan," "Playing
3,000 Golf Courses in Fourteen Lands," and "Crickets, Nature's Expert Fid-
dlers." But the real story was what was happening within the *Geographic*—or,
more to the point, what was *not* happening, especially in the Illustrations Division where Franklin L. Fisher remained all-powerful and either unwilling or
unable to see the need for change.

The *National Geographic's* increasing isolation from the mainstream of
both journalistic and photojournalistic efforts during this decade was due to
several factors, among them the fact that the *Geographic* photographers rarely
entered national photojournalism contests. Therefore, they had none of the
healthy stimulation of seeing what other, perhaps more innovative photographers were doing. Another reason was that the *Geographic* had for so long been

the leading photographic outlet; its staff had grown complacent, confident that they were the best, that outsiders were technically deficient and, therefore, not in the same league as the *Geographic* staff. After all, had not Gilbert H. Grosvenor, on March 6, 1951, received *U.S. Camera*'s Golden Achievement Award for "leadership in the editorial use of color photography"? And had not that citation honored the *Geographic* for having provided "a priceless documentary chapter in the history of color photographic processes"?

There was also the celebration of the *Geographic* by Frank Luther Mott, historian and dean emeritus of the University of Missouri School of Journalism. In his *History of American Magazines*, Mott wrote what Gilbert H. Grosvenor's histories of the *National Geographic* were quick to quote over and over again:

> There is really nothing like it in the world. For more than half a century the *National Geographic Magazine* has not published a single monthly number that has not been interesting and informative, with some measure of value. If it has seemed to some critics too much of a picture book, even they have to admit that in this it is in harmony with its times, and that its pictures are educational to a high degree. By the middle of the twentieth century it had attained the largest monthly circulation in the world at its price, and it had an assured position among the top ten monthly circulations at any price. This is a fabulous record of success, especially since the magazine is founded upon an editorial conviction that rates the intelligence of the popular audience fairly high. The *National Geographic Magazine* has long represented an achievement in editorship and management outstanding in the history of periodicals.

The *U.S. Camera* citation and Mott's laudatory summation were both, of course, the result of the Society's commendable early achievements. The problem with the Society in the 1950s was, however, that its upper echelon of officers was equally entrenched in the past. *Geographic* editors still believed color photographs were interesting as technical achievements, that the act of even being able to take such a photograph provided an end in itself. Therefore, the *Geographic* photographers were still taking color for color's sake, still taking perfectly composed and lighted, professionally challenging photographs—of often boring subjects.

"*National Geographic*'s pictures, with rare exception," Ed Hannigan wrote of that period at the Magazine in a 1962 *U.S. Camera* article, "were all pretty much of the picture postcard type of idealistic beauty, rather than photojournalism." And that picture postcard era reached its culmination in what critics outside the *Geographic* called the "Red Shirt School of Photography"— a reference to the constant use of red shirts, red caps, and red sweaters as props by photographers to brighten up their pictures. (That technique was nothing new; Gervais Courtellement, whose Autochromes had appeared in the *Geographic*s of the 1920s, traveled through Europe, Africa, and Asia with colorful scarves to drape over his models or on nearby fences and walls.)

In fairness to the *Geographic*'s photographers, the emphasis on color came from above: Gilbert H. Grosvenor believed in color; therefore senior editors demanded it, and illustrations editors, reflecting the Chief's views, saw color as being what separated them from lesser, conventional magazines. Color, having become easier to take, was, however, easier to overdo.

Even though Kodachrome was already unnaturally bright, photographers, because of existing imperfect color reproduction methods, splashed the strongest possible colors in their pictures so that they would be more effective in print. One result was that the staff photographers—who were constantly being sent to colorful places to slake what was seen as the public's unquenchable thirst for colorful scenes—would often find themselves needing more color to take advantage of the color film and would resort to placing the people in costume in the photographs. In a *Geographic* article on the modern South, for example, a couple might be dressed in colonial clothing or a mansion's graceful portico might be crowded with Southern belles in antebellum attire. The resulting overposed and artificial photographs were rendered even more ineffective by being jammed together, separated only by hair-thin lines in the Magazine's clumsy, unwieldy layouts.

Former National Geographic Book Service editor Edwards Park relates:

> They tell the story of old Muench, the southwestern photographer, who would go out and set up a tripod in Monument Valley. He'd take a picture with a piece of driftwood in the foreground and those huge monoliths in the background, a lovely sky, the whole thing. He'd shoot that and sell it to *Arizona Highways*. Then he'd snap his fingers and a beautiful girl with a large bosom and something red on would ride into the foreground on a pony and "smile and point at the mountain." Muench would shoot a second picture and sell that photograph to the *Geographic*.

But if one takes into account the context of the times and what the *Geographic* was doing, then the evolution of the "Red Shirt School" is not surprising. The philosophy of the Magazine had always been to increase and diffuse geographic knowledge—and to do so with the brightest possible face. The *Geographic* had not shown the world suffering through the Depression; and only rarely, in their coverage of World War Two, had battle scenes appeared. The photographs we remember from those desperate times were not the *Geographic*'s color photographs because it was never the *Geographic*'s role to provide them. What we do remember are the black-and-white documentary images shot by free-lance photographers with newspaper backgrounds or those who went to work for *Life* magazine; and these photographs were in black and white because that was the way they were shot and appeared out of practical necessity.

Unfortunately for the *Geographic*, during the post-war years and thereafter, black-and-white photographs were increasingly considered the more artistic and creative, while color was considered frivolous and shallow; and that perception also contributed to the *Geographic*'s isolation from the journalistic mainstream. But even if the "Red Shirt School" seems a bit silly *now*, given the kind of magazine the *National Geographic* was *then*, and the sort of article its color photographs were to illustrate, the "Red Shirts" as Melville Grosvenor explained, made perfect sense:

> What happened was that they'd go on these expeditions, and everybody would be in khaki, because that's the color of field uniforms. And they'd come back with the dullest bunch of pictures you ever saw! You couldn't use them editorially because they had no color. So, we decided to have people wear colorful shirts. But some of them went crazy, went to the other extreme for a while. The idea was, if you're going to buy a shirt to wear in the field, why don't you get a lumberman's shirt with plaid colors or something like that? Pictures were just plain monotonous. You might as well be in black-and-white if you're all in khaki or dull green. There's no pep to it. But when a fellow had a red cap on—just a red cap—it would add a little color to the picture.

Although during the 1950s the Geographic's *photographers took advantage of the faster Kodachrome speeds to break away from the more rigidly posed pictures required by the earlier, slower film processes, photographic clichés continued to occur. One particularly popular variant became known as "Smile, and point at the mountain."*

Overleaf: *The "picture-postcard" type of* Geographic *photograph prevalent during the 1950s reached its culmination in what critics outside the Society called the "Red Shirt School of Photography"—a reference to the frequent use by photographers of red shirts, caps, sweaters, or scarfs as props to brighten their pictures.*

295

At the same time, it is important to keep in mind that what the older generation, today, remembers as the traditional *Geographic* photograph—and by "traditional" one really means photographs from the immediate post-war decade—was not the traditional *Geographic* photograph at all. It was simply the reflection of the inexorable aging of its editors, their growing conservatism, complacency, and in Gilbert Hovey Grosvenor's case particularly, his inability for the first time to recognize or act upon the changes in contemporary tastes. The editorship of the Magazine demanded a younger man at the helm, and, perhaps understandably, the one man who had been so responsible for creating the astounding success the *Geographic* had enjoyed, the man who had seen the circulation rise from 1,000 when he became Editor to more than 2,000,000 in 1952, that man was unwilling to let go.

On Saturday night, August 8, 1953, the Magazine's Illustrations Chief, Franklin L. Fisher, was in Los Angeles, California, accepting on behalf of the National Geographic the Photographic Society of America's La Belle Award for outstanding contributions over the years to the development of color photography for magazine use. The following Wednesday, as Fisher was preparing to return to Washington, he was stricken by a heart attack and died. Fisher, who had been hired by Gilbert Grosvenor in 1915, would have been sixty-eight years old two weeks later. His absence was felt, but perhaps more with a sense of relief than of mourning. Some staff members felt that Fisher had been too embedded in the hierarchy to remove, too stubborn to perceive a need for modernization.

During the following year Walter (Toppy) Edwards was eased into the Illustrations Chief post vacated by Fisher's death; but Toppy Edwards, who had been with the *Geographic* for twenty years, was never to be given either Fisher's authority or power.

Herbert S. Wilburn, Jr., was put in charge of color photography; and Kip Ross, who in 1941 had come to the *Geographic* from the E. Leitz Company, makers of the Leica camera, and, by 1945, was an assistant to Fisher on the Illustrations staff, was in charge of black-and-white. The long uninterrupted years of the Illustrations Division's one-man autocratic rule were over.

And then on May 5, 1954, Gilbert Hovey Grosvenor stepped down as Editor of the Magazine and as President of the Society.

By then Gilbert H. Grosvenor had had four babies named after him, a fish, a seashell, an island, a glacier, a trail, a mountain range, a lane, a mountain peak, a lake, and a plant.

John Oliver La Gorce took his place.

The announcement appeared in the July 1954 issue:

> Dr. Gilbert Grosvenor, Editor of the *NATIONAL GEOGRAPHIC MAGAZINE* for 55 years and President of the National Geographic Society since 1920, requested permission to retire in a letter read to the Board of Trustees at its meeting, May 5, 1954. The Trustees voted to accept his retirement, on condition that he become Chairman of the Board. Thus he will continue actively his lifelong interest in geography and the National Geographic Society.
>
> Dr. John Oliver La Gorce, for many years Associate Editor of The Magazine and Vice President of The Society, was elected by the Trustees to succeed Dr. Grosvenor as Editor and President. Dr. Melville Bell Grosvenor, Senior Assistant Editor, was elected Vice President and Associate Editor.
>
> Dr. Robert V. Fleming continues as Treasurer of The Society, and Dr. Thomas W. McKnew as Secretary.

By the time Gilbert H. Grosvenor stepped down as President of the Society, he had been honored by having many entities named for him:

Gilbert Grosvenor II, a grandson; Gilbert Grosvenor La Gorce, son of John Oliver La Gorce; John Grosvenor Hutchison, son of George W. Hutchison, the long-time Secretary of the Society; and Gilbert Grosvenor Edson, grandson of John Jay Edson, an early Trustee and Treasurer of the Society.

Bryconamericus grosvenori eigenmann, a blind fish found in the Urubamba River in southern Peru.

Margarites grosvenori, a pearly white seashell dredged up in the waters west of Greenland.

Grosvenor Island, off the coast of Alaska.

Grosvenor Glacier in Peru, sighted and named by Hiram Bingham, the discoverer of Machu Picchu.

Grosvenor Trail and Gilbert Grosvenor Mountain Range, both in the Antarctic, named in 1929 by Richard E. Byrd.

Grosvenor Lane, the road upon which his Bethesda, Md., house was situated, named after GHG by Elsie May Bell Grosvenor, his wife.

Mount Grosvenor in China near the border of Tibet; named by Joseph F. Rock.

Grosvenor Lake, in Alaska, named by Dr. Robert F. Griggs, expedition leader to the Valley of Ten Thousand Smokes.

Momordica grosvenori, a plant discovered in 1937 in Kwangsi Province, China, by botanist Walter T. Swingle. The Chinese believe the plant to contain qualities of an aphrodisiac.

The Geographic *editors' attempts to appease what they perceived as the membership's unquenchable thirst for colorful scenes resulted in photographers being sent to cover festivals such as the one Jamestown, Virginia, celebrated in 1957 marking the 350th anniversary of its founding (shown above).*

In his acceptance speech, Dr. La Gorce said, "Gentlemen, with your continued help I shall do my best to carry forward the high standards of the educational and scientific work of the National Geographic Society and its Magazine." Little more than a year later La Gorce was one of five journalists and editors from the Western Hemisphere to receive an award from Columbia University for furthering friendship between the peoples of the Americas.

For nearly fifty years La Gorce, formerly an expert press telegrapher and one of the Society's original three employees, had worked in the shadow of his mentor, Gilbert H. Grosvenor. A courtly, six-foot-tall, seventy-five-year-old with an erect military bearing, "JOL"—as he was referred to in the office, although he referred to himself as Grosvenor's "man Friday"—was handicapped by a stutter which miraculously disappeared when he made a public address.

Although advertising and promotion were his chief responsibilities, La Gorce occasionally wrote articles for the Magazine—fishing and sea life were his favorite subjects—and he edited the Society's 1952 *Book of Fishes.*

Gilbert Grosvenor was not the only Society officer to have had land features named after him; La Gorce had had a mountain range and a peak named after him in Antarctica and a meteorological station, too. There was, in addition, a Mount La Gorce, La Gorce Lake and La Gorce Glacier in Alaska, an island and a country club and golf course in Miami Beach, Florida, and La Gorce Arch, a rock formation along the Escalante River in Utah.

Visitors to La Gorce's pine-paneled third-floor office would remark on how dark it was, how it was dominated by La Gorce's weapons collection—which included a thirteenth-century Crusader's sword, a Stone Age axe and a World War Two commando dagger.

La Gorce was a man of simple loyalties—and appalling prejudices.

Gilbert H. Grosvenor's loyalties were to the membership of the Society; John Oliver La Gorce's correspondence reveals, as in this telegram sent to GHG on September 21, 1921, that his loyalties were to Gilbert Grosvenor first of all:

SIXTEEN YEARS AGO TODAY WAS THE MOST FORTUNATE AND HAPPY MILE-STONE OF MY LIFE WHEN I JOINED UP WITH YOU THE BEST LEADER AND FRIEND A MAN COULD HAVE.

J O LAGORCE

Second, his loyalties were to the Society to which he had devoted nearly fifty years; and third, his loyalties were to the membership (so long as it didn't attempt to interfere with that "nonprofit scientific and educational institution's" business of increasing and diffusing geographic knowledge). A 1920 handwritten memorandum from La Gorce to Gilbert H. Grosvenor provides an insight into JOL's attitude toward those members who dared criticize the Magazine's "nude" pictures:

GHG: In considering the letters received now and then from people who believe it their mission on earth to save man from himself and in spite of himself—in the use of tobacco for example—let us not forget that we get as many letters telling us that the Geographic is corrupting the morals of decent people by publishing so many pictures of nude women. Naturally each is entitled to his own opinion but should the mock modesty of 1/1000 of 1 percent of geographic readers upset the educational work of the book—if not why should the same percentage of hopeless asses force tobacco from our ad pages. I know how broad you are but I want to get a line through on the subject.

JOL

Just as La Gorce's and Grosvenor's loyalties differed, so did their prejudices. Gilbert H. Grosvenor, as he once told an interviewer, had "a great sympathy with the idea of combating race prejudice...."

Gilbert H. Grosvenor always said what he thought. John Oliver La Gorce, on the other hand, was more circuitous. It is only through reading his correspondence and memos that one becomes aware that, unless they kept their place, JOL did not like women, blacks, or Jews—though not necessarily in that order.

GHG had always respected women's contributions to science and the arts. There were, after all, precedents: Eliza Scidmore had been on the Magazine's masthead before him, and Ida Tarbell was an Associate Editor in 1901.

"As you know, I have felt for a long time that we should endeavor to develop more women writers for the Magazine," Dr. Grosvenor had written his Assistant Editor, J. R. Hildebrand, in 1949. "Women often see things about the life and ways of people which a man would not notice....Among the educated young women Miss Strider [Personnel Director] has brought to our staff, I am sure there must be some hidden talent. Men are more forward in presenting and asking for assignments than women; perhaps that is one reason why the ladies of our staff have not received as many assignments as the men." Furthermore, GHG's wife, Elsie, an active suffragist, seems to have had enormous influence on him. So La Gorce's attitude toward women in general and lack of appreciation for women on the work force, in particular, seems to represent his views and certainly not those of his Chief who, faced with JOL's intransigence, learned it was easier on occasion simply to bypass La Gorce—an example of which is seen in the parenthetical caveat to the following memo sent on December 2, 1938, by GHG to the editorial staff:

> IT IS INTERESTING TO NOTE THAT OF THE SIX BEST SELLING BOOKS OF FICTION RECORDED IN THE NEW YORK HERALD TRIBUNE OF NOVEMBER 27, THE FIRST, SECOND, THIRD, FIFTH AND SIXTH ARE BY WOMEN; AND OF THE NON FICTION BEST SELLERS, THE TWO FIRST IN POPULARITY ARE WRITTEN BY WOMEN.
>
> G.H.G.
>
> (TO BE SHOWN TO EACH MEMBER OF THE EDITORIAL STAFF EXCEPTING DR. LA GORCE.)

Perhaps no better evidence of La Gorce's attitude toward women can be found than in a letter written by him to GHG occasioned by what La Gorce perceived as a woman staffer's betrayal.

In the fall of 1919, Jessie L. Burrall, a female staffer who had earlier that year contributed the article "Sight-Seeing in School: Taking Twenty Million Children on a Picture Tour of the World," had been made Chief of the Geographic's then-fledgling School Service Division. In September, Miss Burrall had aroused La Gorce's ire by writing him that she was very interested in religious work, and that while traveling for the Society she had been offered the deanship of a college where she would teach religion and would receive, in addition to a $5,000 salary, room, board, and the use of a car.

In a special delivery letter La Gorce sent to Gilbert Grosvenor at his Baddeck summer home, JOL wrote, "I immediately realized why I had always mistrusted her and why, in spite of your constant urging of her qualities, they were always discounted by my somewhat uncanny gift of reading human nature and judging people by what I find...."

"Blooms, Wooden Shoes, and Windmill Say It's Tulip Time in Holland, Michigan" was the caption for this March 1952 photograph from the Geographic. *Not all the members were pleased.* "Your photographers have a mania for photographing costumes," *one wrote.* "I am a nudist and do not care what people wear. I am nevertheless an epicure, and I enjoy reading about the foods of other people."

Hired by Gilbert Hovey Grosvenor in 1905, John Oliver La Gorce remained at the Society for the next fifty-two years. The photograph at top, showing the Georgia pine paneled and raftered "OFFICE OF JOHN OLIVER LA GORCE, THE ASSOCIATE EDITOR" *was published in 1914. Twenty-two years of accumulated clutter and memorabilia later the camera angle was duplicated in this photograph at bottom captioned* "IN THIS ROOM JOHN OLIVER LA GORCE 'HAS LABORED WITH LOVE AND CEASELESS ENERGY'" *when it appeared in 1936.*
"On the office walls...are hung many strange weapons and trophies collected on travels about the world," *the caption explained. "Old ships' running lights suspended above his desk lend a tang of the sea, the elephant's foot at the right brings to mind African treks, and many primitive knives, arrowheads, spears, and shields always interest visitors."*

JOL pointed out that had not GHG hired her, Miss Burrall "would still be a teacher of geography at $1,500, or $1,800, a year in the back-waters of the Middle West..." and that for her to be "even considering leaving this work..." was "disloyalty pure and simple, showing a very selfish spirit...."

La Gorce's letter continued:

> I will not go into the details of my presentation to her of the fact that instead of being the teacher of 400 pupils, in her present work, she was a teacher of 200,000 teachers and 20,000,000 children and she had her Sunday School Class here in Washington which could occupy her from a religious viewpoint. Further, that there was as much comparison between the countrywide prominence of Chief of the School Service of the National Geographic Society and the Dean of a Jerk-Water College as was between the occupancy of the White House and the management of the Washington Baseball Team.

La Gorce explained that he had told Miss Burrall that GHG was aware that her traveling on behalf of the Society necessitated extra expenses, that he would probably increase her salary, and that she should think it over for a few days. Three days later, continued La Gorce's letter, "she came in, looked me in the eye, and said that she had made her absolute decision on the conviction that her work was *here*, that she would stay, and would not consider for a moment any other proposition made to her."

However, La Gorce wrote, Miss Burrall was next offered a job by "some Baptist Association in New York," and despite JOL's "request and demand that she not disturb you in your rest period" she had telegraphed GHG in Baddeck.

JOL urged GHG to dismiss Miss Burrall, assuring him that there would be no difficulty in finding a replacement. "By this," he added, "I do not mean that we can find a person with all her ability as a worker, a talker, as a social climber, an educational parasite, but we can find a teacher of experience...."

La Gorce's letter ends:

> I know that, with your gentle disposition and unwillingness to hurt anybody's feelings, and Mrs. Grosvenor's interest in the lady, you will probably not wish to take the drastic step I suggest....

JOL was right in his assumption: GHG ignored his request and Miss Burrall remained at the Society until 1920.

JOL's attitude toward Jews is revealed in a 1920 memo he wrote to GHG containing his response to a proposal that the National Geographic Society lend its name to the distribution of films with which the Society had no previous association and which were made and developed primarily for commercial reasons:

> G.H.G.: No one has a keener enjoyment or a more fulsome appreciation of what motion pictures have done for humanity than I, but at the same time the industry is known to be practically in the hands of some of the most unscrupulous men and associations in the country,—mainly Jews who have by a process of elimination and massacre absorbed or done away with the smaller fry, and their business methods have been an unpleasant odor in the commercial nostrils for a long time....

But the most revealing of the La Gorce memos was the one he sent GHG on March 3, 1926, from which the following is extracted:

G.H.G.: May I give you a thought in connection with campaigning for new members in Southern states?

In the matter of the Southern states we have, of course, the danger of negro nominations, but after all that is merely a question of applying thought to the method and at first it will be a trifle more expensive per member, but every other organization in the country is confronted with the same thing and are not afraid to enter the field, so why should we be? I would suggest that we put Boutwell on the job of securing from the N.E.A. or from the Bureau of Education (if we do not have such lists in our files) the names of all negro schools and colleges in the Southern states, and to obtain a list of the faculties. These names could be checked against our membership lists as though they were all nominations, and in the mailing out of nomination letters such names could be omitted by the simple process of putting a tab on the Addressograph card that would trip it through without printing, as in the case of dues.

A second precautionary measure would be to make a special list of all nominations coming from towns where these schools and colleges are and, without giving the name of the nominator, submit the list to some member who we know is white, asking their cooperation in the matter of glancing over the list enclosed and to indicate any names that they know are not of the white race. It will be unnecessary to go into any further detail with any white Southerner, for he, or she, will know why we are making this request and will be only too glad to cooperate. There are few large cities in the South comparatively speaking, and in places of even 30 to 50,000 all white people either know each other, or know who the other fellow is, for that is peculiar to the South....

There is no indication that GHG ever acknowledged JOL's recommendation or that the intricate procedure was ever adopted. Sixteen years later, La Gorce was urging GHG not to run a proposed article on black members of the military because:

I fear it would promptly bring insistent demands from the group of educated Negro agitators who for political reasons have been encouraged and aided by a wellknown source to strike now for social equality in this country....

I do not discount the importance of negro soldiers in the present all-out manpower necessity, but accounts of the Spanish-American and First World Wars do not credit negro soldiers with the sort of courage needed to win battles.

Personally I have always had a kindly feeling for the race, and get along with them because I understand them. We could go as far as to include now and then a picture of Negro troops in training when illustrating appropriate articles, but I'd vote against an article about them.

JOL's prejudices apparently had little effect on GHG and little impact on the Society's policies. His strong personal feelings aside, JOL had made significant contributions to the Society during his tenure: His promotion efforts had increased membership from 10,000 in 1905 to more than two million by the time he became Editor. Advertising, too, had flourished in the Magazine under his direction. In fact, no one since has equaled his advertising record.

Considering that his strength was primarily in promotion and advertising and that he had somewhat limited editorial experience, why was La Gorce made Editor of the Magazine and President of the Society? For several reasons. Chief among them, one suspects, was the feeling that La Gorce *deserved* the recognition as a reward for his lengthy service. A second reason was that he

La Gorce poses at the foot of the stairs beneath his portrait in the clubhouse of Miami Beach's La Gorce Country Club, named in his honor.

La Gorce's Editorship—beginning with Gilbert H. Grosvenor's retirement in 1954, and ending with his own retirement in 1957—was perceived by most of the staff as an interregnum between Grosvenors.

was expected to be an interregnum leader, a custodian, a keeper of the flame. Under GHG's reign, the Society had prospered, and La Gorce was determined to see that during his three years of tenure nothing was changed. The third reason was that in the context of the times La Gorce's prejudices really were not all that surprising. He, like GHG, was ultra-conservative, a product of the Victorian era; but unlike GHG, La Gorce reflected the biases of that earlier time. La Gorce's wartime memo, one should note, was written twelve years before the Supreme Court, in *Brown* v. *Board of Education of Topeka*, would unanimously ban segregation in the public schools.

La Gorce retired on January 8, 1957, after serving less than three years as Editor and President. At the time, GHG presented a resolution to the Board that reflected the close bond and mutual loyalty he and JOL had shared for half a century. In part, GHG said, "With everything the Society has done...he has been identified. Many of our most useful and interesting projects he originated. He labored...to help develop the organization and bring it to the dignified position it now holds in the life of our country."

Just as photography at the *Geographic* was stagnating during the 1950s, so was its writing. The Magazine was still running stories like Jean and Franc Shor's "From Sea to Sahara in French Morocco" with such lines as:

> Morocco is today a neighbor with trouble in its own house. The night we landed in Casablanca terrorists shot up a French cafe a block from our hotel, killing two Frenchmen and wounding three. In Marrakech, as we watched a parade honoring the retiring Resident General, a bomb shattered a French Army unit 50 yards from where we stood. Three soldiers died, 28 were injured. And as we drove from Meknès to Rabat, through Morocco's breadbasket, the horizon was black with smoke from burning grainfields fired by arsonists.

Despite its acknowledgment of the civil disruptions then occurring in French Morocco, the piece swiftly becomes upbeat and positive. A friend at the American Consulate General in Casablanca tells the Shors he thinks their trip from the sea to the Sahara will be no more dangerous than "driving an equal number of miles in American traffic. You and Jean aren't concerned with politics," he says. "You're interested in the land and the people...."

Franc Shor, a member of the *Geographic* staff for twenty-one years and one of its two Associate Editors at the time of his death in 1974, was a legend at the Society. Supreme Court Justice William O. Douglas called Shor "the most traveled man I have ever met." Shor had traveled around the world at least a dozen times; and once said he had been to every country in the world, including some that weren't legally countries when he visited them. Witty, erudite, charming, Shor was the sort of man who, while in Athens for a January 1956 "Athens to Istanbul" *Geographic* piece, would be photographed by Queen Frederika of Greece—who would then ask for and receive a credit line on the photograph when it was published in the Magazine.

Once Shor was forced to bail out of an airplane over China—and took

On assignment in what was then Formosa in 1949, staff writer Frederick G. Vosburgh and companions rode a push car across crude log bridges spanning mountain ravines. "Many people get killed on this railroad?" Vosburgh asked. "Not so many," was the answer.

302

Although legendary Associate Editor Franc Shor, in fact, came from Dodge City, Kansas, could do rope tricks like Will Rogers, was a magnificent horseman, and, although his mother, in fact, broke horses for the U.S. Cavalry in World War One, many of the *Geographic*'s staff still found some of Shor's personal accounts hard to believe.

"So many of Franc Shor's stories you just assumed were lies," Bart McDowell recalled. "Franc would say, 'of course, I speak fluent Chinese'—one of those throwaway lines—and Garrett would say he didn't believe it, that Shor was deluded. And then Shor and Garrett went to China to do the Quemoy-Matsu story ["Life Under Shellfire on Quemoy," March 1959] and some taxi-driver on the street in Hong Kong began to discuss god-knows-what with Shor, and Franc responded in such fluent Chinese the guide said, 'Oh, you come from Chungking!' and suddenly it would turn out that Shor's crazy stories were true!"

nothing with him but his typewriter. He was an "old China hand" who, in order not to be caught short of cash, wore cufflinks made of the gold bars Oriental money-changers recognized as currency wherever one traveled in the days before Chairman Mao.

On one occasion Shor spent five weeks tracing, partly on foot, the path of the first Crusade from France to Jerusalem. On another occasion he trekked across the Gobi desert retracing Marco Polo's journey from Venice to China—"a trip," noted *The New York Times* in Shor's obituary, "that exposed him to fever, bandits, wolves, sandstorms and nearly ended in his death."

"Franc Shor was a great character," Luis Marden recalls. "A great lady's man, a great wine connoisseur—he had a private wine cellar at the Ritz in Paris—a great intellect." It is not inappropriate that Marden speaks of Shor with the fondness one club member feels for another, for the National Geographic in the 1950s was still run very much like a private men's club—although photographer Dean Conger, who joined the Geographic staff in 1959, thought the Society of that period more akin to a Southern plantation. And with reason.

A woman hired by the Society about that time recently said, "One of the first things I noticed when I came here was that all the chauffeurs and the elevator operators were black. They called everybody 'Mr. So-and-so,' and everybody called them by their first names. There was this *noblesse oblige*, a hierarchy that was catered to. I was born and raised in the Midwest and had had no experience with people or places like this. And one day I woke up and suddenly I realized I was living in The South."

There were quite a few women then working at the *Geographic*, but nearly all of them were in secretarial positions. As Anne Chamberlin in her 1963 *Esquire* magazine piece would note, the Society was "considered by many of Washington's best families as a fine, safe place for a post-debutante daughter to put in a few years dabbling in secretarial work before marriage."

However, not all the young women hired in those days were post-debutantes. Legends Editor Carolyn Patterson, who retired in 1986 after nearly thirty-seven years with the Society, came to the National Geographic in 1949 after having served as the first woman police reporter on the *New Orleans State*. Although Miss Strider, the maiden lady Personnel Director then in her mid-fifties, thought Carolyn Patterson overqualified, she hired her as a file clerk but later rebuked her for unladylike behavior—Patterson had been walking too fast in the corridors.

"There was a very, very distinct idea of what ladies did and what ladies did not do here," Carolyn Patterson explained. "In those days we had two dining rooms, one for the men and one for the women and [in] the women's dining room we would lunch in large groups. And in my luncheon group I once asked the single question, 'Has anyone thought of putting a picture on the cover of the *National Geographic*?' Everyone looked terribly shocked that I would even ask the question, and someone reported me to Miss Strider....She called me into her office and said to me, 'The National Geographic is a traditional organization, Mrs. Patterson, and if you don't understand that means keeping the Magazine's cover the way it is, you perhaps do not belong here.' "*

*In July 1942 the American flag had appeared for the first time on the Magazine's cover; cover photographs began appearing regularly in September 1959.

It was precisely because, by the 1950s, the Society had become such a "traditional organization" that the Magazine began to lose the creative impetus that had lifted its circulation from 2,200 in 1900 when Gilbert Hovey Grosvenor became Managing Editor to 2,100,009 in 1954 when he retired. Grosvenor's daring, so instrumental in creating a popular magazine during the previous fifty years, had dissipated with age until all that remained were his Victorian sensibilities. And it was those sensibilities, now entrenched—and, indeed, celebrated—as traditions by his successor, John Oliver La Gorce, that brought about the stagnation of the Magazine, a stagnation evident in its sluggish circulation which increased by but 53,201 during La Gorce's three-year tenure, an average gain of only 17,794 per year.

One long-time Geographic staffer, comparing Gilbert Hovey Grosvenor and John Oliver La Gorce, recalled GHG as "being an enthusiast rather than a professional journalist, and sort of 'inventing the wheel' as he went along," the kind of editor, he continued, who "would lose interest after the first blush of enthusiasm." JOL, on the other hand, was "the phlegmatic ballast of the operation, [the man who] made sure things were spelled correctly, that there weren't any 'typos,' that there was a sense of follow-through and thoroughness.... He was always a good cleaner-upper, but not an innovator in any way."

"Poor old JOL," the staffer said, "stood around like Anthony Eden to GHG's Winston Churchill for all those years, crown prince until he got senile, and then when he finally did get [to be editor], he was never able to do anything with it."

Certainly La Gorce *could* have done things as editor and didn't, not only because, as has been mentioned, he was determined to see that nothing was changed, but also because, if anything, he was even more conservative than his predecessor.

Gilbert M. Grosvenor, who was hired by La Gorce (and told by JOL, "G-Gilbert? the only advice I c-can g-give you is d-don't d-dip your p-pen in the c-company ink well,") recalls one telling incident from that era: "GHG knew what he was doing when he published semi-nude pictures. He 'showed people like they were,'" Gil said, "but he also knew that this was the kind of 'trade-mark' you made. JOL, particularly in the latter years, was a little prudish about that.... There was a case where GHG approved a set of pictures and there was a bare-breasted lady [who]...was apparently very beautiful...Dee Andella [then Printing Production Division chief] had the proofs made. GHG said, 'Yes, these proofs are fine,' and went away, left JOL running the show. JOL looked at the proof and said, 'That's awful!' and he covered her up. So Andella had to go back and put her breasts in shadow or something...GHG came back to town, looked at these final proofs, asked, 'What's going on here? You have ruined this photograph!'...It went back. They made the plate all over again... and that's the way it ran."

If the Magazine was treading water during the La Gorce interregnum, the fact that it did not drown was due to La Gorce's Associate Editor and heir apparent, Melville Bell Grosvenor. In turn brilliantly and imaginatively tutored as a boy by his ebullient grandfather, then strictly dominated and suppressed as an adult by his somewhat distant father, Melville Grosvenor—charged with the kind of enthusiasm and vision one might expect of a man emotionally more Bell than Grosvenor—was impatiently awaiting his turn.

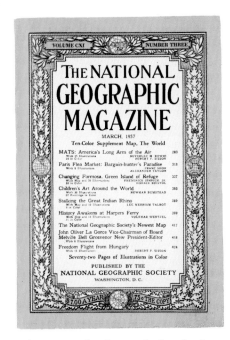

The Geographic languished under La Gorce. Between January 1954 and his retirement in January 1957 only 53,201 new members joined the Society, bringing its total to 2,144,704. Melville Bell Grosvenor's long-awaited Editorship would change all that.

"Because it's there."

"'Why did you want to climb Mount Everest?' This question was asked of George Leigh Mallory, who was with both expeditions toward the summit of the world's highest mountain in 1921 and 1922, and who is now in New York. He plans to go again in 1924, and he gave as the reason for persisting in these repeated attempts to reach the top, 'Because it's there.'"
—The New York Times, March 18, 1923

"Well, we knocked the blighter off!"
—Sir Edmund Hillary, May 29, 1953, returning with Tenzing Norgay from their successful struggle to Everest's 29,028-foot summit, to expedition member George Lowe. As reported in Sir Edmund Hillary's July 1954 *National Geographic* article, "The Conquest of the Summit"

For the April 1891 issue of the Magazine, geologist Israel C. Russell, a founder of the Society, wrote in his "Summary of Reports on the Mt. St. Elias Expedition" that the purpose of the National Geographic Society's first sponsored expedition* was "to study and map...the vicinity of Mt. St. Elias...and, if practicable, to ascend it."

En route to Mount St. Elias, Russell found "a wonderful panorama of snow-covered mountains, glaciers, and icebergs...hoary mountain peaks, each a monarch robed in ermine and bidding defiance to the ceaseless war of the elements."

Russell survived "torrential rains," "blinding snowstorms," "roaring avalanches," "huge grizzly bears," "yawning chasms," "treacherous passes," and was—despite a bout of snowblindness—the first to see and name 19,524-foot-high Mount Logan, the continent's second highest mountain. Russell failed, however, in his attempt to reach Mount St. Elias' 18,008-foot peak.

Since Russell's report, *National Geographic* has been filled with accounts of the world's great mountains scaled or attempted, climbers' triumphs or tragic deaths, expeditions' successes or defeats. Seven years after Russell's attempt, Eliza Ruhamah Scidmore reported on the Italian Duke of the Abruzzi's successful ascent of Mount St. Elias in 1897; and in 1913, Walter Woodburn Hyde wrote this pithy description of his feelings upon having reached the 15,771-foot summit of Mont Blanc:

> Do you open your eyes wide in astonishment at the wonderful sight? By no means! You shut them as tight as you can and throw yourself down on the

The Mount St. Elias expedition was sponsored jointly by the National Geographic Society and the U.S. Geological Survey.

Opposite: *In 1890, a founder of the Society Israel C. Russell and his party crossed the moraines of the Malaspina Glacier in Alaska while on their expedition "to study and map...the vicinity of Mount St. Elias" for the Society's first expedition, sponsored jointly with the U.S. Geological Survey.*

Preceding spread: *First conquest of Antarctica's highest peaks. In December 1966 and January 1967 the unclimbed Sentinel Range of the Ellsworth Mountains, the highest mountains in Antarctica, fell to members of the American Antarctic Mountaineering Expedition. Here expedition leader Nicholas B. Clinch struggles to the top of the Vinson Massif—at 16,860 feet, Antarctica's highest peak.*

"I recall flying around and over the peak...sitting on a gas can by the open door and snapping pictures with a bulky camera—while a rope around my waist let me lean out far enough, and no farther," wrote Bradford Washburn (above) *posing with the forty-eight-pound aerial camera with which, in 1936, he photographed Alaska's Mount McKinley on Society-sponsored flights that "resulted in the first complete photographic record of North America's ice-encrusted roof-top."—From* Great Adventures with National Geographic

Opposite: *Landing on McKinley's flanks at 10,000 feet, higher than any previous plane, Terris Moore taxis up to Bradford Washburn's 1951 expedition base camp with supplies and mail.*

snow in utter weariness of mind and body, resenting the impertinence of your guides, who urge you to look about.

In 1924, the year after his American lecture tour, George Leigh Mallory, with his partner Andrew Irvine, vanished into the mists on Everest's northeast ridge during his third expedition to the mountain. Before his disappearance, Mallory had expressed a sentiment that all mountain climbers can understand: "Have we vanquished an enemy? None but ourselves."

In 1929, G. H. H. Tate was appalled by the silence on the 9,092-foot summit of South America's Mount Roraima. "One feels oppressed, dwarfed," he noted, "...almost as if one were a trespasser."

That same year explorer-biologist Joseph F. Rock was in China attempting to reach a vantage point from which he might view Minya Konka, "one of the loftiest peaks of western China." Rock wrote in "The Glories of the Minya Konka," for the October 1930 Magazine:

> Our descent [from Chirpin Pass] was very difficult. Men and beasts and loads were many times catapulted...into a snow bank up to our necks....
>
> As the storm continued, to go on the next day was impossible. [My men] were snowblind and were suffering terribly. Fortunately, I had cocaine, with which I made a solution, and this I dropped into their eyes to relieve their pain....To the north, it seemed as if one had entered a different world. There not a tree could be seen: the entire landscape was a bleak waste....
>
> When we reached 15,000 feet we found the snow lay deep...but, fortunately for us, it was frozen, and, as it bore our weight, we advanced cautiously....
>
> And then suddenly, like a white promontory of clouds, we beheld the long-hidden Minya Konka rising 25,600 feet in sublime majesty.
>
> I could not help exclaiming for joy. I marveled at the scenery which I, the first white man ever to stand here, was privileged to see.

In 1933, nine years after Mallory disappeared on Everest's slopes, two especially redesigned British camera-equipped biplanes took off from a landing field near Purnea in Bihar, India, for the 154-mile flight north to Mount Everest in Nepal. In his *National Geographic* account of the journey, "The Aërial Conquest of Everest: Flying Over the World's Highest Mountain Realizes the Objective of Many Heroic Explorers," Lt. Col. L. V. S. Blacker, O.B.E., wrote of clearing Everest's amazing, immaculate crest: "I looked down through the open floor and saw what no man since time began had ever seen before. No words can tell the awfulness of that vision...."

One year later the *Geographic* published Miriam O'Brien Underhill's "Manless Alpine Climbing: The First Woman to Scale the Grépon, the Matterhorn, and Other Famous Peaks Without Masculine Support Relates Her Adventures."

"Non-climbers often ask me how a woman can be strong enough for exertion that would tax an 'athletic young man,'" she wrote. Her answer was "A certain muscular vigor is indispensable, of course...but technique, knowing how to use strength to the best advantage is more important. The greater the technique, the less the power required"—a reply that would appear as relevant today to any climber, male or female.

In 1948, Maynard M. Miller reported on the first American ascent of Mount St. Elias of 1946; Bradford Washburn's account of the first ascent of 20,320-foot Mount McKinley's west buttress, "Mount McKinley Conquered by New Route," appeared in August 1953. (Society staffer Barry C. Bishop,

then a college student, was on that climb, too.) "New Mount McKinley Challenge—Trekking Around the Continent's Highest Peak" by Ned Gillette was his story in the July 1979 issue of the circuit of McKinley.

Asia's Minya Konka (24,900 feet high), Muztagata (24,757), K2 (28,250), Annapurna (26,504), Anyemaqen (20,610), Ama Dablam (22,494) all succumbed to American climbers—but not without loss of life. Despite one of the worst avalanche seasons ever, and despite American Women's Himalayan Expedition members Vera Watson and Alison Chadwick-Onyszkiewicz, both careful and skilled mountaineers, having fallen, roped together, to their deaths, 1978 saw American women reach the summit of Annapurna—the first Americans and the first women to do so.

Still there was Everest.

Between 1921 and 1953 seven major expeditions had attempted to conquer Everest—and failed with awesome loss of life. And then, on March 10, 1953, a British party led by Brigadier Sir John Hunt sallied forth for another attempt. "What manner of mountain is this which for so many years so easily shrugged off all assaults and claimed the lives of at least 16 men?" Hunt asked in the pages of the July 1954 *Geographic*. His explanation followed:

> Other peaks demand more actual climbing. Alaska's Mount McKinley, for example, measures 19,000 feet from its lowland base, while Everest rises only about 12,000 above the 17,000-foot Tibetan plateau. Himalayan winds are fierce, but the Scottish Highlands, battered by the North Atlantic's hurricanes, endure gales as terrible. Everest's crags and crevasses test any man's ability, but half a dozen Alpine peaks offer technical problems of greater severity. Everest can chill a man to the marrow with summer temperatures down to −40° F. at night; yet on the Greenland icecap and elsewhere explorers have lived through cold worse by 30 or 40 degrees.
>
> What makes Everest murderous is the fact that its cold, its wind, and its climbing difficulties converge upon the mountaineer at altitudes which have already robbed him of resistance. At 28,000 feet a given volume of air breathed contains only a third as much oxygen as at sea level....Above 25,000 feet the climber's heavy legs seem riveted to the ground, his pulse races, his vision blurs, his ice ax sags in his hand like a crowbar. To scoop up snow in a pan for melting looms as a monumental undertaking. In the words of a Himalayan veteran, Frank Smythe: "On Everest it is an effort to cook, an effort to talk, an effort to think, almost too much of an effort to live."

The first assault team of Tom Bourdillon and Charles Evans reached the 28,700-foot South Peak, a mountaineering record, but were forced to descend when Evans' oxygen set failed, leaving them with neither time, oxygen, nor strength to tackle the final ridge. "Now they handed over the expedition's hopes to Hillary and to Tenzing," wrote expedition leader Hunt.

Hillary and Tenzing Norgay reached the summit on May 23, 1953. Almost exactly ten years later the American Mount Everest Expedition, sponsored by the National Geographic Society, with staff man Barry C. Bishop, achieved its historic twin assaults on the summit from the South Col and West Ridge but not before climber Jake Breitenbach died on the mountain. Bishop, Lute Jerstad, Tom Hornbein, and Willi Unsoeld were caught at night on the descent at 28,000 feet with neither oxygen, tents, nor sleeping bags. Bishop and Unsoeld suffered severe frostbite. Bishop lost his toes and parts of

the little fingers of both hands,* Unsoeld lost all but one of his toes.

But of all the mountaineers there is none whose accomplishments inspire more awe among fellow climbers than Reinhold Messner, who, with companion Peter Habeler, first achieved Everest's summit—*without* bottled oxygen, but with the help of Sherpa porters—on the 1978 Austrian expedition.

Two years later Messner attempted Everest without high-altitude porters, without climbers, without bottled oxygen, without a radio, aluminum ladder, or rope.

Messner's only climbing equipment were ski poles, an ice ax, an ice screw and a rock piton "for holding my tent, or even my body, to the ground in case of a severe storm." In "At My Limit," his October 1981 *Geographic* article, Messner wrote, "I was attempting the greatest challenge, to me, in mountaineering—to climb the highest mountain on earth completely on my own."

Before dawn on August 18, 1980, Messner set out from his High Base Camp at 6,500 meters (21,325 feet) and minutes later, he was crossing a snow-bridge over a crevasse when,

> Suddenly it went, crumbling into powder and chunks of ice.
>
> I was falling—falling into the deep. It felt like eternity in slow motion as I bounced back and forth off the crevasse walls. In the next moment I came to a sudden stop. Or had it been minutes?...Blackness surrounded me. "My God! Perhaps I will die down here!"...
>
> I fumbled with my headlamp. As it flashed on, the walls of ice shimmered a dark blue-green. They were two meters apart where I stood, but nearly joined at the top. The snow platform that had halted my fall was no more than a meter square....I became acutely aware that if the platform collapsed beneath my weight, I would hurtle into the abyss....

With an ice ax in one hand and a ski pole in the other, Messner kicked step after step into the ice until he could escape, then pushed on. At 7 A.M., Messner reached 6,990 meters (22,930 feet) and the saddle of the North Col. Two hours later he was at 7,360 meters:

> "Making good time," I thought, as I climbed over the rolls and bulges that form the lower part of the North Ridge....Snow swirled about my head. Gusts of wind began to sap my energy.
>
> At 7,500 meters I could feel myself slowing considerably. I must not become exhausted, I told myself. The next two days would be far more strenuous.

Finding a piece of rope left behind by an earlier Japanese ascent, Messner was led to compare large expeditions with multiple climbers and Sherpa high-altitude porters to his own attempt:

> My method...differed immensely. The solo climber, alone against the elements on a Himalayan giant, is like a snail. He carries his home on his back, and moves slowly but steadily upward. No relays. No ferrying up of supplies. No setting up a series of camps higher and higher. No assault team kept well rested for the final push.
>
> Everything I needed—tent, sleeping bag, stove, fuel, food, climbing equipment—had to go on my back. I would put up my tent for sleep, and then pack it and take it with me the next day. There was no one to carry a second tent, and I was not using oxygen, which could have given me more

The first American to attain Everest's 29,028-foot summit was James W. Whittaker, who reached it with Sherpa Nawang Gombu on May 1, 1963. Four other members of America's first Everest expedition made it to the top as well: Thomas F. Hornbein, William F. Unsoeld, Luther G. Jerstad, and the Society's Barry C. Bishop (above), shown bracing himself against the summit's 70-mph winds while holding the Society flag he carried with him. "If you have to crawl on your hands and knees, you're going to get there," he had told himself.

Opposite: In 1963, climbers of America's first Everest expedition approach Everest's snow-plumed south summit from Lhotse's slanted face.

Following spread: Its back in Tibet, Mount Everest (left), at 29,028 feet, joins Lhotse (center), at 27,890 feet, and Nuptse (right), at 25,850 feet, to form a cradle for the Khumbu Glacier, traditional southern route to Everest's summit.

---

*While on a post-Everest speaking tour, which had brought him to a large midwestern university, Bishop was momentarily at a loss during the question period when a young woman asked, "Other than the loss of the tips of your little fingers and your toes, were any of your other extremities permanently damaged by frostbite?"

strength. The oxygen apparatus is too heavy for an Alpine-style ascent, a single push for the summit. Also my own theories rule out its use. I want to experience the mountain as it really is, and truly understand how my own body and psyche relate to its natural forces. By using an artificial oxygen supply, I feel I would no longer be climbing the mountain towering above me. I would simply be bringing its summit down to me.

Messner made camp the first night at 7,800 meters (25,580 feet). Although his native tongue was German, he found he was talking to himself in Italian, urging himself to eat. "I shoved dried meat, cheese, and bread into my mouth. Just those small movements were exhausting. 'I must begin the cooking,' I told myself. I needed to drink at least four liters of water a day; to dehydrate could be fatal. The tent was fluttering wildly in the wind." Messner scooped snow into his cooking pot to melt and mix with dried tomato soup; he drank that, then two pots of Tibetan herbal salt tea. He dozed off fully clothed, then awoke not knowing "whether it was evening or morning. I didn't want to look at my watch," Messner later wrote. "Deep inside I was frightened. Not only of the present situation. My fear encompassed all my 30 years of climbing mountains: the exhaustion and desperation, and the thundering avalanches. These sensations spread over me and merged into a deepening fear."

The morning of August 19, Messner lightened his load as much as he dared and started off. Within an hour he was wading through knee-deep snow as he closed in on the North Face's steeper slope, which rises toward the Northeast Ridge 455 meters (1,500 feet) below the summit.

Convinced I would be forced to abandon my attempt soon if I had to climb in the deep snow, I searched for an alternate route. The vast snow area of the North Face extended to my right. Several avalanches recently had poured down its flank. With the fresh snow swept away, perhaps the surface would be hard. It was my only chance. Climbing gradually with each step, I began the long crossing to the Great Couloir....Concentrating on each step, I failed to notice the weather turning bad.

By three that afternoon Messner had reached 8,220 meters (26,962 feet) and was exhausted and frustrated by his lack of progress.

I wanted desperately to find a bivouac site. But I could see none.

One hour later, on a snow-covered rock ledge, I managed to pitch my tent. I wanted to photograph myself there. But I hadn't the strength to screw the camera onto the ice ax, put it on automatic, go back ten steps, and wait for the click. Far more important was to prepare something to drink....I measured my pulse while I was melting snow; it was racing—far more than a hundred beats a minute.

What if the fog did not lift by morning? Should I wait? No, that was senseless. At this height there is no recuperation. By the day after tomorrow I would be so weak that I could never advance toward the peak. Tomorrow I had either to go up or go down. There was no other choice.

The morning of August 20 was clear, but clouds were closing in. I strapped my crampons to my boots and took my camera over my shoulder and my ice ax in one hand. Everything else I left in the tent.

Climber Vera Komarkova is reflected in Sherpa Mingma's sunglasses atop the 26,504-foot summit of Annapurna with Sherpa Chewang (at left) and *Irene Miller. This first ascent by Americans and women of the world's tenth highest mountain was reported in the March 1979 issue.*

At that elevation, over 27,000 feet, Messner was oxygen-starved and hallucinating: "After a short time I missed my rucksack. It was my friend, my partner," he wrote. "I had conversed with it. It had edged me on when I was exhausted. Without it, however, the journey was easier—much easier. Besides, my second companion, the ice ax, was still with me."

Messner's familiarity with the reports of previous British expeditions helped him locate the best route into and up the Great Couloir, a "physically taxing but technically not too difficult" climb until he came to "a snow gully leading to a steep step interspersed with rocks. After a while," he wrote, "soft snow slowed my pace."

> I climbed on hands and knees, like a four-legged animal, sluggish and apathetic. A dark rock wall blocked the path. Something pulled me to the left. Making a small loop, I bypassed the obstacle.
>
> I now stood just below the peak. The fog was thick, and I could hardly orient myself. The next three hours seemed to pass without notice. I climbed instinctively, not consciously. The clouds opened for brief moments, giving fleeting glimpses of the peak against the blue sky.
>
> Suddenly I saw the aluminum tripod! There it was—the blessing of proof, the curse of desecration, on that supreme place of solitude—barely peeking out of the snow, a piece of cloth frozen around the top. The Chinese had anchored it at the highest point in 1975 to make exact measurements.
>
> I sat there like a stone. I had spent every bit of strength to get there. I was empty of feeling. I needed to take several pictures. Each required monumental effort. Patches of blue sky graced me briefly, then clouds closed in once more and swirled about as if the whole earth was pulsating. For the second time I had reached the highest point on earth, and once again I couldn't see anything. This time it simply didn't matter.
>
> I still do not know how I managed to achieve the summit. I only know that I couldn't have gone on any longer. Slowly I rose and began the descent.

In Reinhold Messner's *Geographic* article a final photograph shows him exhausted and dehydrated at his High Base Camp. It was accompanied by the following caption: "First came tears, then the story of the ordeal, told in a flood of memory. Along with Messner's story came the realization that he had pushed himself as far as he could go, learning, as British poet C. Day Lewis wrote,

> *"Those Himalayas of the mind*
> *Are not so easily possessed:*
> *There's more than precipice and storm*
> *Between you and your Everest."*

In October 1986, six years after his triumphant conquest of Mount Everest, Reinhold Messner reached the summit of neighboring Lhotse, the world's fourth highest mountain. He thereby became the first person in history to have reached the summit of all fourteen of the world's mountains whose altitudes are over 8,000 meters (26,246 feet). Known to climbers as the "eight-thousanders," the fourteen are all in the Himalayas. Messner achieved mountaineering's grand slam over a period of sixteen years. He had climbed Lhotse as he had Everest—and the world's twelve other highest peaks—without the use of bottled oxygen.

"Have we vanquished an enemy?" Mallory had asked. His answer, "None but ourselves," is one Reinhold Messner clearly understood.

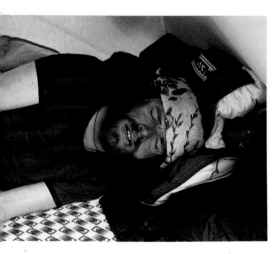

Above: *Alpinist Reinhold Messner, shown recovering from the ordeal, set off from his High Base Camp at 21,325 feet on August 18, 1980, and two and a half days later reached Everest's summit. "I was attempting,"* he wrote for the October 1981 Geographic, *"the greatest challenge, to me, in mountaineering—to climb the highest mountain on earth completely on my own." His accomplishment—without porters, without climbing companions, without a radio, and without bottled oxygen— has been hailed as mountaineering's supreme achievement.*

Right: *Messner had "spent every bit of strength" to get to the top. But now with the aluminum tripod—"the blessing of proof, the curse of desecration"—left anchored to Everest's highest point in 1975 by the Chinese, Messner expended the "monumental effort" to set up his camera and self-timer, then stumbled, dizzy with exhaustion, to the peak.*

Following spread: *"Looking back at Cerro Torre. Photo from just below summit of Fitz Roy after bivouacking on summit snowfield without sleeping bags, standing up most of the night. Before dawn we started moving the last half-hour to the true summit. As we were nearing the top, the first rays of 'alpenglow' hit the peaks....Cerro Torre, considered the world's most difficult rock spire, is the highest point in the foreground at left.... Behind is the Patagonian Icecap, 200 miles of continuous ice on the border of Chile and Argentina...."—From photographer Galen Rowell's notes to describe this never before published photograph*

"Like a Cluster of Rockets on a Quiet Night"

"Change and resistance, one mellows the other. But back in Jack La Gorce's day he wouldn't change anything. His whole editorial policy was: Steady as you go, don't upset the apple cart. And I went along with it because that's what he wanted. But I kept all those ideas tucked up in the back of my head—for the day I became editor."

—Melville Bell Grosvenor, as quoted by Priit Vesilind in his Master of Arts thesis, "National Geographic and Color Photography," December 1977

During the 1954–57 interregnum, there was never any doubt about who would succeed La Gorce. Melville Bell Grosvenor had, in effect, been running the Magazine during those years. So in 1957 when MBG, chafing at La Gorce's conservatism, was finally appointed Editor, one *Geographic* staffer compared Melville's arrival to "a cluster of Fourth of July rockets on a quiet night."

The fireworks were understandable; in one way or another, Melville Grosvenor had been preparing for command of the Geographic ever since 1902, when the five-month-old child, with his father guiding his hand, had affixed a shaky "X" to the dedicatory document which was then sealed and deposited inside the cornerstone of Hubbard Memorial Hall, the Society's new permanent headquarters. Later, with his grandfather Alexander Graham Bell's huge hand gripping his tiny fist, the infant Melville troweled a bit of mortar to help set the cornerstone in place.

Melville's relationship with his grandfather Bell was special to them both. Melville spent his boyhood summer vacations at the Bell home in Baddeck, Nova Scotia, where young Melville basked in his grandfather's love. "Those were wonderful years," he said. "...I learned more from Grandfather than anyone...I was his first grandson and his two sons had died in infancy, so he concentrated all his desire for training a son on me...."

Alexander Graham Bell died in 1922; Melville was then twenty-one years old and a midshipman at the United States Naval Academy. Melville graduated from Annapolis in 1923, served a year as an ensign aboard the battleships *Delaware* and *West Virginia*, resigned his commission in 1924, and began work at the National Geographic.

For the next eighteen years, from 1924 to 1942, Melville wrote and edited picture captions (or "legends" in Geographic parlance); for the following eight years he edited pictures; and from 1950 to 1957 served as an Assistant Editor and later Associate Editor of the Magazine. When he finally became Editor, Melville was fifty-five years old and brimming with ideas—most of which he had presented earlier to his father and La Gorce, only to have them turned down. Still, in the three years following his father's retirement in 1954, Melville had prepared the way for his editorship by hiring young people as full of ideas as he was—Wilbur E. Garrett, Robert L. Breeden, Thomas J. Abercrombie, and Thomas R. Smith among them.

Almost immediately upon assuming the leadership of the *Geographic*, Melville had to contend with a "palace revolt" within the Illustrations Division. Walter Meayers (Toppy) Edwards, who had started at the *Geographic* as Franklin Fisher's secretary, had simply assumed the Illustrations Editor post vacated in 1953 by Fisher's death; Toppy Edwards reported directly to Melville. Herbert S. Wilburn, Jr., was Assistant Illustrations Editor in charge of color, and Kip Ross was Assistant Illustrations Editor in charge of black-and-white photographs. Under Wilburn and Ross were the picture editors, including Garrett and Breeden.

Although nominally their supervisor, Edwards had little to say about who worked for him. "All of us who were hired in those days were hired by Melville," Garrett explains, "and put under Toppy whether he liked it or not."

Edwards was primarily a picture man. Although he had occasionally written for the Magazine, his chief responsibility for some fifteen years was selecting most of the color pictures for it. Kip Ross's background had been with the Associated Press. He knew journalism, and had done a book on photography, and because he resented having to work under Edwards, constantly undercut and subverted his Illustrations Chief.

Bob Breeden and Bill Garrett were working in black-and-white illustrations under Kip Ross, and urging a modernizing of the Magazine's photographic layout and style. Because Ross was amenable to their tastes, the Magazine's black-and-white pages began to contain more photographs with natural lighting and greater use of white space to create a stronger design. The color sections, however, remained a jumble of miter cuts, hairline rules, and carefully posed set-up shots. And so, as Garrett says, there was a "schism in the Illustrations Division," and

> what made the whole thing intolerable is that there was no layout department. The picture editor was the layout person even if he had no talent at all. It was a totally impossible situation for everybody because it became competitive between the black-and-white picture editors and the color. The color editors would finish the color dummy, so to speak, and then there would be some holes left where the page make-ups went, in which we would then put some black-and-white pictures. And even though Toppy was the head of the department, the black-and-white layouts would go straight to Melville because Kip was much looser. Melville was kind of using the black-and-white department as his experimental lab for make-up and design. It was wonderful because we were allowed freedom. The guys who were on the other side weren't....

The issue came to a head not because of the actions of anyone in the Illustrations Division, but because outsiders from the newly created Book Service Division were brought in: Merle Severy, who had been moved from the staff of the *School Bulletin** to head Book Service, and Howard Paine, a young designer hired by Melville Grosvenor to do layouts for Severy.

Severy was delighted to be away from the School Bulletin. *He had written a piece for it that had opened, "If you have a touch of the smuggler in you, then Andorra is the place for you." La Gorce had called Severy into his office and said, "Harumph, Mr. Severy, there are no smugglers among the membership."*

Melville Grosvenor's Leica camera was always at the ready as in this 1959 photograph of the Editor in Cambodia at the great Khmer temple of Angkor Wat. His interest in photography had been encouraged by both his father, Gilbert Hovey Grosvenor, who started Melville taking pictures when he was eleven or twelve years old, and then by his grandfather Alexander Graham Bell, who appointed Melville, at age thirteen, his official photographer of experiments.

BOOK SERVICE

Since 1957, the year MBG created the Book Service Division, nearly fifty separate titles have been mailed out to millions of purchasers. It is difficult to gauge how widespread the influence of these books might be, but Book Service Associate Director Ross S. Bennett tells the following story: In 1962, sophomore Ronald M. Nowak, at Tulane University in New Orleans, read in a Book Service volume titled *Wild Animals of North America* that the red wolf, found in the bayou area of Louisiana and Texas, was fast disappearing. But other books he had read made no mention of the decline. So Nowak decided to investigate. He began writing letters

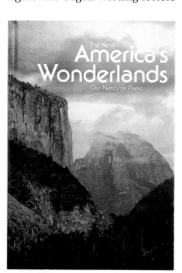

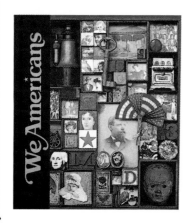

National Geographic Society

asking for information on the red wolf population; he then contacted the United States Fish and Wildlife Service. When his own study confirmed the red wolf's decrease, Nowak helped launch a national campaign that led to a federal effort to save the animal. Nowak went on to become a biologist.

"Years later, when Book Service editors decided to publish a new edition of *Wild Animals* and sought a chief consultant," Bennett reports, "they found their expert right in their own backyard—Ron Nowak, staff mammalogist in Fish and Wildlife's Office of Endangered Species."

Perhaps the Book Service volume closest to Melville Grosvenor's heart was *Men, Ships, and*

the Sea. Melville wanted the book written by someone who had "lived the sea" and wrote of it with passion. He was able to get Alan Villiers, author of many of the most popular *Geographic* seafaring articles, to write the text. Not only was *Men, Ships, and the Sea* one of the first Book Service items that contained material not primarily picked up from the Magazine, it was also an oversized volume—a fact that, Book Service Director Charles O. Hyman recalls, "caused all kinds of hell...Melville was angry because it didn't fit in his bookshelf. He had his bookshelves designed so that they were exactly the height of the Magazine. Merle Severy [the first Director of Book Service] had to sneak into Melville's office with carpenters while MBG was away and raise the bookshelves so *Men, Ships, and the Sea* would fit."

Since 1962, Book Service subjects have "ranged across the spectrum of human interest, from Indians to insects, from national parks to historic shrines, from birds to dogs," says Associate Director Bennett. "Readers have marveled at animal behavior, explored our living planet, discovered Britain and Ireland, journeyed into China, diagrammed the human body, and followed the adventure of archeology. They have watched history come alive in the rise of Western civilization through volumes emphasizing everyday life. ...They have sampled spiritual tolerance in *Great Religions of the World* and rediscovered themselves in *We Americans.*"

Although their painstakingly researched volumes have received wide critical acclaim, the reviews that please Book Service staffers most are ones from young students such as the Brooklyn boy who wrote, "I think your books are wonderful. ...I am 13 years of age and someday I hope to work for the Society along with all of you."

Although Book Service, established in 1957, was new, the Society, of course, had been publishing books for years: *Alaskan Glacier Studies* and *A Book of Monsters* (about insects) had both, for example, been published in 1914, and there had been books on horses (1923), birds (1915, 1918, 1951), and dogs (1919) as well. In 1951, the Society had published *Everyday Life in Ancient Times* with a text by the scholar Edith Hamilton, among others, and *Stalking Birds With Color Camera*. In 1954, Severy, along with Andrew Poggenpohl, had put together *Indians of the Americas*, with a text by Matthew W. Stirling and others. That book, published in 1955, sold 379,152 copies and was such an enormous success that Melville, who wanted to increase the Society's participation in book publishing, felt encouraged enough to establish the Book Service Division.

Until *Indians of the Americas*, the Society's books had been simply a collection of magazine pieces with plates. But when Howard Paine started designing for Severy, the Society's books suddenly began to have a different look—the sort of open-layout-with-lots-of-white-space look that the younger members of the Magazine's Illustrations Division had been clamoring for and couldn't get. The result: resentment.

Severy explains, "The Magazine couldn't get away with the kind of stuff the Book Service did. You couldn't tamper with the Magazine. Psychologically, however, the books were not sacred ground. We'd been allowed to do things with the books because nobody took us seriously—except Melville."

The confrontation occurred after Merle Severy showed Toppy Edwards a copy of his latest book while they were on an airplane en route to a Chicago convention and recommended that Edwards have Howard Paine help with the Magazine layouts. Upon arrival in Chicago, Edwards phoned Garrett in Washington and told him to give the dummy he was working on to Howard Paine to re-do with white space. Garrett had already completed it.

"Is that an order?" Garrett asked.

"Yes," Edwards said.

Garrett said, "Okay," and went down to Personnel to resign. Edwards' phone call had also enraged the other members of the Illustrations Division. All of them had been trying to update the design of the Magazine, to open up the layouts, to use some white space, and Edwards hadn't let them. "Suddenly," Garrett recalls, "here was an outsider named Howard Paine who was going to be allowed to do it!"

When Edwards returned, he summoned the entire Illustrations staff to his office for a meeting with Merle Severy, Howard Paine, and a feltboard. Paine began to instruct the Illustrations Editors on how to lay out the Magazine pages.

"Toppy was a fairly small man, and he sat behind this huge desk," Garrett remembers. "And here was Howard Paine, nicest guy in the world, who'd been brought in with a feltboard to tell us how to lay out pictures. You could just cut the tension with a knife."

Paine, new to the *Geographic* and unaware of the conflicts within the Illustrations Division, was only trying to do what he had been told. Staff members who, like Breeden, had been doing layouts and suggesting the very changes Paine was proposing, felt they hadn't been heard, resented it, and made their feelings known.

Garrett recalls:

> Breeden was getting tight-lipped and intense, and after a while he asked between clenched teeth, "Are *you* trying to tell *me* how to crop a picture?"
>
> I didn't say a word. I kept my mouth shut because as far as I was concerned, I was through anyway. I was gone. But others—like Brooks Honeycutt who was a former Navy [pilot] and wasn't going to take this—they were more vocal....
>
> Finally, after hearing enough grumbling and resentment, Toppy jumped to his feet and pounded on his desk and said, "You're all acting like a bunch of..." and then, reaching for a word—I think he meant to say "anarchists," but what he finally said was, "a bunch of *communists*!" And Andy Poggenpohl, who was a very benign, jolly-faced, slightly portly older guy, just exploded.
>
> Poggenpohl's father had been an Admiral in the Czar's navy and I suspect he was murdered by the communists. I'd never seen him get so mad before. He jumped to his feet and said, "I don't have to take this from you!" and stomped out the door.

The rest of the staff soon followed, leaving Toppy Edwards, Howard Paine, Merle Severy, and the feltboard behind. The meeting was over.

Ironically, twenty years later, Howard Paine was Art Director and Merle Severy was an Assistant Editor on the Magazine. Looking back on that incident—known still within the Illustrations Division as "Black Friday"—Paine recalled having been so excited by his work that he had not been aware of what he had done. "I was so young," he said. "I wasn't sophisticated enough to know that I was treading heavily on toes."

Almost immediately the look of the Magazine changed, reflecting acceptance of new layout and design ideas of the younger men.

His Illustrations Division in chaos, his leadership compromised, Toppy Edwards became ill. Shortly afterward, Melville sent him on a cruise to the Caribbean, and, on January 28, 1958, while Edwards was still recovering, Melville notified him that Herb Wilburn was replacing him as Illustrations Editor. That same year, Kip Ross suffered a stroke and retired as head of the black-and-white section. The last vestige of the old "Red Shirt School of Photography" was gone, and the way was opened for the new breed. Melville transferred Toppy Edwards to the Foreign Editorial staff, where he covered assignments as a writer and photographer; he retired in 1973 after a forty-year career. Herb Wilburn, the sole survivor of the Illustrations triumvirate in place when MBG took office, quickly consolidated his position by combining the black-and-white and color sections. Although Melville had, in a sense, stood back and let the upheaval in the Illustrations Division sort itself out, there was never any question about who ran the *Geographic*: Melville Bell Grosvenor did—and with an enthusiasm his father had never shown.

Over and over again, older staffers speak of Melville's childlike enthusiasm. "MBG's eternal boyishness made him a wonderful guy to work with," Severy said. "He was approachable, salable on new ideas. He was warm, generous; he could be reasoned with. He woke up with stars in his eyes believing the world was a wonderful, exciting place."

Once Severy had a project he wanted to do and, anticipating MBG's resistance, thought up ten reasons why it should be done. After the second reason, Melville said, "Yes." Severy continued to provide reasons until Melville interrupted, saying, "Merle, you must learn to take 'Yes' for an answer."

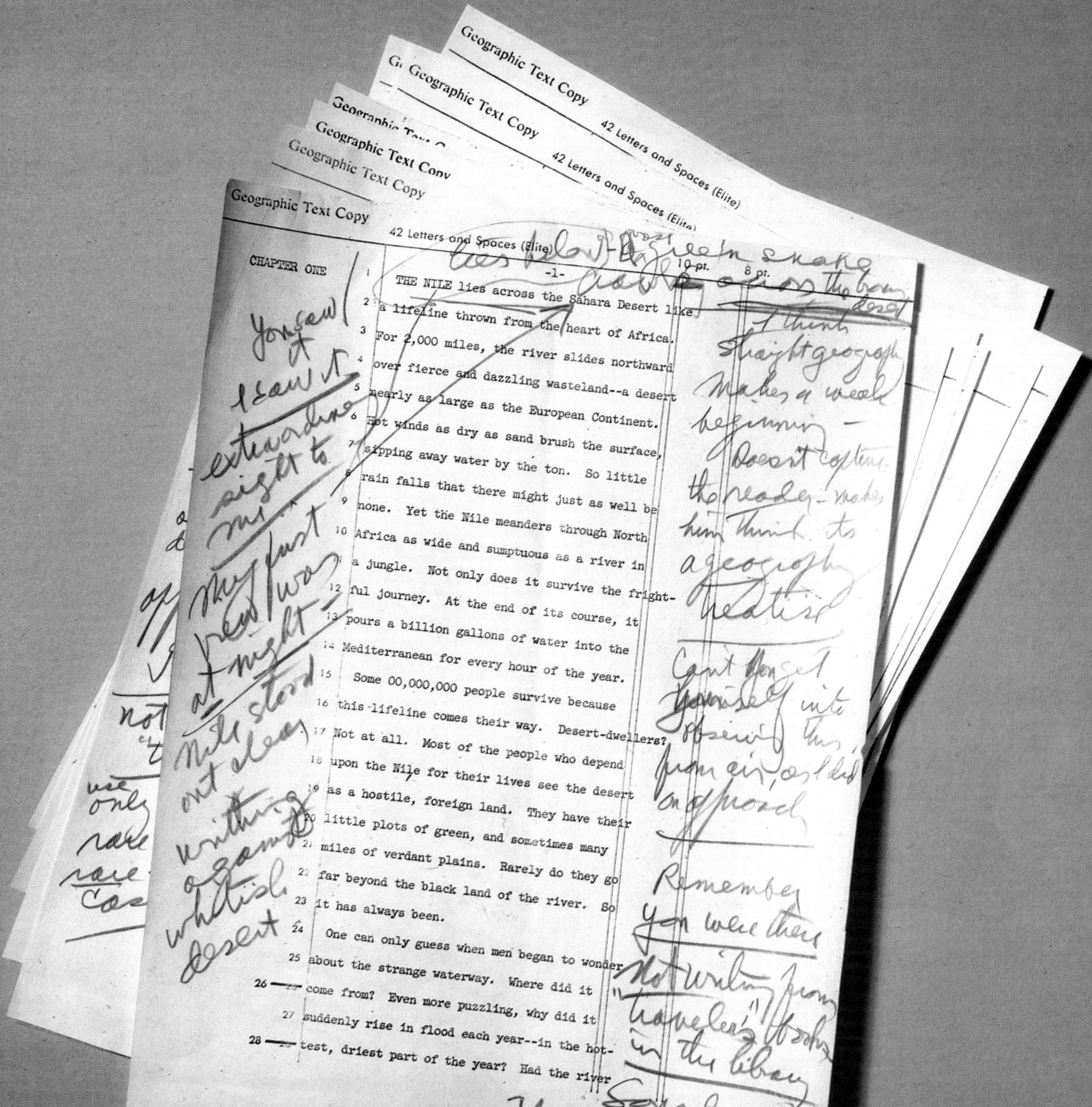

Geographic Text Copy
Geographic Text Copy
G. Geographic Text Copy
Geographic Text Copy
Geographic Text Copy
Geographic Text Copy
42 Letters and Spaces (Elite)
42 Letters and Spaces (Elite)
42 Letters and Spaces (Elite)

CHAPTER ONE

-1-   10 pt.   8 pt.

1 THE NILE lies across the Sahara Desert like
2 a lifeline thrown from the heart of Africa.
3 For 2,000 miles, the river slides northward
4 over fierce and dazzling wasteland--a desert
5 nearly as large as the European Continent.
6 Hot winds as dry as sand brush the surface,
7 sipping away water by the ton. So little
8 rain falls that there might just as well be
9 none. Yet the Nile meanders through North
10 Africa as wide and sumptuous as a river in
11 a jungle. Not only does it survive the fright-
12 ful journey. At the end of its course, it
13 pours a billion gallons of water into the
14 Mediterranean for every hour of the year.
15 Some 00,000,000 people survive because
16 this lifeline comes their way. Desert-dwellers?
17 Not at all. Most of the people who depend
18 upon the Nile for their lives see the desert
19 as a hostile, foreign land. They have their
20 little plots of green, and sometimes many
21 miles of verdant plains. Rarely do they go
22 far beyond the black land of the river. So
23 it has always been.
24 One can only guess when men began to wonder
25 about the strange waterway. Where did it
26 come from? Even more puzzling, why did it
27 suddenly rise in flood each year--in the hot-
28 test, driest part of the year? Had the river

Samples of the sort of jumbled color layouts that led to a "palace revolution" within the Magazine's Illustrations Division during the early months of Melville's Editorship. The confrontation occurred when young picture editors grew frustrated because they were not permitted to modernize.

Photographer Luis Marden remembers MBG as "an intuitive editor. Quite often he could not explain to you why he wanted something done, but he damned well knew that this was the way to do it, or that it was worth doing. And he intuitively knew what was going to be well received by the readers of the Magazine....MBG never lost the sense of wonder—'wonder' in the best sense of the word....If you came to him with an idea that had any merit, that was a good idea, he'd catch fire from it. He'd get up from his desk and start pacing. He'd say, 'That's great! Do you think we can do this?' You'd find yourself running to keep up with him. 'Yes, that's great!' he'd say. 'Go ahead and do it!' "

When Melville Grosvenor was hired at the *Geographic* the number of members was 905,700; when he became Editor in 1957 membership was 2,175,000. It was 5,500,000 when he resigned ten years later. The decade during which Melville Grosvenor was President of the National Geographic Society and Editor of its Magazine "is remembered by old-timers on the staff," Luis Marden says, "as a golden age." What were Melville's innovations?

Melville initiated printing the subjects of Magazine articles on its spine and changed the name of the Magazine: In December 1959 *The National Geographic Magazine* became *The National Geographic*; three months later Melville further shortened it to *National Geographic*.

He changed the face of the Magazine, narrowed the yellow border, and pruned the oak leaves one by one until, not long after MBG retired as Editor, the last leaf had fallen.

Aware that many readers simply looked at the *Geographic*'s photographs and ignored the articles, MBG structured photograph captions to provide distillations of the text they illustrated. The emphasis was on liveliness. The verb "to be" was not allowed. "In those days," one staffer remembers, "the style was never to use the word 'said.' No one ever *said* anything. You'd use a quote, and the person chuckled, or he grimaced, he murmured, or grunted, and so on. You had to maintain a high level of excitement in the writing. Every verb had to be an active verb; every sentence active...."

By the time Melville became Editor, the Geographic was beginning to run out of unexplored South Pacific Islands and undiscovered mountain ranges. With the encouragement of by-then Trustees Melvin Payne and Walter Myers, Melville expanded the Society's contributions to research and exploration and directed its energy and funds toward the investigation of the next two great unknowns: the sea and space.

"He was full of ideas," said Melville's writer-colleague Nat Kenney. "Wild ideas to get behind some Cousteau expedition, find out how NASA was simulating weightlessness in space, talk to this woman Goodall with the chimpanzees. Well, at least four out of five times, the readers were just wild for what MBG was wild about."

Under Melville's direction, the Magazine published numerous articles by Cousteau; readers learned from Barry Bishop "How We Climbed Everest" and watched our nation's fledgling space program reach out and touch the moon.

Readers lived with Harald Schultz's "Brazil's Big-lipped Indians" and L.S.B. Leakey's early hominid, Zinjanthropus. They shared a young man's adventure in "A Teen-Ager Sails the World Alone" and survived disasters in "Avalanche!" and "Earthquake!" and "Florence Rises From the Flood."

They dove in the Yucatán a thousand years into the past with Luis Marden

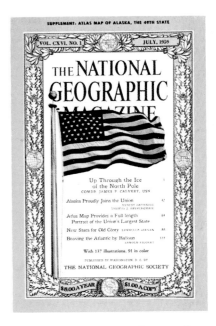

"We all opposed him at first," Gilbert M. Grosvenor recalled of the staff's reaction to Melville's 1959 proposal to regularly put color photographs on the cover of the National Geographic. *The July 1959 cover (above) displayed the then-new forty-nine-star flag. Other cover photographs soon followed and began appearing regularly with the September 1959 issue. One member quipped, "If the Lord had meant the* Geographic *to have a picture on the cover, He would have put one there in the first place!"*

Under Melville's direction, the Magazine's oak-leaf border was pruned, its yellow border narrowed, and even its name changed. Five months after the flag cover, The National Geographic Magazine became The National Geographic; *and three months later, in March 1960, simply* National Geographic.

Top: *With painstaking precision an artist adds a place name to the nomenclature sheet of the Society's 1963* Atlas of the World; *and,* above, *a cartographer bends over a sheet of translucent plastic to ink a maze of North American rivers and shorelines onto one of the eight separate images that, combined in printing, will make up a single map.*

to bring Maya artifacts back "Up from the Well of Time." And in "Mystery of the Monarch Butterfly" discovered with staff natural scientist Dr. Paul A. Zahl where those lovely, fragile creatures went on migration.

A great atlas had long been a dream of Melville's; and because the geographic changes brought about by World War Two and its aftermath demanded nothing less than a complete remapping of the world, MBG felt now was the time to make his dream come true. The cartographic staff was doubled, and new techniques for drawing maps on film were developed. As each map sheet was completed, it was issued as a Magazine supplement.

Bart McDowell, now an Assistant Editor, as a "very, very junior staff member" was brought in to assist with the first atlas, and recalls:

> These atlas-sized maps were loose in the Magazine, and the idea was that we would sell the members a kind of package that was essentially just a chunk of cardboard and glued paper that would be a receptacle for all these free maps as they came along. An announcement needed to be prepared pointing out that this "atlas folio" was available—but what could you say about it? It was just glue and a piece of cardboard.
>
> Out of the blue I thought the whole "sell" was *the lore of maps!* The marvelous things that happen to people who work with maps: Christopher Columbus, that man in Antwerp who compiled the first Atlas, Robert Louis Stevenson and *Treasure Island.* What great escape reading maps are: wondering about latitudes, what kind of people inhabit such a place....
>
> Well, Melville saw what I had written and was thrilled by it. "This really deserves a byline," he said. He got to thinking even more, became even more enthusiastic, and then he said, "I think what we'll do is give it *my* byline!"

It was the ultimate accolade.

The National Geographic Society's 300-page *Atlas of the World* was published in 1963. Melville called it "the Society's largest, handsomest, and most ambitious contribution to geographic knowledge." It contained an index with 125,000 place-names, a description of every nation, its flag, and information ranging from "Great Moments in Geography" to "Temperatures Around the World."

The *Atlas* was followed by a free-floating globe, an idea that came about as a result of then Associate Editor Frederick G. Vosburgh's telling MBG he had been invited by the U.S. Navy to visit Antarctica, into which, for the first time, big C-130 aircraft were to fly from New Zealand.

Melville had said, "Great! How far is Antarctica from New Zealand?"

"About 1,600 miles, I think," Vosburgh answered. "Let's measure it on the globe."

Melville's office globe was a floor model that spun on a fixed axis. To measure the distance at the bottom of the world Melville and Vosburgh had to get down on their hands and knees.

In a sanitized version of what happened next, Melville wrote, "I exclaimed that we should have a globe you could pick up in your hands and turn to any spot. Mr. Vosburgh agreed. We called in Chief Cartographer [James M.] Darley and his associate Wellman Chamberlin, and our globe program was born...."

It was Chamberlin who suggested adding the geometer, a plastic cap to facilitate measurements of distance and area. Inevitably the globe became known as the "globe with a thinking cap."

"A CHARACTER OUT OF FICTION": THE SOCIETY'S LUIS MARDEN

In an April 1978 Editor's page salute to Luis Marden, written by the grandson of the man who had hired him, Gilbert M. Grosvenor called Marden's forty-two-year career as a staff member "so remarkable that he sometimes seemed a character out of fiction," and went on to say "he knows Tahiti, Fiji, and Tonga as well as his native Boston, and is at home in five languages, in any one of which he can deliver anecdotes with a consummate actor's skill."

Seven years later, *Washington Post* staff writer James Conaway described Marden as "one of the last old time adventurers, the epitome of a phenomenon once known as 'the Geographic man.' He traveled first class and carried an expense-account book, containing the entry 'Gifts to Natives.' "

Elsewhere in this book's pages Luis Marden's foresight in the role that 35mm cameras and Kodachrome film would play at the *Geographic* has been noted, as has his pioneer work in underwater photography with Jacques-Yves Cousteau, but the extraordinary range of his assignments demands mention too.

In 1936, on his first foreign assignment, Marden (*above, left*) photographed the Maya city Chichén Itzá—and himself. It was Luis Marden who found and photographed in the waters off Pitcairn Island in 1957 the remains of Captain Bligh's *Bounty*, sunk by mutineers more than a century and a half before. A new species of orchid found by Marden in Brazil was later named *Epistephium mardeni* for him; and so was the new species of sand flea, *Dolobrotus mardeni*, he had plucked from the Atlantic's depths.

In the Yucatán for an article appearing in the January 1959 issue, "Dzibilchaltun: Up from the Well of Time," Marden dove into a Maya natural well and photographed or recovered from the murky ooze eighty feet beneath the surface centuries-old artifacts—among them the unbroken 1,000-year-old earthenware jar shown *below*.

Luis Marden's most recent adventure was to sail with his wife, Ethel, a mathematician, from Spain's Canary Islands to the Bahamas to determine the actual track Columbus followed to the New World. Using daily plotting sheets, trigonometry, and small navigation computers, the Mardens' track led to Samana Cay and the history-making November 1986 articles by Luis Marden ("The First Landfall of Columbus") and Joseph Judge ("Where Columbus Found the New World").

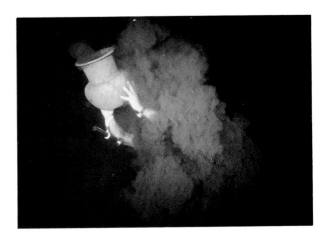

"Even as the new headquarters opened [in 1964]," Melville Grosvenor later recalled, "our 40-year-old membership records building in another section of Washington was bulging at the seams." In 1965, the Board of Trustees authorized the purchase of a hundred-acre site near Gaithersburg, Maryland, for the construction of new Membership Center Building.

A few days before the groundbreaking, Melville went out to the site with Boris V. Timchenko, the landscape architect. Disturbed by the planned location of the building in relation to the eleven-acre spring-fed lake and the rest of the terrain, MBG asked Timchenko, "Will the lake be used to best advantage?"

"[Not] the way the building is staked out now," the landscape architect replied.

"Then move the building," Melville said.

Timchenko realigned the stakes, literally turning the 420-foot building twenty-five degrees on its axis. The architects worked in a frenzy to finish revised drawings before the giant earth-moving machines arrived.

MCB—as the new Membership Center Building's name inevitably came to be abridged—opened in 1968 housing more than 1,100 employees and two giant computers to help keep track of the Society's nearly eleven million members in 170 of the world's 174 nations and to maintain all phases of communication with them.

Operations at MCB fall under the direction of Society Executive Vice President Owen R. Anderson, whose firm supervisory control led editorial wags to dub the Gaithersburg center "Andersonville."

"We were pretty well regimented in the membership fulfillment operation compared with the editorial world," Anderson laughingly admits.

A certain amount of regimentation could be expected of an organization that takes care of the complete membership operation. The membership files are kept at MCB; its staff copes with membership renewals, the mailing of bills, the receipt of all remittances. "And since we have the files, we're responsible for addressing all the promotion pieces, all the billing operations, etc., for *all* the magazines and books," Anderson says. "Poor service is a great problem in various fulfillment houses. If you're not getting your magazine to the members, people will get disgruntled and they won't want it."

In 1986 alone, MCB received 21,595,224 pieces of mail, an average of 85,695 pieces of mail *per day*! Even though a largely automated computer operation provides immediate answers to many of the members' inquiries, many prefer instead to telephone, so in addition to the vast volume of mail, the Center receives an average 1,209 calls a day.

Anderson suggests that "maybe part of the National Geographic Society's high renewal rate [an astounding eighty-six percent in 1986] is due to the service that MCB offers."

On a "big day" more than half a million pieces of mail have arrived at MCB. The heaviest mail day is generally a result of the summer remittance operation mailing. "We send out a promotion called a voluntary summer remittance plan six months in advance," Anderson explains. "Memberships expire in December, and we ask the members to renew six months in advance so we can hold down renewal costs. As a result almost fifty percent of the members renew at that time. Therefore, when this huge mailing goes out in June a peak mail hits us the last of June or early July." In one day in July 1986 582,614 pieces of mail reached MCB.

Melville pushed for perfection and innovation in photography and showed that he was willing to pay for it by tripling the photographic budget. At Melville's direction, Illustrations Editor Wilburn increased the size of photographs printed in the Magazine and initiated the practice of having every article begin with a double-page picture. And each issue contained as much color as possible.

As dramatic as the changes were inside the Magazine, the outside was due for changes, too. Melville had grown tired of looking at the Magazine's same old conservative covers. So one day he asked several of his associates—among them his son Gilbert M. Grosvenor—to come to his office. On his desk MBG had a big pile of *Geographics*. As soon as the door was closed he picked up the Magazines and threw them on the floor, face up. "For a moment we wondered what was wrong with the Editor," Ted Vosburgh recalled. "Then he challenged us. 'Look at those Magazines,' he said, 'and tell me which is the latest issue.' Nobody could.

" 'You see!' declared Melville triumphantly. 'They all look alike.'

"Then he produced a dozen Magazines on which striking color photographs had been pasted," Vosburgh's recollection continues.

" 'Now look at these,' he said, holding up one with a picture of a geisha on the cover. 'If you saw this on a coffee table in a friend's home, you'd think, oh yes, there's last month's *Geographic* with a story on Japan. Or you'd think, I haven't seen that one yet; it must be the new issue.'

" 'Hereafter,' declared the Editor, almost belligerently, 'there's going to be a color picture on the cover.' "

A photograph of the new forty-nine-star American flag appeared on the cover of the July 1959 issue; there was no photograph in August; in September a small Navy jet fighter appeared, and color pictures have been on the cover of the *Geographic* ever since.

The Society next changed printing plants and processes. When Melville took over the Magazine, he wrote, "stood at a crossroads. With a rising membership of 2,175,000, we faced the fact that two and a half million would be the most that could practically be printed by the method then in use. No longer could we hope to meet the needs of our members with presses that printed one sheet at a time, stacked the sheets, waited for the ink to dry, then printed the other side."

Thomas W. McKnew, then Executive Vice President of the Society and Secretary of a special Printing Methods Committee, and other members of the Committee debated whether to continue printing in Washington or to change processes and go elsewhere. They decided a move from Washington was the only solution. McKnew supervised moving the printing out of Washington, where it had been (with the exception of a brief period in 1901 when it had been printed in New York), taking it to R. R. Donnelley & Sons' high-speed color presses in Chicago, where, as MBG noted, "instead of the slow inserting of sheets, there would be a continuous web of paper running off a roll and through the presses at a rate of 1,200 feet a minute. Inks would be 'heat-set' (dried instantly) as the web passed over a hot drum....

"Although enormously complicated, the transfer of our printing to Chicago was carried out without a hitch, and by mid-1960 the move was complete. No longer would we have to limit our color illustrations to a few sections of 4, 8, or 16 pages. Now we could have an all-color magazine—color photographs on every page—and press capacity to meet every foreseeable need."

Above: *The bathyscaph* Trieste *surfaces after its January 23, 1960, record dive of nearly seven miles into the abyss of the Mariana Trench. Jacques Piccard reported the descent that he and Lt. Don Walsh made to a depth of 35,800 feet in the August 1960 article "Man's Deepest Dive."*

Right: *Seven months later, on August 16, 1960, U. S. Air Force Captain Joseph W. Kittinger paused in the door of an open gondola at an altitude of 102,800 feet, prayed, "Lord, take care of me now," and stepped into space. Kittinger's account, "The Long, Lonely Leap," in the December 1960 issue, told of his falling more than sixteen miles in four and a half minutes before his parachute opened and he landed safely.*

In 1965, Melville launched the Special Publications Division with Robert L. Breeden as its chief. Twenty years later Breeden had, as writer William Hoffer was to note in a *Regardies* magazine cover story on the National Geographic, "after a fortuitous encounter with Jacqueline Kennedy in 1961, built a somewhat separate empire within the National Geographic Society that now accounts for 40 per cent of its sales."

"I've told many people," Bob Breeden has said, "that if Jack Kennedy hadn't been elected President, I wouldn't be in the book business; I'd still be on the Magazine staff."

When Jacqueline Kennedy entered the White House, the only guide available for the thousands of tourists who daily toured its historic rooms was a skimpy mimeographed sheet published by the National Park Service. The First Lady, who, eleven months after moving into the White House, had formed the White House Historical Association "to enhance understanding, appreciation, and enjoyment of the Executive Mansion," let it be known that a mimeographed sheet would not do. A member of the Association contacted Melville and asked whether the Society could produce a proper, accurate, inexpensive, color-illustrated guidebook. Melville agreed, pending Board approval, to put the Geographic at the First Lady's disposal to take pictures and put out whatever kind of book she wanted. He then called a group of editors into his office, among them Bob Breeden, and told them the Society was going to do this book as a public service—and that the layout had to be done in three weeks. Associate Editor Franc Shor was assigned to supervise the project.

Sometimes working in twenty-four-hour stretches, Breeden and Donald J. Crump, also a picture editor, completed a layout of the book in the allotted three weeks. When MBG, Shor, and Breeden took it to Mrs. Kennedy, she didn't like what she saw; the book, following *Geographic* style, had too many human interest pictures in it for her taste. So Breeden and Crump went back to work. With researcher Pat Rosenborg and two staff photographers—Bates Littlehales and George Mobley—they produced what was essentially a catalogue of rooms and items in the Executive Mansion. A White House curator wrote the text; the Society donated the remaining editorial labor, typesetting, and design. That book, *The White House: An Historic Guide*, went on sale July 4, 1962. The entire first printing of 250,000 copies sold out in less than ninety days.

"The book was instantly successful," Crump, now Director of the Special Publications and School Services Division, recalls. "We had to go back to press again and again. And it led to other books."

Breeden and Crump were still doing their regular Magazine jobs then. "Don Crump and I used to joke about it," Breeden recalls. "We'd say, 'Well, we'll finish this and just do our regular jobs.' And then we'd say, 'No, the next thing you know we'll be doing a Capitol book.'"

Two weeks after publication of the White House guide, the newly formed U.S. Capitol Historical Society asked the Geographic to do a book on the Capitol. And while work was proceeding on that, the Federal Bar Association asked that the Society do a book on the Supreme Court. And then Melville thought the Washington Monument deserved a book...

Breeden and Crump were continuing to work an inordinate number of hours—nights, weekends, holidays—at the office trying both to put out these books and to keep their Illustrations Chief happy. Along with their regular picture editing jobs on the Magazine, Bob Breeden and his staff did, in addition to

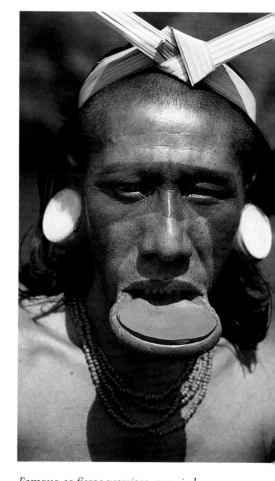

Famous as fierce warriors, married men of the Suyá distort their lower lips with wooden disks and their earlobes with plugs of twisted palm leaves. Only sixty-five Suyá were left when five small boatloads of them paddled down the Xingu River, a tributary of the Amazon, seeking medical treatment for dysentery and other benefits of civilization. Anthropologist Harald Schultz returned with them to report on life in their jungle home for the January 1962 article "Brazil's Big-lipped Indians."

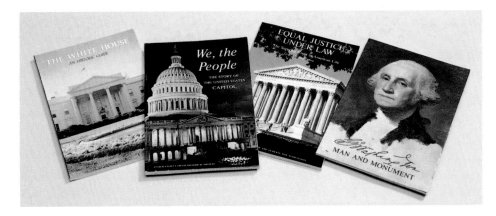

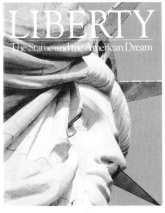

"A beautiful book," President John F. Kennedy commented when MBG handed him a copy of The White House: An Historic Guide, *produced by the Society at the request of the First Lady. The White House guide was first in a series of public service books prepared on such topics as the Capitol, the Supreme Court, the Washington Monument, and Presidents. More recently,* Liberty *has generated more than $4 million for the Statue of Liberty Foundation.*

the White House book, the Capitol's *We, the People; Equal Justice Under Law* on the Supreme Court; *G. Washington, Man and Monument;* and a book on the Presidents. All of these books were done as a public service by the Geographic (and a significantly expensive one), but it finally got to be too much. Breeden wrote a confidential memo to Melville Grosvenor suggesting, in effect, that the National Geographic should be publishing these small books "not just for others," but for itself and that a Department of Special Services be formed to do them. Breeden's memo continued:

> With a staff you could count on one hand, I am sure I could produce at least three such books a year. The Society, I estimate, could net $250,000 on its books and many times over this amount on free publicity on public service books.
>
> All of this would be done apart from our present Book Service, which already has an expanded program.

Breeden insisted in his memo that such a department "would not involve hiring new personnel" and concluded:

> When you stop to think about it, a Department of Special Services and a small book program is not anything new for the Geographic. We've been doing it for two years!

Melville called Breeden into his office and asked, "But Bob, you wouldn't want to leave the Magazine, would you?"

"Well no, not really," Breeden said, "but I would like to get to the point where I would not have to spend so many hours down here and could be with my family more."

Melville said he would think about it. Melvin Payne, President of the Society, appointed a task force to study the feasibility of publishing a series of low-cost books. The committee came back with a positive report: the Society could publish annually four high-quality, color-illustrated books averaging 200 pages each. With relatively low promotional costs the members could have them at a reasonable price; moreover, test mailings proved conclusively that the members would buy such books.

In 1965, the Special Publications Division was born. (Twenty years later, as a result of the creation of many new products, Breeden's original staff that "you could count on one hand" had grown so large it practically needed a small building of its own.)

By then Melville had also moved the Society into television—a medium his father had avoided, but an inevitable outgrowth of the Society's im-

344

SPECIAL PUBLICATIONS

"Why don't you print these articles as a separate book? My son used them for his homework and got an A." Letters such as this from members played a part in Melville Grosvenor's decision to launch the Special Publications Division in 1965 with Robert L. Breeden as its chief.

"An outgrowth of the public service books about the White House, the Capitol, and the Supreme Court, Special Pubs were intended to be small, magazine-page-size books," Breeden points out, "not the large format type of books done by Book Service.

"Melville had a great deal of respect for Merle Severy who was then chief of Book Service. One day Melville asked me to come in, and he said, 'Bob, I know you're going to be doing these books. Do you think they'll interfere with Book Service?'

"I told him I didn't think there would be any problem at all, and I asked, 'Would you like for me to go talk with Merle about this?'

"He grinned, obviously relieved, and asked, 'Bob, would you do that?'

"I went to see Merle," Breeden said, "and he was, of course, confident these small books wouldn't hurt the big ones. He saw no problem with it—although, I will have to say, there certainly developed a very strong feeling of competition between the two book services."

Special Publications and School Services, now under Director Donald J. Crump, publishes books that are generally more narrow-focused and tightly defined than Book Service's large-format "encyclopedic" books.

Special Pubs quickly moved into publishing books on a wide variety of popular topics such as U. S. history, wildlife, earth sciences, undersea and space developments, and New World

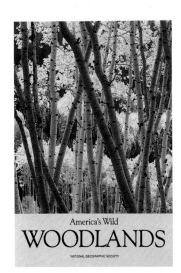

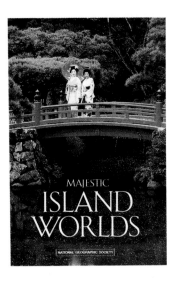

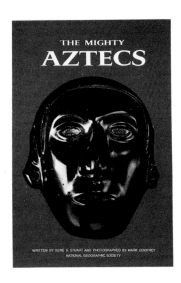

archeology—some of which were big enough to further blur to an outsider the already fuzzy distinction between Book Service and Special Pubs.

There is one distinction, however, that any publisher would envy. From an original core of 28,000, there are now more than 200,000 readers who purchase each Special Pubs volume without even seeing the announcement of the year's four new titles. "Forever Pubs," as these loyal members are known, helped push sales of Special Pubs' now eighty-seven titles to more than 34,000,000 by the end of 1986.

345

mensely popular lecture* series, which, since 1888, had thrilled Society members, first with lantern slides and later with color films. As Melville wrote in 1970:

> Many early members saw the North Pole meet the South Pole in 1913 when Robert E. Peary introduced Roald Amundsen; both described their discoveries.
>
> Many listened spellbound to Wilbur Wright telling of the birth of powered aviation. Later audiences rode vicariously with Richard E. Byrd on his pioneer flights over the Poles, ascended with balloonists Albert Stevens and Orvil Anderson into the stratosphere, took the first successful deep-ocean dive with William Beebe in his bathysphere.
>
> In the fifties and sixties, we presented other dramatic film lectures by modern trailblazers who devoted, and often gambled their lives to win new geographic knowledge: Sir Edmund Hillary, who, with Sherpa guide Tenzing Norgay, first reached the top of Mt. Everest; Commander [James F.] Calvert, who surfaced his atomic submarine *Skate* at the North Pole; Captain Cousteau and Dr. Leakey....
>
> The popularity of our lectures encouraged us to feel that a far wider audience awaited the color films that the Society's experience and talent could produce.

Perhaps even more encouraging than the lectures was the guest appearance Melville made with Luis Marden in January 1958 on NBC's television program "Omnibus." Marden, the previous year, had found, while diving off Pitcairn Island in the South Pacific, the remains of Captain William Bligh's ship H.M.S. *Bounty*, burned and sunk by mutineers 167 years before. Marden's discovery had made world headlines, and a Society lecture film, *The Bones of the Bounty*, was so successful that Melville began to believe the Society should be producing its own television shows.

In 1961, a documentary film department was organized by the Society's first Director of Television, Robert C. Doyle, who had the responsibility of producing four TV documentaries a year—and coming up with outside commercial sponsorship for the films.

In 1963, funds were made available to film the three-team assault on Mount Everest by Americans, one of whom was Barry C. Bishop, National Geographic staff geographer. This expedition, partially sponsored by the Society, was, in addition to being the first to ascend the summit by the West Ridge, the first ever on which motion pictures were taken from the peak:

> "The wind whipped and tore at us as we perched precariously on the earth's highest pinnacle," wrote an expedition member of how he and his teammate Luther (Lute) Jerstad made the film. "The American flag chattered in the gusts....Lute stuck his ice ax in the hard snow, anchored a motion-picture camera to its head...the ax shuddered in the wind. Lute's silk-gloved fingers began to freeze as he turned the camera's metal crank...."

The color motion-picture footage was edited and assembled; expedition member James Ramsey Ullman wrote the script. Encyclopaedia Britannica and Aetna Casualty Insurance agreed to sponsor it; David L. Wolper produced

A National Geographic Society lecture was scheduled the evening of November 22, 1963. That afternoon President John F. Kennedy was shot and killed in Dallas, Texas. Several Geographic staff members were sent to Constitution Hall to turn away Society members who might not have heard that the lecture was cancelled. An elderly couple, intercepted on their way to the hall, were told there would not be a lecture that night because the President had been assassinated. Shocked by the news, one of them asked, "Our Dr. Grosvenor?"

LEAKEY AND PROTÉGÉES:

Among the Magazine's "stars" who became television celebrities were several recipients of financial support from the Society's Committee for Research and Exploration such as the Leakeys ("The Legacy of L.S.B. Leakey," broadcast in 1978), Jane Goodall ("Miss Goodall and the Wild Chimpanzees," 1965), Dian Fossey ("Gorilla," 1981), and Dr. Biruté M.F. Galdikas (with Dian Fossey in "Search for the Great Apes," 1976).

Since 1959, the year Dr. and Mrs. Louis S. B. Leakey discovered the 1.75-million-year-old skull from a manlike creature they called *Zinjanthropus*, the Society supported their work in East Africa. In five decades of unearthing bones of hominids and other animals in Tanzania's Olduvai Gorge, the late paleoanthropologist Louis Leakey greatly expanded the record of early man. In the picture *near right top* Louis and his son, Richard, study the fossil skull of a man- and ape-like creature.

Leakey believed that by studying the great apes, insight could be gained into the behavior of early man. As a result, a direct link exists between the Leakeys' work and that of Jane Goodall *(near right bottom)*, Dian Fossey *(far right top)*, and Biruté Galdikas *(far right bottom)*—all Leakey protégées.

Beginning in July 1960, when Jane Goodall (shown here with Research and Exploration Committee members Leonard Carmichael, Melvin Payne, and T. Dale Stewart) began to live among the chimpanzees of Tanzania's Gombe Stream National Park, Society grants have helped support her continuing study of their behavior in the wild—she discovered that usually sociable chimpanzees can, on occasion, become killers and cannibals. She proved that they fashion and use tools—an observation of ob-

GOODALL, FOSSEY, AND GALDIKAS

vious importance for the scientific world. "Tool using always used to be considered a hallmark of the human species," Jane Goodall later said.

In December 1986 the Society—and the world—was shocked and saddened by the brutal murder of Dian Fossey in the forested mountains of Rwanda. She had made her home there since 1967, when she began her long-term study of the mountain gorilla. Fossey's years of painstaking observation of these, the largest of the great apes, helped change our perception of this endangered species from that of a fearsome jungle giant to that of a shy and unaggressive fellow creature.

Anthropologist Biruté Galdikas, who raised young orangutans at her camp in the Tanjung Puting Reserve study area in the Bornean rain forest, has also spent years trying to protect an endangered primate species.

BOOKS FOR CHILDREN

The Society's children's book program began in 1972 with publication of the first set of Books for Young Explorers, a series designed for children four through eight years old.

The inspiration behind the series was then *National Geographic* Editor Gilbert M. Grosvenor, who was looking for non-fiction books designed for durability, illustrated with quality color photographs, and of interest to both the child listener and the parent reading aloud—books that he could share with his own young son. Books for Young Explorers were developed by the Special Publications Division under the leadership of Robert L. Breeden and his associates Donald J. Crump and Philip B. Silcott. Their goal was to stimulate in young children a love of reading by making books with texts that children themselves could read and that would sustain the interest of young listeners. The staff, working with educators and librarians, scientists and children's book authors, produced the first of the series of thirty-two-page books: *Dinosaurs, Dogs Working for People, Lion Cubs: Growing Up in the Wild,* and *Treasures in the Sea.* Packaged in sets of four, the Books for Young Explorers series are offered annually to Society members in time for Christmas.

In school and at home, the opinions of children in the four to eight age group were solicited in the selection of titles. Under the guidance of a reading specialist, texts were tested on children to help set an appropriate readability level. Children were also shown photographic layouts to determine if they had any difficulty in "reading" the pictures.

In 1978, however, Books for Young Explorers were temporarily "put on hold" when the demographic trend of lower birth rates resulted in a lower demand from members. Three years later, in 1981, the program was reinstated. And the press run for its four-book set that year was 200,000 copies. By 1986, with the publication of Set 13—*Baby Bears and How They Grow, Saving Our Animal Friends, Animals That Live in Trees,* and *Animals and Their Hiding Places*—the press run had jumped to 410,000 copies. By the spring of 1987, more than twenty-one million individual Books for Young Explorers were in print.

In the late 1970s, the Society turned its attention to slightly older readers. Children who had avidly read Books for Young Explorers were growing up; and it was decided that the children's book program should grow up with them. The result was a series of Books for World Explorers for children ages eight to thirteen designed to be both educational and fun. The first book, *Secrets From the Past,* rolled off the press in October 1979, and amazingly—as in *Amazing Animals of the Sea, Amazing Mysteries of the World, Amazing Animals of Australia,* and *Computers: Those Amazing Machines*—one 104-page book after another has followed every three months since then.

The most recent children's book venture has been the Action Book. Each "pop-up book" contains twelve pages with dimensional and moving parts. Ingenious paper engineering enables a butterfly to take wing in one book, while monkeys dangle from three-dimensional trees in another. Introduced in 1985, the new program received instant acclaim.

it, and on the evening of September 10, 1965, *Americans on Everest*, narrated by Orson Welles, was seen on television.

This was no blurry, strung-together sequence of still pictures showing distant specks clinging to the side of the world's tallest mountain. It was superb close-up color motion-picture photography of the expedition members' torturous struggle against mind- and bone-chilling 20° below zero temperatures and 60 mph winds, against collapsing ice walls and treacherous snow cornices, against numbing fatigue and oxygen deprivation. It was about courage and determination, risk and achievement, and it took its millions of viewers along with Barry Bishop (who had sworn to himself: "If you have to crawl on hands and knees you're going to get there") up 29,028 feet to Everest's summit.

The day after *Americans on Everest* was shown, an ecstatic CBS reported to the Society that the film had received the highest ratings of any documentary ever telecast. Melville Grosvenor's "far wider audience" had been reached; and a formidable new era of increasing and diffusing geographic knowledge had begun.

During Melville Grosvenor's decade at the helm, the Society turned out one stunning documentary after another. *Americans on Everest* was followed by *Miss Goodall and the Wild Chimpanzees, Dr. Leakey and the Dawn of Man, The World of Jacques-Yves Cousteau*. The Magazine's "stars" became television stars. And then, in 1968, the Geographic special *Amazon* made television history when its lush river and jungle scenes were watched by thirty-five million people. It was the first television documentary to top all other shows in a two-week rating period.

Under Melville's leadership, massive new buildings rose to house the Society. He commissioned one of America's best-known architects to create a new headquarters to be linked with the older headquarters by a tunnel beneath a magnolia-shaded parking lot—and paid for it in cash. "The concrete capstone of Grosvenor's contributions," the *Washington Post* reported on September 13, 1974, "is represented by the white marble headquarters building at 17th and M Streets, NW, designed by Edward Durell Stone. It is proof that modern architecture can have grace and beauty."

The new headquarters was the first Washington building designed by Stone. On January 18, 1964, President Lyndon B. Johnson dedicated the gleaming, marble-clad ten-story building, saying, "Today in this house of exploration, let us invite exploration by all nations, for all nations." President Johnson was an appropriate choice to have dedicated the building. He had once told Melville, "My mother brought me up with the Bible in my right hand and the *Geographic* in my left."

With color photographs on the cover, vivid all-color Magazines, new printing techniques, expanded support of research and exploration, the advance into television and book publishing, Melville Grosvenor had taken what many in the senior echelons of the Society considered a "fussy, stagnant empire" and directed it through a decade of exciting, explosive expansion.

"However," as *The New York Times* noted, "Dr. Grosvenor did not dramatically modify the magazine's traditional tone of gentlemanly detachment from the ugliness, misery and strife in the world."

Publishing, outside the Society, continued to deride the Magazine's see-no-evil attitude. Calling the National Geographic Society "the least exclusive, farthest flung and most improbable nonprofit publishing corporation in the world," *Time* magazine in a June 15, 1959, article, characterized Gilbert H.

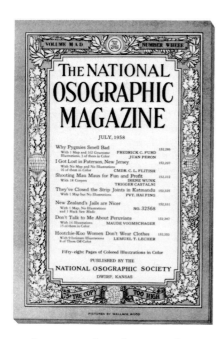

Look again. At first glance, Mad *magazine's 1958* National Osographic Magazine *so mimics the* National Geographic *style that it risks being passed by. However,* Mad's *parody lead story, "Why Pygmies Smell Bad," inadvertently foreshadowed the* Geographic's *September 1986 article "The Intimate Sense of Smell," with its startling photograph of laboratory-jacketed women sniffing bare-chested men's armpits.*

Grosvenor's "I was always taught not to criticize other people" attitude as being the basis for the "unrealistic hue to the *Geographic*'s rose-colored world."

The National Geographic "thrives on a policy of daring serenity and a 'dear Aunt Sally style,' as staff members refer to it," *Newsweek* wrote in its November 11, 1963, issue, remarking on the Society's seventy-fifth anniversary in that year. "The world of National Geographic is usually a sunlit Kodachrome world of altruistic human achievement in settings of natural beauty, a world without commercial blemish or political disturbance, even, it sometimes seems, a world without germs or sin."

Melville Grosvenor's response to those criticisms was always to admit they were true, but, as he told one reporter, "We've always tried not to point out the sores. What the hell, there's so much wonderful stuff in the world, why get into the sordid too much? I mean, that's not our job."

It was with that attitude in mind that three years after Melville Grosvenor's resignation, *The New York Times Magazine* titled its piece on Gilbert M. Grosvenor's accession to the editorship "With the National Geographic On Its Endless, Cloudless Voyage." For *Esquire*'s "Two Cheers for the Geographic," Anne Chamberlin wrote, "The Magazine has mastered the art of admitting there are bad smells but not dwelling on them."

Even *Mad* magazine took on the *Geographic* in a 1958 parody called *The National Osographic Magazine* that contained such lead articles as "Why Pygmies Smell Bad" and "Don't Talk to Me About Peruvians" and a piece called "Africa is for the Birds," which said:

> Far up the trail, I could make out the form of a tall princely Buktuktu, his k'kkkaty glistening in the sunlight. Obviously, the Jdu-Jdu drums had heralded our arrival and he was a royal welcoming committee.
>
> He came forward with a smile that disclosed the sharply filed teeth of the Gwam'mbmba aristocracy. He bowed low, showing us his wumbt'tu, and I shielded Evelyn from the sight as best I could.

Chamberlin, in *Esquire*, rose to the *Geographic*'s defense, quoting one devotee: "Laughing at the *National Geographic* is like laughing at Harvard. No matter how hearty and well-deserved the laughter, it is still a great institution."

Had it not been, the Society might not have so readily withstood the chaos within its Illustrations Division and what Melville's son Gilbert M. Grosvenor would refer to as the "meandering oxbows of communications" and decisions "dependent upon who was the last person to get to MBG and his enthusiasm." And just as the Society had withstood the chaos within, it withstood the chaos without: the troubled 1960s—that decade of brutal and heartrending political assassinations and imprisonments, of escalating "protective reaction" raids against Southeast Asian villagers, that decade of racial and generational polarization which culminated so despicably in senseless campus killings and in mind-bending and life-ending experimentation with drugs, that decade of monolithic megacorporational indifference toward the environmental and ecological havoc they had wrought, that decade of Presidential and Congressional chicanery, that decade of alarming and confounding economic and spiritual crises....

The *Geographic* was able to withstand the storms primarily because of the leadership of Melville Bell Grosvenor: the right man, in the right place, at the right time.

On February 4, 1966, Gilbert Hovey Grosvenor, ninety years old, lay down for an afternoon nap at his Baddeck summer home. He had spent his first Christmas without his beloved wife and companion of so many years. Elsie May Bell Grosvenor, First Lady of the Geographic, had died the day after Christmas, 1964.

Dr. Grosvenor had grown increasingly frail as the New Year 1966 had commenced and, knowing the end was near, his family had gathered around him to say good-bye. Just before his nap that February afternoon, a daughter heard Dr. Grosvenor mention a recurring dream, one in which he was either just setting out on a voyage, or returning to port.

Looking back on his "long and happy tenure" as Editor of the Magazine and President of the Society, he had written, "Those golden years of my editorship bring to mind a fragment from Tennyson's beautiful poem *Ulysses*,"

> For always roaming with a hungry heart,
> Much have I seen and known—cities of men
> And manners, climates, councils, governments.

Gilbert Hovey Grosvenor never awoke from that nap.

That evening, President Lyndon B. Johnson and Mrs. Johnson telephoned their condolences, and the next day sent the following wire to Melville:

> As the Nation grieves the loss of your father, inquiring minds everywhere grieve the loss of a leader. Through words and pictures and through his own unswerving dedication, Gilbert Grosvenor opened the wonders of the world we live in to three generations of Americans. No mountain was too high, no sea too deep, no climate too forbidding for the teams of the National Geographic Society.
>
> If a great national power has an obligation to know and to understand the nature of the world around it, then it can truly be said that Gilbert Grosvenor played a vital role in America's coming of age.

More tributes, from former Presidents and other Chiefs of State, poured in, but among the most touching was the memory of a Society staffer who recalled that as "a shy cub from New England" he had been invited to spend a weekend with the Grosvenor family and their guests at their Florida winter home. As Vice President and Associate Editor Frederick G. Vosburgh wrote in his October 1966 *National Geographic* tribute to Gilbert Hovey Grosvenor:

> White linen suits were in style then and, on the advice of friends, he had brought one. But nothing had been said about white shoes, and when he went down to dinner he saw to his horror that of all the men present he alone was wearing black shoes. He had no others; there was nothing he could do in his embarrassment but try to lose himself among the guests.
>
> In a few minutes Dr. Grosvenor quietly excused himself and disappeared upstairs, to return shortly. He was wearing black shoes.

In 1967, one year after his father's death, Melville Bell Grosvenor, age sixty-five, fulfilled his vow not to deny his successors the opportunity to lead that had for so long been denied him. He stepped down from active editorship. Frederick G. Vosburgh was appointed Editor to replace him—a position that, like the editorship of John Oliver La Gorce, would serve as an interregnum between Grosvenors. Gilbert M. Grosvenor would succeed Vosburgh in 1971.

Melville Grosvenor's turbulent decade had been one of unquestioned success. While major mass-market magazines such as *Life, Saturday Evening Post,* and *Colliers* were folding all around him, the *Geographic*'s revered skipper had steered the Society into the twentieth century—and not a moment too soon.

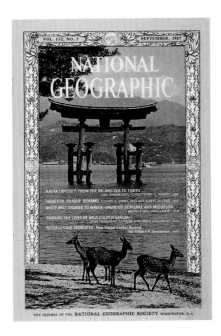

In 1967, at the close of Melville Grosvenor's decade as Editor, membership had more than doubled from 2,178,040, at the end of 1957, to 5,607,457.

"The Choice Is the Universe —or Nothing"

"There is no way back into the past: the choice, as [H.G.] Wells once said, is the Universe—or nothing. Though men and civilizations may yearn for rest, for the Elysian dream of the Lotus Eaters, that is a desire that merges imperceptibly into death. The challenge of the great spaces between the worlds is a stupendous one, but if we fail to meet it, the story of our race will be drawing to its close. Humanity will have turned its back upon the still untrodden heights and will be descending again the long slope that stretches, across a thousand million years of time, down to the shores of the primeval sea."

—From Arthur C. Clarke, *Interplanetary Flight: An Introduction to Astronautics*, as quoted by Carl Sagan in his article "Mars: A New World to Explore," in the December 1967 Magazine

Above: *The first* Explorer *balloon tugs at its tethers prior to its near disastrous July 28, 1934, ascent.*

Opposite: *November 11, 1935. Wearing football helmets borrowed from a high school squad, Capt. Orvil A. Anderson (left) and Capt. Albert W. Stevens (right) pause outside the gondola of the stratosphere balloon* Explorer II *before their ascent (sponsored by the Society and the U.S. Army Air Corps). They reached 72,395 feet—a record that stood for twenty-one years.*

Preceding spread: *Designed to bridge the gap between manned flight in the atmosphere and manned flight in space, the X-15 rocket plane thunders at five times the speed of sound on its April 30, 1962, run, carrying test pilot Joseph A. Walker to a record altitude of 246,700 feet—above 99.996 percent of the earth's atmosphere.*

Since its inception, the National Geographic Society has been in the forefront of those supporting aviation and space exploration. In 1911, at a time when American interest in aviation seemed to be faltering, Wilbur Wright took advantage of his opportunity as speaker at a Society banquet to warn the membership and President Taft, the Society's guest, that "the leading nations of the earth are taking up the subject, our own nation being the first of all to begin it. But unfortunately, there seems to be some hesitation at present."

In 1914, Alexander Graham Bell, who had sketched a rocket plane in 1893, predicted, "...heavier-than-air machines...of a different construction from anything yet conceived of, will be driven over the earth's surface at enormous velocity, hundreds of miles an hour, by new methods of propulsion....Think of the enormous energy locked up in high explosives! What if we could control that energy and utilize it in some form of projectile flight!"

Airship pioneers Hugo Eckener, Amelia Earhart, and Charles Lindbergh each lectured before the Society. So did stratosphere balloonists Auguste Piccard and Albert W. Stevens, to whom the altitude records then belonged.

The March 1933 *Geographic* carried Piccard's account, "Ballooning in the Stratosphere: Two Balloon Ascents to Ten-Mile Altitudes Presage New Mode of Aërial Travel," and told of his record-breaking ascents in hydrogen-filled balloons to altitudes of 51,775 feet on May 27, 1931, and to 53,152.8 feet on August 18, 1932. "The stratosphere is the superhighway of future intercontinental transport," Piccard proclaimed.

In October 1934, the *Geographic* contained "Exploring the Stratosphere," Capt. Albert W. Stevens' account of the National Geographic Society–U.S.

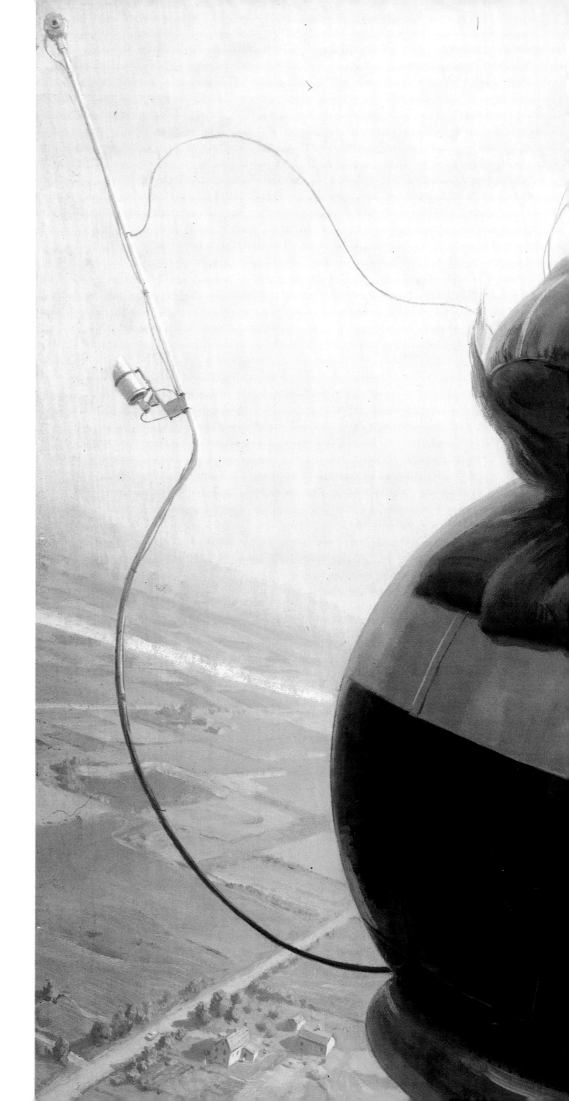

"Then things happened fast. Anderson disappeared. The balloon exploded. The gondola dropped like a stone...." wrote Capt. Albert W. Stevens in the October 1934 National Geographic.

Falling at a mile a minute, with only 800 feet to go, Maj. William E. Kepner, atop Explorer's *gondola, saw Stevens wedged in the hatch opening by the bulk of his parachute and frantically kicked him free as Capt. Orvil A. Anderson, sucked out of the hatch by the force of the wind, parachuted safely down in the distance. Kepner jumped with but 300 feet remaining before the gondola "hit the earth with a tremendous thud."—Painting by Tom Lovell, 1963*

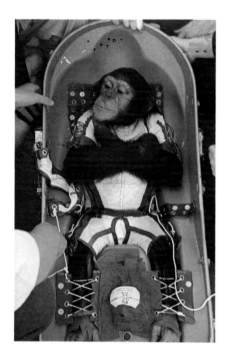

Above: *"Ham Greets Rescuers with Nonchalance. Arms folded jauntily, every freckle standing out, the chimpanzee awaits release from his couch aboard the recovery [ship] LSD* Donner. *Long hours of training paid off: Ham quickly recovered from his eight-hour ordeal. His reward: a big red apple."—From "Countdown For Space" in the May 1961 issue*

Opposite: *May 5, 1961. Spectators crowding Cocoa Beach cheer as America's first astronaut, Alan B. Shepard, Jr., in* Freedom 7 *atop an eighty-three-foot thirty-three-ton Redstone rocket, rises from Cape Canaveral for a fifteen-minute suborbital flight. Three weeks earlier, the Soviet Union's Maj. Yuri Gagarin had completed the first manned spaceflight, a one-orbit mission aboard the Soviet spacecraft* Vostok.

Army Air Corps Stratosphere Flight in *Explorer*, until then the largest free balloon ever constructed.

As Stevens related, in the calm, windless hours shortly before dawn on the morning of July 28, 1934, "Captain [Orvil A.] Anderson and I climbed into the gondola; Major [William E.] Kepner to its rope-enclosed top, the better to direct the take-off." Then, at 5:45, cast-off! The ground dropped away as *Explorer*, carrying three men and over four tons of scientific instruments, rose rapidly into the clear morning air.

At 15,000 feet the airtight manholes were closed; at 40,000 feet the balloon's ascent was halted for nearly two hours to test the scientific instruments, and then the balloon was permitted to rise again until one in the afternoon; at an altitude of nearly 60,000 feet, gas was valved out and the balloon slowed.

It was at that point that disaster struck. Stevens wrote:

> Suddenly a clattering noise was heard…caused by part of…a small rope—falling on the roof of the gondola.
>
> What had caused the cord to drop? Looking still higher, we were startled to see a large rip in the balloon's lower surface…Through the overhead glass porthole we watched the rent in the fabric gradually become larger and larger. The minutes passed slowly by….
>
> Little…talking was done, for our ears were strained to get warning sounds from above us. Soft swishing noises came through the roof of the gondola from time to time. Each of these sounds meant a new rent, or an increase in length of a rip already there….

Outside the gondola the temperature was nearly −80° F. A narrow band of ice had formed inside the gondola two feet below the top. Three quarters of an hour later, *Explorer* had descended to 40,000 feet and its speed was increasing; a half hour later it had reached 20,000 feet.

> Major Kepner and Captain Anderson each forced open a hatch, and for the first time we felt we were free….
>
> We all climbed out on top and took a good look at the balloon. It was pretty badly torn. Many more tears and rips had appeared in it. The question was, How long would it hold together?…
>
> Suddenly the entire bottom of the bag dropped out…I climbed back into the gondola and started discharging ballast….
>
> At 10,000 feet we really should have left the balloon, but we did not wish to abandon the scientific apparatus. So we stayed on. At 6,000 feet we …decided we had better leave. The last altimeter reading I gave was [3,000 feet above the ground]….
>
> In the meantime Captain Anderson, atop the gondola, had been having difficulty with his parachute. The release handle had caught on something and the parachute pack had come open….

Anderson tucked the folds of parachute under his arm and stepped down into the hatch, blocking the opening through which Stevens planned to jump. Stevens shouted at Anderson to get out of the way.

> Things started to happen fast.
>
> [Anderson]…disappeared and I knew he had leaped. As he jumped the balloon exploded….
>
> The gondola dropped like a stone….

Anderson had been sucked out of the gondola by the force of the wind. The gondola was now falling at the rate of a mile a minute. Kepner, still riding atop

the balloon, saw that Stevens was wedged by the bulk of his parachute in the tiny hatch opening and struggling to get out. With only 800 feet remaining before the speeding gondola smashed into the ground, Kepner frantically kicked Stevens free. And then, with but 300 feet left, Kepner himself jumped.

Stevens jerked his ripcord, and his parachute fluttered free:

> The folds of white silk opened in a large circle—and then a portion of the balloon fabric…fell on top of my parachute…luckily the parachute slid out from under and worked itself free.
>
> How about Kepner and Anderson? I looked around and saw the other two parachutes in the air and knew they were safe. Directly below me, I heard the gondola hit with a tremendous thud, and saw a huge ring of dust shoot out. Forty seconds later I hit—fortunately with a much lighter thud…

Stevens, Kepner, and Anderson rolled up their parachutes and hurried to the cornfield near Holdrege, Nebraska, in which their gondola had fallen.

> [We] found…our beautiful black and white globe had been crushed like an eggshell…Inside, the instruments…were a heart-breaking mass of wreckage…
>
> [But] the two sealed barographs, hung outside the gondola…suffered scarcely any damage, [and showed] we reached an altitude…of 60,613 feet above sea level…only a trifle more than two of our balloon lengths short of the official world record of 61,237 feet, [attained] by Commander T.G.W. Settle and Major Chester Fordney [the year before].

Stevens, Anderson, and Kepner were confident that another balloon could be built that would break the altitude record easily.

Approximately sixteen months later, on November 11, 1935, Captains Anderson and Stevens in *Explorer II* lifted off at 7:01 A.M. from their Strato-bowl launching site near Rapid City, South Dakota, and at 11:40 ascended to their maximum ceiling at the very fringe of space, setting an altitude record of 72,395 feet that was to stand for the next twenty-one years.

From nearly 14 miles above sea level, Stevens wrote,

> The earth…was a vast expanse of brown, apparently flat, stretching on and on…a foreign and lifeless world.…The horizon itself was a band of white haze. Above it the sky was light blue, and…at the highest angle that we could see it…black with the merest suspicion of very dark blue.

Stevens and Anderson remained at that altitude for an hour and a half conducting experiments, their instruments busy. Then with millions of listeners as far away as Europe and South America able to listen to their voices over a short-wave broadcast direct from the gondola, Stevens and Anderson in *Explorer II* started back down.

> …10 miles above the earth…I was amused when I overheard the instructions, given on short wave, of an eastern announcer to his fellow announcers.

Tethered by a cord, Maj. Edward H. White II moves freely 100 miles above earth. Orbiting at 17,500 mph, he had little feeling of speed and none of falling, only "of accomplishment."
—From "America's 6000-Mile Walk in Space" in the September 1965 issue

"Don't play up this record business, boys, until we are sure that they have gotten down safely," he suggested. "There is still plenty of chance for them to crash and they have to come down alive to make it a record...."

In 1949, in conjunction with the California Institute of Technology, the National Geographic Society launched the most ambitious astronomical project undertaken to that time: a monumental Sky Atlas, prepared at Palomar Observatory, that took more than seven years to complete. The National Geographic Society–Palomar Observatory Sky Survey's 1,758 photographic plates covered a volume of space at least twenty-five times as large as any before charted, mapped the heavens to a depth of one billion light-years, revealed millions of galaxies containing billions of stars, and changed our whole concept of the universe.

In May 1959, the Magazine published the "First Color Portraits of the Heavens," which showed such things as the Crab Nebula glowing "like veined marble," the Andromeda constellation's "whirling blue pinwheel," and a red and gold "cosmic smoke ring." But photographs can tell only so much—even about our closest companion, the moon.

Nobody really knew what would happen if someone set foot on the lunar surface until June 1966 when *Surveyor I* landed on a gray, shallow crater in the Ocean of Storms and began transmitting back to earth the first of its more than 11,000 photographs from the moon. In the October 1966 issue, Homer E. Newell, then Associate Administrator, NASA, reported that *Surveyor*'s photographs

> gave us a remarkably clear and intimate view of the lunar face, so close that we can measure and count particles only a fiftieth of an inch across. It even provided us with a glimpse beneath the moon's surface.
>
> For the first time, because of Surveyor, Project Apollo officials feel real assurance that an astronaut can safely set foot on the moon, that the moon's surface will support him, and that he will not be swallowed up in a thick sea of dust.

Three years later Kenneth F. Weaver was writing in his May 1969 article "And Now to Touch the Moon's Forbidding Face" for the *Geographic*:

> Goddess of the Night they called her: Selene, Artemis, Cynthia, Luna. Men worshiped and feared her, believing that her mystical powers influenced life on earth. They closely marked her inconstant face, measured time by her waxing and waning. They sang of her splendor, named her "bright wanderer," "fair coquette of heaven," "sweet regent of the sky," "Mother Moon."
>
> Once she was unreachably remote—the province of poets, of shepherds and nomads, of lonely astronomers and not-so-lonely lovers.
>
> Today, in the twelfth year of the Space Age, earth's natural satellite has become the concern of everyman, and the object of the most intensive scientific and technological effort in history. Hundreds of thousands of people, and industrial firms by the thousands, have turned their energies toward realizing the goal of putting men on the moon.

Apollo 17 astronaut Harrison Schmitt investigates a fifty-foot boulder in the moon's Taurus-Littrow Valley.

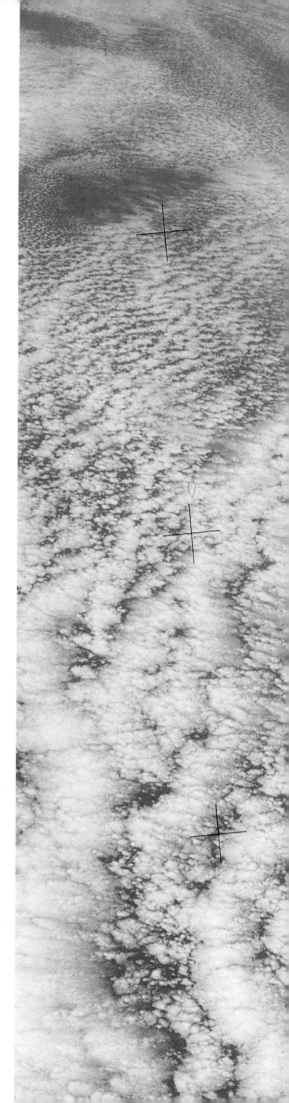

Man has already probed at the moon with some 45 spacecraft bearing names like Luna and Zond, Ranger, Surveyor, and Orbiter. From afar we have poked and scratched its surface, hammered on its rocks, assayed its chemistry, measured its temperatures. We have tested it with radio waves, the exhaust of rocket engines, and magnets. We have photographed from close up all but the merest smidgen of its tortured face.

And now, as this is written, a few chosen men, astronauts and cosmonauts, train with monastic zeal for the fantastic attempt that will bring alive the tales of Jules Verne and H.G. Wells.

The list of *Geographic* manned spaceflight articles is a condensed history of this country's space program en route to the moon. The first manned Mercury launch atop a Redstone rocket was covered in "The Flight of *Freedom 7*," by Carmault B. Jackson, Jr., and "The Pilot's Story: Astronaut Shepard's First-hand Account of His Flight," which appeared in September 1961, four months after Alan B. Shepard, Jr., became the first American to ride a pure rocket into space. Shepard's launch was followed by Gus Grissom's in *Liberty Bell 7*, the last launch atop a Redstone rocket. Now that the Mercury capsule had been proven safe, the next stage of manned flight, atop the more powerful Atlas booster, was to test the performance of pilots in a more extended weightless condition. Robert B. Voas' "John Glenn's Three Orbits in *Friendship 7*: A Minute-by-Minute Account of America's First Orbital Space Flight" covered this stage. Voas was the Training Officer of Project Mercury; his article appeared in June 1962, four months after Glenn's launch, and recalls those pre-man-on-the-moon days when every space flight was breaking new ground. It also reminds us that such flights were dangerous—a lesson reinforced when the *Challenger* space shuttle exploded in January 1986.

Reading Voas' article today, one sees that even with Glenn's flight there were indications that NASA felt there were certain things it was better for the astronauts not to know. At the end of Glenn's first of three earth orbits, as Voas wrote, "a radio signal from the spacecraft indicated that the landing bag, which would act as a cushion when the capsule hit the water, had been deployed prematurely. If this signal proved valid, it would mean that the heat shield, which is attached to the landing bag, had also come free and would not protect the spacecraft from the fiery heat of re-entry." In other words, the capsule's heat shield would have sheared off during re-entry, and *Friendship 7*, with Glenn inside it, would have incinerated in seconds.

Glenn's first indication that there might be a problem occurred at 02:26:36 [2 hours, 26 minutes, and 36 seconds] into the flight, when capsule communicator Gordon Cooper in Muchea, Australia, asked Glenn, "*Friendship 7, will you confirm the landing bag switch is in the off position?*"

Voas' piece continues:

Affirmative, Glenn answers. *Landing bag switch is in the center off position.*

Cooper asks: *You haven't had any banging noises or anything of this type?*

Negative.

▶

Right: *With one solar panel lost en route to orbit, 100-ton Skylab orbits silently 270 miles above the cloud-shrouded earth.*

Overleaf: *The space shuttle* Challenger, *inverted high above Baja California in 1984, launches a satellite for testing materials in space.*

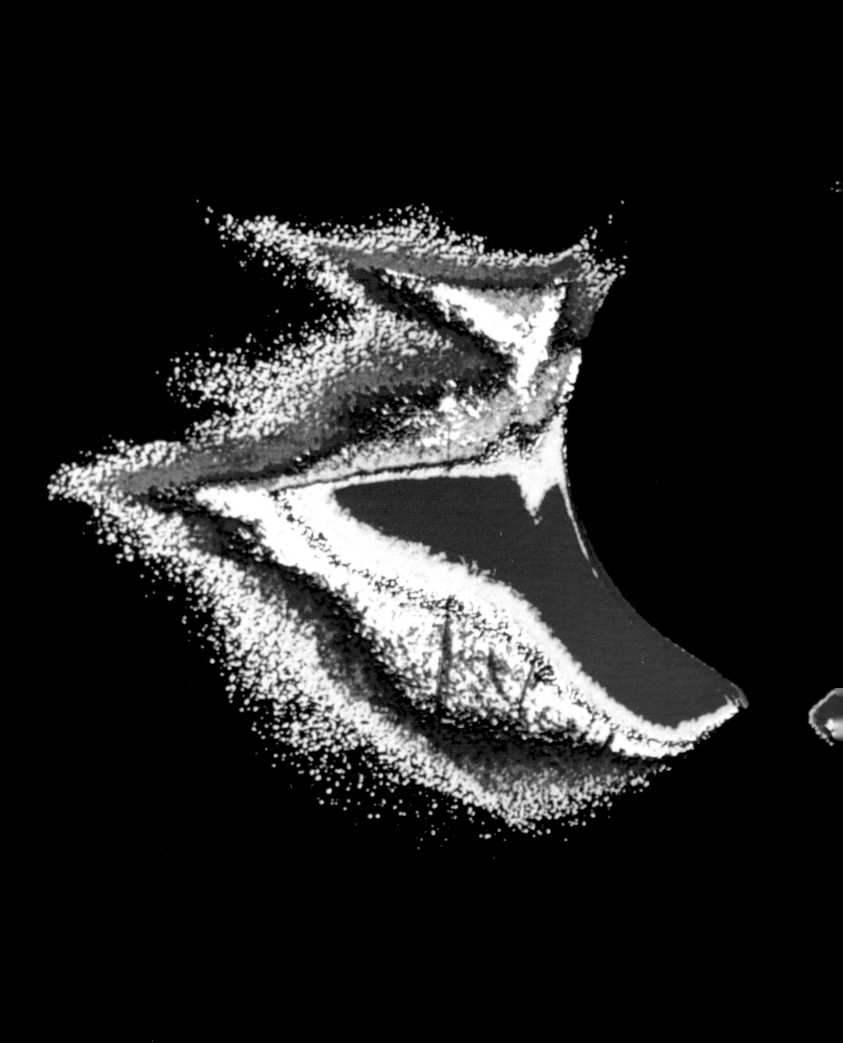

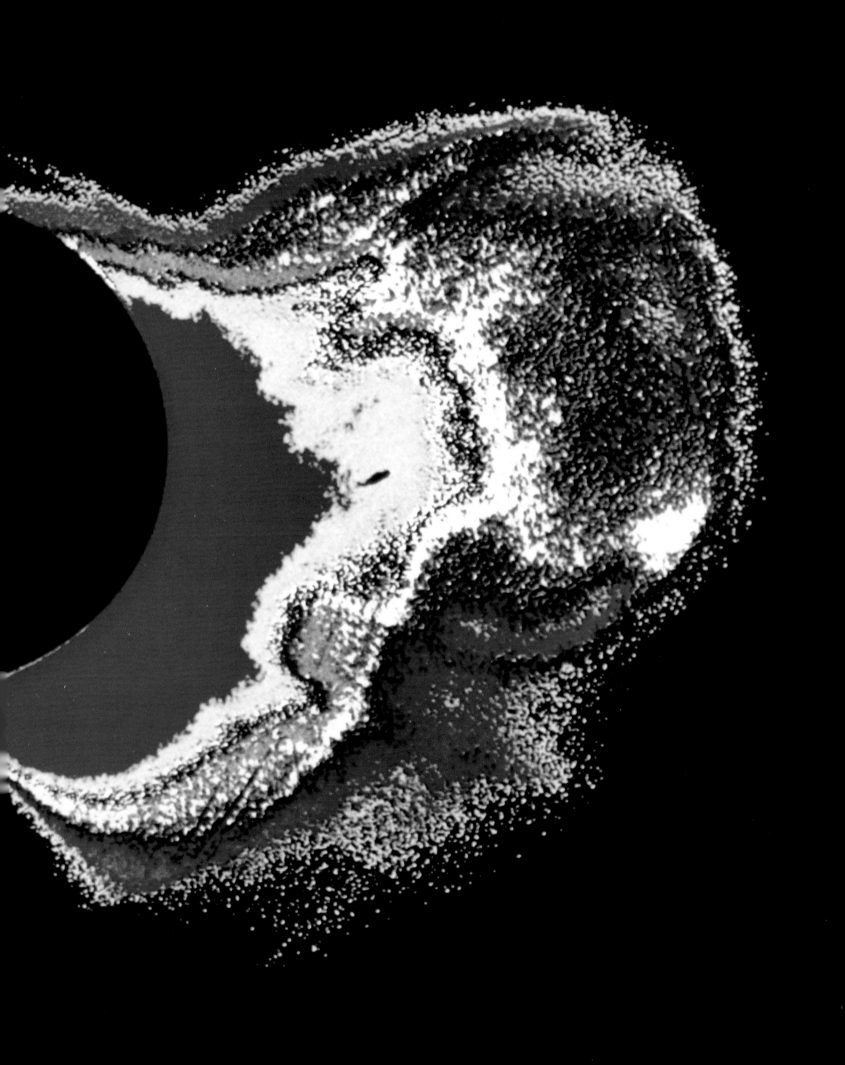

Above: *The Viking 2 lander touched down on Mars' Plains of Utopia in 1976, and its cameras and sensors commenced gathering data on the Red Planet's weather, atmosphere, chemistry, and biology.*

Opposite: *A bispectral mosaic made up from Viking Orbiter I images of Mars covers nearly an entire hemisphere of that planet.*

Preceding spread: *The sun's fiery corona glows in this color-coded photograph taken from the Skylab space station.*

Nearly four hours into the flight the autopilot is behaving erratically and Glenn bypasses it to take control of the spacecraft himself. At 04:38:25 into the flight the Texas capsule communicator in Corpus Christi identifies itself and tells Glenn:

> *We are recommending that you leave the retropackage on through the entire re-entry....*
> *This is Friendship 7. What is the reason for this? Do you have any reason?*
> *Not at this time. This is the judgment of Cape Flight.*

Glenn has still not been told that NASA is worried his heat shield has slipped. Glenn is busy bringing the nose of the Mercury capsule up until instruments indicate he is flying straight and level with the earth and ready for re-entry. Robert B. Voas' narration continues:

> 04:40:23: *Friendship 7, this is Cape...Recommend you...retract the [peri]scope manually.*
> *Roger. Retracting scope manually.*
> *While you're doing that* [Cape tells Glenn], *we are not sure whether your landing bag has deployed. We feel that it is possible to re-enter with the retropackage on. We see no difficulty at this time in that type of re-entry.*

During re-entry the heat shield, as its name implies, protects the pilot from the friction-heated air that slams against the capsule with a force of 8 g's. [Eight times gravity] The shock wave not three feet from his back incandesces with a temperature nearly five times that of the sun's surface.
04:42:52: *Seven, this is Cape....We recommend you....*

Shepard's voice fades away as the communications blackout begins. Glenn wonders what his message was.

Manually he starts the capsule rolling. Like a rifle bullet, the capsule must revolve slowly during re-entry for maximum accuracy in hitting the landing area.

Suddenly he hears a report; one of the stainless-steel retropack straps is hanging directly in front of the window.
This is Friendship 7. I think the pack just let go. His message goes unheard.

The needles on the indicators begin to move back and forth slightly. Glenn counters with the control stick to keep the capsule from oscillating too much.

Outside the window he sees an orange glow. Its brilliance grows.

Now the orange color intensifies. Suddenly large flaming pieces of metal come rushing back past the window. What can they be? He thought the retropack was gone. For a moment he feels that the capsule itself must be burning and breaking up. Deceleration holds him, pressing him back into the seat, or he would involuntarily rise up, expecting at any moment to feel the first warmth as the heat burns through the capsule to his back. But it doesn't come.

He keeps moving the control handle, damping oscillation. He peers out through the window from the center of a fireball. All around him glows the brilliant orange color. Behind, visible through the center of the window, is a bright yellow circle. He sees that it is the long trail of flowing ablation material from the heat shield, stretching out behind him and flowing together.
04:43:39: *This is Friendship 7. A real fireball outside.*
04:47:22: The fireball is fading. He hears the Cape calling, slightly garbled,
How are you doing? He answers, *Oh, pretty good.*

Glenn splashed down at 04:55:24. Not quite nine and a half minutes later the U.S. Navy destroyer *Noa* began hoisting Glenn and his capsule on board.

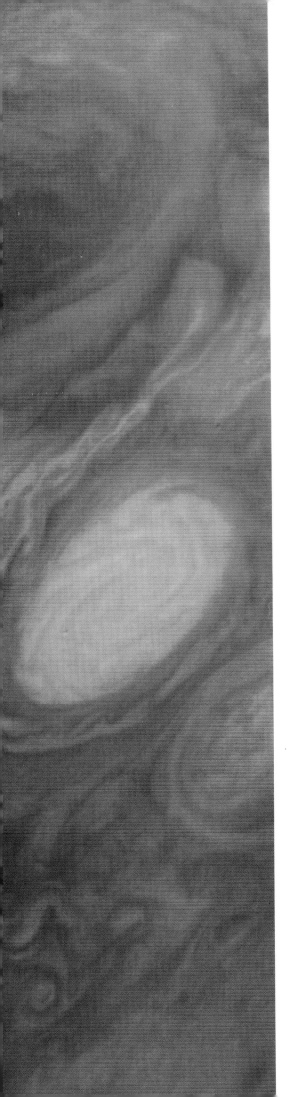

The Mercury program gave way to the Gemini program. The larger two-man Gemini spacecraft proved, among other things, that man could stay in space long enough to go to the moon and back, that a man could work effectively outside his spacecraft, and that the sort of rendezvous and docking techniques a moon landing would require could be carried out.

When the success of the Gemini program cleared the way for Apollo and the first landing on the moon, the National Geographic was there to provide more striking coverage. "'A Most Fantastic Voyage': The Story of Apollo 8's Rendezvous with the Moon," by Lt. Gen. Sam C. Phillips, appeared with Ken Weaver's "And Now to Touch the Moon's Forbidding Face," in May 1969.

Man's first steps on the moon, "First Explorers on the Moon: The Incredible Story of Apollo 11" took up sixty-two pages of the December 1969 Magazine—an issue that included a record containing sounds of the space age, accompanied by Astronaut Frank Borman's narration.

Understandably, the stories of the historic moon walk overshadowed the launches that had gone before. But consider Apollo 8—launched from Cape Kennedy on December 21, 1968, seven months before Armstrong and Aldrin of Apollo 11 stepped out onto the moon—and the risk that mission entailed.

The Apollo space capsule in which Astronauts Frank Borman, James A. Lovell, Jr., and William A. Anders were launched sat atop a giant 364-foot-high, 6.2-million-pound Saturn V rocket, *which had not yet been used for a manned flight*. The flight demanded such absolute accuracy of navigation that Apollo 8 would miss the moon by only 80 miles after a launch that took place 230,000 miles away. The capsule containing Borman, Lovell, and Anders was then required to orbit the moon ten times before returning. And although Apollo 8's voyage might, as NASA's safety chief Jerry Lederer pointed out, pose fewer unknowns than had Christopher Columbus', the mission still "would involve risks of great magnitude and probably risks that have not been foreseen. Apollo 8 has 5,600,000 parts and one and one-half million systems, subsystems, and assemblies. Even if all functioned with 99.9% reliability, we could expect fifty-six hundred defects."

But, as Apollo 11 Command Module pilot and National Geographic Society Trustee Michael Collins wrote in his book *Carrying the Fire*, what really made Apollo 8 so different from all previous spaceflights was that for the first time man

> ...was going to propel himself past escape velocity, breaking the clutch of our earth's gravitational field and coasting into outer space as he had never done before. After TLI [translunar injection] there would be three men in the solar system who would have to be counted apart from all the other billions, three who were in a different place, whose motion obeyed different rules, and whose habitat had to be considered a separate planet. The three could examine the earth and the earth could examine them, and each would see the other for the first time. This the people in Mission Control knew; yet there were no immortal words on the wall proclaiming the fact, only a thin green line, representing Apollo 8 climbing, speeding, vanishing—leaving us stranded behind on this planet, awed by the fact that we humans had finally had an option to stay or to leave—and had chosen to leave.

◄ ───

Jupiter's Great Red Spot—at upper right in this photograph of the planet's highly reflective cloud cover—a persistent feature of its atmosphere, is an anticyclone large enough to swallow two earths.

373

The Apollo program ended in December 1972 with the return of Apollo 17 and the last of the twelve American astronauts who had walked on the moon and brought back 843 pounds of rocks. Even while the Apollo program was underway unmanned spacecraft had been sent out to explore the planets. Mariner 9, the first spacecraft to survey the planet Mars from orbit had been launched in May 1971 and revealed a Martian landscape cut by a 20,000-foot-deep, up to 150-mile-wide equatorial canyon stretching nearly a fifth of the way around the red planet, and massive volcanoes—among them the nearly 15-mile-high *Olympus Mons,* which towers more than 50,000 feet above Everest. Pioneer 10, the first spacecraft to visit the outer planets, was launched in March 1972. In January 1977 the Magazine's cover was a photograph taken from a Viking lander *on Mars!* Mariner 10's voyages to Mercury and Venus were chronicled in June 1975. Four and a half years later, the January 1980 Magazine contained the Voyager spacecraft's detailed photographs of "Jupiter's Dazzling Realm"; July 1981 carried Voyager's photographs of Saturn; and the August 1986 issue, "Uranus: Voyager Visits a Dark Planet."

As manned and unmanned spacecraft reach farther and farther out, mankind's knowledge of the universe expands with them. But there will always be those who would suggest that the lives lost and the monies spent in exploration might have been better used. Journalist and explorer Walter Wellman's response to those who questioned a previous century's efforts to attain an equally debatable goal, is as relevant to manned space flight to Mars today as it was to that former "most cherished of all geographical prizes." In the December 1899 *National Geographic Magazine* Wellman wrote:

> We may have differences of opinion as to the value of reaching the Pole. If we apply the utilitarian test, it is of small moment; but so is a poem. And what is polar exploration but an epic of endeavor, in which all sordidness is left behind, and in which a man, knowing the risks and the chances of failure, ventures his life and his all in a combat against the forces of ignorance? For I deem it beneath the dignity of man, having once set out to reach that mathematical point which marks the northern termination of the axis of our earth, which stands as a sign of his failure to dominate those millions of square miles of unknown country, to give it up because the night is dark and the road is long. He will not give it up. The polar explorer typifies that outdoor spirit of the race which has led conquering man across all seas and through all lands, of that thirst for knowing all that is to be known, which has led him to the depths of the ocean, to the tops of mountains, to dig in musty caves, to analyze the rays of light from distant worlds, to delve in the geologic records of past times. It will carry him to the North Pole, too, and that before many years shall have passed. Any one who supposes anything else of man doesn't know man. His acquaintance with human nature— with the nature of the adventurous races of our zone and times—is limited.

Anyone who supposes that the National Geographic Society will not continue to report man's explorations of space with the same attention it has paid exploration of our own planet for the past one hundred years, doesn't know the Geographic.

James Herzat's painting from the January 1985 article "The Planets: Between Fire and Ice" depicts the now-cancelled Galileo probe, sched- *uled to have orbited Jupiter in 1988 where it would have released scientific instruments designed to help understand that planet's atmosphere.*

"Holding Up the Torch, Not Applying It"

"The Geographic's way is to hold up the torch, not to apply it."

—Frederick G. Vosburgh, Editor, *National Geographic Magazine*, 1967–1970

"He is one of the great editors of our time. No magazine has ever had a steadier, more reliable helmsman than Ted Vosburgh. This ship will miss his hand on the tiller, although I know the crew he has gotten together will do a great job."

—Melville Bell Grosvenor, on Ted Vosburgh's retirement in 1970

On August 1, 1967, ten years after assuming the positions of Editor of the Magazine and President of the Society, Melville Bell Grosvenor retired from both offices to become Chairman of the Board of Trustees and accepted appointment to the newly created position of Editor-in-Chief.

Despite MBG's opposition, the offices of Editor and President, which had been combined at Gilbert H. Grosvenor's insistence in 1920, were separated again—this time at the insistence of the Trustees, who felt that the organization had grown too big to be run by one man. Splitting the two offices, they believed, would enable the Editor to concentrate on running the Magazine and the President to direct his attention to the business of the Society.

Dr. Melvin M. Payne was elected Society President. He had started thirty-five years earlier as a secretary to the Society's then Assistant Secretary Thomas McKnew and had worked his way up through the ranks.

Frederick G. (Ted) Vosburgh, a Vice President and Associate Editor, the number two man under MBG on the Magazine, was appointed Editor—only its eighth since the Society's founding and the fourth in this century.

Robert E. Doyle, a Vice President and Associate Secretary, became Secretary, and Gilbert Melville Grosvenor was moved up from his former position as a Vice President and Senior Assistant Editor of the Magazine to become, along with Franc Shor, one of the Magazine's two Associate Editors.

At a staff meeting held before the formal appointment, Melville Grosvenor explained that separating the offices of President and Editor would "give others up and down the line a chance to grow and develop, and will help keep the Society alive and vibrant and in the hands of young, imaginative people, as it should be."

But Ted Vosburgh, into whose hands the Magazine had been placed, was only three years younger than Melville.

Vosburgh had joined the *Geographic* in 1933, spent eighteen years on the staff before becoming an Assistant Editor, and another five before becoming

Preceding spread: *"The seal of the Society shall consist of... the western hemisphere, from 0° to 180° west from Greenwich, with the legend 'National Geographic Society' above and 'Incorporated A.D. 1888' below, as in the design herewith." —From Article IX of the by-laws of the National Geographic Society, adopted April 17, 1896.*

The bronze seal shown here is approximately seven feet in diameter and gleams in the marble floor of Explorers Hall. It and its twin were laid at the main entrances to the Society's Edward Durell Stone–designed headquarters in 1964 when the building was completed.

"Pointing out deadlines, Vice President and Associate Editor Gilbert M. Grosvenor briefs the [Executive Editorial] council [in the Control Center] on the status of forthcoming issues."—From the National Geographic October 1967 issue

"Stories were being scheduled in and out all the time and day to day and nobody knew who was responsible for what," Gilbert M. Grosvenor recalls. "Stories would go off to the engraver months late that had been scheduled months before…I felt very strongly that the Magazine was going through utter chaos. The Control Center was my idea. I built it in my basement and I ran it." Once Gil Grosvenor had the endorsements of Editor Vosburgh and Editor Emeritus Melville Grosvenor, the Control Center—unpopular in the beginning with the staff—"became a fait accompli."

Senior Assistant Editor in 1956. The following year, Melville's first act upon being named Editor had been to appoint Vosburgh his Associate Editor.

When Vosburgh became Editor ten years later, in 1967, he was already sixty-three years old. His reputation among staff members was for inflexibility and conservatism, rather than for imagination.

Melville was the one with the imagination, the big ideas, the broad strokes. Vosburgh was responsible for keeping MBG's excesses in check. "MBG was the first to say he could not have run the Magazine without Ted Vosburgh," Melville's son Gil is quick to point out. "Ted was his right-hand man. When MBG took off for a couple of months at a time, it was FGV who stayed home and picked up the pieces and put the Magazine together and made damn sure the Magazine was accurate and factual and came out—if not on time, at least that it came out."

Vosburgh was the "stickler for accuracy," the man who would remain late night after night working over the proofs. On one occasion, while the July 1964 issue was being printed, Vosburgh had the presses stopped when he discovered the engravers had eliminated a restrictive comma from a sentence in Peter T. White's "The World in New York City." At a cost of $30,000, Vosburgh had the comma restored and with it the *Geographic*'s reputation for accuracy.*

Vosburgh's caution as an Associate Editor was not necessarily an advantage as Editor. Although now Senior Associate Editor Joseph Judge observed that "under Vosburgh words like 'threatened,' 'polluted,' and 'imperiled' began appearing in the Magazine's titles with regularity," Vosburgh had to be talked into doing those stories. "He would have been the one to avoid that sort of topic very carefully," Edwards Park explained, "because showing the bad side of things, he felt, wasn't our job. Our job was to show the good side of things."

*The sentence had read: "Just the two boroughs on Long Island, Brooklyn and Queens, house 4,660,000 people more than either Chicago or Los Angeles." The $30,000 comma was inserted after "people."

Ted Park, who had come to the *Geographic* from newspaper work, and left to become Editor of *Smithsonian* magazine, remembers Vosburgh fondly. "I'd been on the Features desk...of the old Boston *Globe*," Park recalled, "and Vosburgh liked something I'd tried for the *Geographic*. He didn't use it, but he liked my style. He was very sort of old-fashioned about the way he told you, 'That's a rattling good story!' He used wonderful words right out of the 1920s—in fact, everything Ted did was right out of the '20s.

"Vosburgh was even nuts about birds, just as old Gilbert Grosvenor had been," Park continued. "We had Alexander Wetmore from the Smithsonian as chief author on our bird book, and Wetmore had come over to our office to look at some layouts that Merle Severy had put together for him. Merle showed him this perfectly beautiful picture of a full moon with one bird silhouetted against it—a terrific picture. Wetmore glanced at it, just this tiny speck against the moon and said, 'That's a brant goose.'

"There was this silence," Park said, "and we all just looked at Wetmore in awe. Then Vosburgh, tongue-in-cheek, asked, 'Male or female?'"

Another favorite story, recalled by Assistant Editor Bart McDowell, has Vosburgh telling a writer, "Can't we brighten up this lead with a few statistics?"

When Ted Vosburgh retired in 1970 after nearly thirty-seven years with the *Geographic*, including a three-year tenure as Editor, Society President Melvin Payne praised him saying, "The *National Geographic* has lost the services of a man whose talents were particularly suited to its needs, whose voice has been heard and heeded on every major editorial decision in the past 20 years. The *Geographic* during the Vosburgh years came closer to absolute accuracy than any other publication I know, and I think the Society's growth in members over this period from 900,000 to 6,900,000 is in considerable part due to the fact that Ted helped give them a magazine they could not only enjoy, but trust as well."

Ted Vosburgh was the *first* Editor of the *Geographic* who was not—as one of the staff described each of his predecessors—"a charming, wonderful amateur." Vosburgh was an *editor* and a *professional*. He did not interfere with the photographic side of the Magazine; instead, he concentrated on the editorial side and left Herb Wilburn, Chief of Illustrations; Robert Gilka, Director of Photography; Gilbert M. Grosvenor, an Associate Editor; and Wilbur E. Garrett, an Assistant Editor, to dictate the tone and style of the photography.

The result was that pictorially, at least, the Magazine began to portray some of the realities of a world that, during Vosburgh's leadership, included growing domestic and international protests against American pursuit of an unpopular war in Vietnam, the assassinations of Robert F. Kennedy and Martin Luther King, Jr., the Six Day War between Israel and Arab nations, the Soviet Union's invasion of Czechoslovakia, the capture by the North Koreans of the U.S. Navy's electronic intelligence-gathering ship *Pueblo*, the horrors of starvation in Biafra, the My Lai massacre trial of Lt. William Calley, and President Richard M. Nixon's expansion of the war into Cambodia.

If the Magazine still stepped gingerly around controversial issues, it at least began to recognize that controversial issues did exist. One such issue was the damage being done to the environment; and the *Geographic*'s reporting on it in "Our Ecological Crisis" for its December 1970 issue was, for both the conservative Magazine and its conservative Editor, a major turning point.

Gilbert M. Grosvenor interviewing Marshal Tito, with interpreter Mrs. Maté Meštrović, for the February 1962 article "Yugoslavia's Window on the Adriatic."

"I had waited three weeks for my audience with Marshal Tito, President of the Federal People's Republic of Yugoslavia—and an avid amateur photographer. Knowing this, I had taken along far more photographic equipment than I needed. My plan was to soften up the Communist leader with small talk about photography and then bore in with pointed questions about the Yugoslavian economy.

"We met at his island hideaway, where he received me graciously but sidestepped my questions. I then decided to try my photography maneuver.

"I opened my aluminum camera case, and my gunstock tripod fell out. Burly security men materialized all around me. An aide leaped in front of Tito to shield him. Tito, seeing the cause of the panic, roared with laughter. 'You almost ended badly,' he said."—From Gilbert M. Grosvenor's introduction to Images of the World.

More than twenty-five years later, Gil Grosvenor still remembers that incident with Tito's secret service clearly. "When...the gunstock tripod fell out....They came out of the trees and knocked me down," he said. "I was flattened!*"*

Publication of that piece marked the Magazine's return to journalistic advocacy, a practice it had all but abandoned after 1916, when the National Geographic Society guided by Gilbert H. Grosvenor, responded to Congress' request for additional financial support with a significant monetary contribution to ensure the survival of California's majestic redwoods in the Giant Forest of Sequoia National Park.

Vosburgh's philosophy, expressed in his phrase "The *Geographic*'s way is to hold up the torch, not to apply it," was to present the facts about a problem and to let the readership decide what, if anything, to do about it. In his October 1967 "Threatened Glories of Everglades National Park," for example, he toured the drought-stricken Everglades, pointing out without editorializing that the diversion of Lake Okeechobee's waters, which previously had flowed southward into the Everglades en route to the sea, posed a threat to the Everglades' ecosystem. Nowhere in the piece did Vosburgh suggest the solution; but because of the way he had presented his facts, he didn't have to.

Frederick G. Vosburgh retired as Editor in October 1970. He was succeeded by thirty-nine-year-old Gilbert Melville Grosvenor.

Like John Oliver La Gorce's appointment, Ted Vosburgh's tenure was perceived by many as merely an interregnum between Grosvenors. But despite the fact that Gil Grosvenor's father, Melville Bell Grosvenor, his grandfather Gilbert Hovey Grosvenor, his great-grandfather Alexander Graham Bell, and his great-great-grandfather Gardiner Greene Hubbard had led the Society before him, there is nothing in the Society's charter that says a Grosvenor has to head—or even be hired by—the National Geographic.

Gil Grosvenor's ascendancy was less a *fait accompli* than one might think. As one senior staff member said, "When Melville retired he made Ted Vosburgh Editor and Mel Payne President with, I think, the understanding that Ted would stay on for three or four years and then Gil would be made Editor and that was the way it was going to be. But there was some resentment."

Gilbert M. Grosvenor had never intended to join the *Geographic*; his plans centered on his being a pre-medical student until, during the summer between his junior and senior years at Yale University, he went to the Netherlands as part of an international work force that was rebuilding dikes ravaged by that previous spring's disastrous floods. "I took a camera along," he has said, "and I guess that's when I was bitten by the bug."

By 1954, Gil Grosvenor had graduated from Yale and was hired by John Oliver La Gorce. With the exception of two years' military duty with the Army's Psychological Warfare Service, Gil Grosvenor has worked at the *Geographic* ever since.

During the next fifteen years GMG—as, of course, he would become known—wrote and/or photographed articles on Bali, Ceylon, the Italian Riviera, Copenhagen, Monaco, "Yugoslavia's Window on the Adriatic," the flight of the *Silver Dart*, a reproduction of Canada's first airplane (the original of which had been built by his great-grandfather Alexander Graham Bell and his fellow members of the Aerial Experiment Association), and President Dwight D. Eisenhower's 1960 visit to Asia, for which he won a first prize award in the nineteenth annual News Pictures of the Year competition.

Soon after GMG came to the *Geographic* he commenced a two-year apprenticeship in the administration, serving with the Secretary, the Treasurer, the membership-fulfillment section, and the correspondence office. He then returned to the editorial side of the Magazine with, as then Society President Melvin Payne said, "a real knowledge of how the magazine relates to our millions of readers." However, to Mel Payne mere possession of that knowledge did not mean that Gil Grosvenor was necessarily qualified to become Editor. Twenty years older than Gil Grosvenor, Payne "was working at the *Geographic*, doing responsible work," he said, "while Gil was still in school."

"Payne thought Gil was an upstart...," one Senior Editor said. "I think he resented Gil as a kid who had had it all his way, the 'silver spoon,' every door open to him, and he just walked into the job."

Doors had not opened easily for Melvin Payne. He was born the son of a railroad freight conductor on May 23, 1911, and grew up in a poor section of Washington, D.C. Orphaned at ten, he was reared by an uncle and aunt and contributed to his keep through earnings from a newspaper route and, later, by writing features for the New York *Sun*. Still later, while working during the day, he attended the National University Law School (now part of George Washington University) at night. In 1932, shortly after he had begun working for Thomas McKnew, Payne transferred to Southeastern University, where he continued to attend night school until he had obtained his law degree. In 1958, Payne had been elected to the post of Vice President serving as Associate Secretary in charge of administration; later that same year he was elected to the Board of Trustees. From that point on, Payne's rise in the hierarchy had been swift, and when Melville retired in 1967, Payne was the Society's newly elected President. He had come up the hard way and was very proud of that.

When the Board announced the appointment of Gil Grosvenor as the Magazine's ninth Editor, Mel Payne called Gil into his office. "He told me he was not in favor of it," Gil said, "but that was the way the Board went and he'd

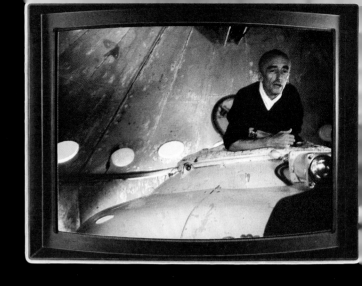

AWARD-WINNING TV

Because of the success of the Society's lecture film *Bones of the Bounty,* shown on NBC's television program *Omnibus* in January 1958, Melville Grosvenor came to believe the Society should be producing its own television shows. And in 1961 a documentary film department was organized under Robert C. Doyle—"T.V. Doyle," as he was known, to distinguish him from then Vice President and Associate Secretary of the Society Robert E. Doyle.

Doyle was given the responsibility of producing four TV documentaries a year; and without any network commitment to show the films, the Society plunged ahead, confident that a market existed for documentaries made to the Geographic's own high standards. But it was not until the Society joined with David Wolper in the mid-1960s that its television productions began to be aired.

The Society's first documentary, *Americans on Everest,* was already in production. Broadcast on September 10, 1965, by CBS, it made television history. Narrated by Orson Welles and sponsored by Encyclopaedia Britannica and Aetna Casualty Insurance, this Geographic Special on the three-team assault of Mt. Everest by Americans—one of whom was National Geographic staffer Barry C. Bishop—received the highest rating of any documentary until then telecast. Although Wolper had nothing to do with the creation of this film, he was given sole credit as producer.

Americans on Everest was followed by *Miss Goodall and the Wild Chimpanzees:*
VIDEO: *Jane Goodall in small boat on Lake Tanganyika. In background, mountains at Gombe in the East African nation of Tanzania (formerly Tanganyika).*
AUDIO: Voice Over (Jane Goodall): "When I first arrived at the

Gombe Stream Reserve, I felt that at long last my childhood ambition was being realized. But when I looked at the wild and rugged mountains where the chimpanzees lived, I knew that my task was not going to be easy."
VIDEO: *Jane in boat. 2 chimps play. Jane observes.*
AUDIO: V.O. (Jane Goodall): "The chimps very gradually came to realize that I was not dangerous after all. I shall never forget the day after about 18 months when, for the first time..."
VIDEO: *Chimp group.*
AUDIO: V.O. (Jane Goodall): "...a small group allowed me to approach and be near them. Finally I had been accepted....it was one of the proudest and most exciting moments of my whole life."
VIDEO: *Jane with notebook near chimps.*
AUDIO: V.O. (Alexander Scourby): "Chimpanzees are as distinct from one another as are human beings, and Jane gave them names..."
VIDEO: *Chimpanzee "Flo."*
AUDIO: V.O. (Alexander Scourby): "...as she came to recognize them. Old Flo, with her bulbous nose and ragged ears, is matriarch of the family Jane would come to know best..."

Opposite: *Top row, left,* Americans on Everest; *right,* The World of Jacques-Yves Cousteau. *Middle row, left,* Rain Forest; *right,* Journey to the Outer Limits. *Bottom row, left,* The Incredible Machine; *right,* The Sharks. Left: Filming the Impossible. Below: Himalayan River Run.

Next came *Dr. Leakey and the Dawn of Man* and *The World of Jacques-Yves Cousteau.* And in 1968 the Society made television history again when its special *Amazon* was watched by over thirty-five million people and became the first television documentary to top all other shows in a two-week rating period.

In 1975 National Geographic Specials entered into partnership with WQED/Pittsburgh, supported by a Gulf Oil Corporation grant, to produce documentary programs for Public Television. Their success can be measured by the statistics: By 1986 nine out of the top thirteen most-watched shows on PBS were National Geographic Specials (with *The Sharks, Land of the Tiger, The Incredible Machine,* and *Great Moments With National Geographic* ranking one, two, three, and four).

FILMS THAT EDUCATE

In the early 1970s the question was, "How do you reach a generation of kids who have grown up with movies and television?" The answer: "With films and videos." And in 1972 the Educational Films Division was born.

The first eight educational films were primarily motion-picture footage from the Society's award-winning television Specials and re-edited for classroom use. But despite highly favorable critical response, sales of the educational films grew slowly—perhaps because they were a new product offered by an organization primarily identified with publishing. However, by the late 1970s, most of the films had become original productions conceived, controlled, and completed by a full-time staff; by then their quality was recognized and sales increased.

With time, the range of Educational Films' subject matter broadened. Ambitious film series on the American Revolution and Shakespeare were produced. There were impressive films in the areas of earth sciences and life sciences. More recently films on economic subjects like capitalism, socialism, Communism, and the stock market have been made. In 1983 a ten-film series on United States geography was released; another ten films on nations of the world soon followed.

Today the list of Educational Films offered has grown from the eight in the 1973 Educational Services Catalog to more than 200.

support me. He was very up-front and honest about it."

Whatever resentment there might have been over Gil Grosvenor's appointment quickly dissipated. Staffers felt released from the "very tight, careful, cautious straitjacket" they felt Ted Vosburgh had put them into. They saw that Gil's strength lay in hiring capable people and in leaving those he had inherited free to do what they did best.

"With Gil as Editor," Senior Assistant Editor William Graves said, "there was a great climate of warmth and tolerance and encouragement....Gil ruled with a much lighter, easier, warmer hand than anybody. He was accessible."

Gilbert M. Grosvenor was not the sort of seat-of-the-pants Editor that his father had been. Shy, sometimes uncomfortable with other people, business-oriented, attentive to detail, dependent upon statistical analysis, GMG was an Editor in many ways more like his grandfather Gilbert Hovey Grosvenor, with whom he had been very close, than his father. He had to be. After Melville's tumultuous decade, Gil Grosvenor's was seen as the calm after the storm. Editor from 1970 to 1980, Gil Grosvenor led the *Geographic* through a period of consolidation, organizational stability, and a careful polling of readership tastes—a management method that caused many of the Magazine's editorial staff to bristle.

"I think it's wonderful to have the attitude that you can be above such things as surveys and testing," Gil has said, "but let me tell you, when you get to the top and you're looking at the report card—and the report card is the renewal rate—statistical analysis is the name of the game."

The *National Geographic* is understandably extremely sensitive to the membership renewal rate. It is what *drives* the Magazine. To replace the loss of one percent of the renewal rate today costs the Society approximately a million and a half dollars.

Magazine subscribers pay in advance. They are buying on faith in the Magazine's reputation and its continuing ability to interest them. And yet, despite what some in the Society perceived as the Editor's caution, his unwillingness to move without statistical support, the Magazine under Gil Grosvenor moved away from the "letter to Aunt Sally" travelogue style of reporting and into articles that both editorially and pictorially reflected the post-Vietnam realities.

Change at the Society never occurs swiftly; senior officers quote the maxim "Evolution, not revolution." So if, as Editor, Gil intended to publish a story he felt was extremely important on a subject that was of only marginal interest to the readers, he would make certain the same issue contained at least two or three articles on proven popular topics. He could not afford to risk losing readers. *Geographic* staples like "I Live With the Eskimos" and "Headhunters of Remotest New Guinea" continued to appear, as did articles on "The Wasp That Plays Cupid to a Fig" and "Shy Monster, the Octopus." But in addition to these conventional pieces, under Gil Grosvenor articles of a sort that had rarely appeared in the Magazine since the end of World War Two were also published: articles like "East Germany: the Struggle to Succeed" that attempted to present an objective look at nations whose political systems were in opposition to that of the United States, and articles such as "Can the World Feed its People?" that, with Steven L. Raymer's photographs of starving people in Bangladesh, the new Editor felt were important and expressive of his social conscience.

"Warmed by the sun, monarchs descend in swarms upon a stretch of soggy ground [in Mexico's Sierra Madre] where they draw water, a signal they are ready to venture north. There, flower nectar and water will be their diet. Monarchs, still sluggish [from near-freezing temperatures], will cling anywhere, as on the coat, hat, and face of guide Juan Sanchez.

"As days lengthen, the butterflies also begin to mate. Then they depart, flying solo at speeds ranging from 10 to 30 miles an hour."—From "Found at Last: the Monarch's Winter Home" in the August 1976 issue

387

Top: *Third graders at the Janney Elementary School in Washington, D.C., respond to a* World *magazine story idea shown them by Barbara Steinwurtzel. Students' reactions to ideas being tested help shape the magazine. Out of every ten ideas presented to children, only two make the grade.*

In 1985 World *celebrated its tenth anniversary and its pre-eminence in the children's magazine field.*

A WORLD OF FUN AND LEARNING

By 1974, Gilbert M. Grosvenor had two pre-school children of his own and was deeply disturbed by reports of how young people were turning to passive television as a substitute for reading and the interactive learning that reading promotes. He wanted the Society to produce something that might help stem that tide.

"I dreamed up *World* magazine in Baddeck," Gil said. "I saw that kids were ignorant of geography and they were getting hooked on this junk on television. There weren't many good kids' magazines out—most of them were comic books."

He returned to Washington and prototypes of the magazine were developed. Staff members consulted with educators and child psychologists. They studied children's periodicals, children's books, children's television programs, and—most important—they studied the children themselves. Hundreds of ideas for stories, projects, and games were tested on children. After a year and a half of intensive research ended—and once Gil had the Board of Trustees' endorsement—the first magazine was produced by the Special Publications Division. Its editor was Ralph Gray, who had been editor and "The Old Explorer" columnist of *World*'s predecessor, *The School Bulletin*, for so many years.

The first issue of *World*—with its name proclaiming a world of interesting information to youngsters eight and older, with its cover photograph of a girl sliding down a snowbank promising the magazine would be fun, and a bold, fresh look inside to capture the attention of the TV generation—went out to 536,000 subscribers in September 1975.

That premier issue contained stories on an adventure in the High Sierra, on archeologists uncovering the ruins of Pompeii, and on smoke jumpers and snake charmers. Readers looked into odd animal eyes and at chimpanzees in a zoo nursery. In addition to things to read and see, there were things to do: tips on backpacking, a maze in which every fork in the path required a decision based on knowledge of safe outdoorsmanship, and a pull-out supersize page to hang on the wall—a feature that continues to be *World*'s most popular "extra."

World magazine has evolved over the years to reflect the changing tastes of its young readers. "Working with this age keeps the staff on its toes," present editor Pat Robbins says. "Children ask a lot of questions. They are interested in everything, and often they don't see things the way adults do." To challenge children's creativity *World* constantly comes up with new ideas, contests, and quizzes. *World* readers have designed stone gargoyles for Washington Cathedral, in Washington, D.C., saved a redwood grove in California, and assembled their own punch-out globes. But *World* staples continue to be stories about outdoor adventure and fun, animals and conservation, new places and faces, and science and technology.

Today *World*, the most widely read general interest magazine in its readers' age group, brings its special kind of educational fun to more than a million readers.

But in 1977, the evolving social awareness of the Magazine resulted in its carrying articles on Cuba, Harlem, Quebec, and South Africa—pieces that precipitated a direct confrontation between *National Geographic* Editor Gil Grosvenor and conservative members of the Board of Trustees led by Melvin M. Payne, the Society's former President and, as of 1976, the new Chairman of its Board of Trustees.

Black Star writer-photographer Fred Ward's "Cuba Today" appeared in the Magazine's January 1977 issue. By any other magazine's standards, the piece would have been considered an objective assessment of Cuba under Castro's Communist regime. But it angered Trustee Crawford H. Greenewalt, Chairman of the Board of E.I. du Pont de Nemours & Co., because one of Du Pont's chemical plants had been seized by Castro.

The month after Ward's "Cuba Today" appeared, the Magazine published "To Live In Harlem" by two black freelancers, Trinidadian writer Frank Hercules and photographer Leroy Woodson. The Harlem article marked a different sort of milestone: sympathetic reporting on the American black—virtually ignored by the Magazine in the past.

No article on any Southern state during the previous seventy years, for example, had mentioned segregation, lynchings, the Ku Klux Klan, sit-ins, freedom rides, or black poverty and unemployment. In the Magazine's 1949 regional survey, "Dixie Spins the Wheels of Industry," blacks were not mentioned, nor did they appear in any photographs.

"To Live in Harlem" was a basically upbeat piece emphasizing how *livable* Harlem was. Hercules had interviewed Harlem politicians and churchmen, social leaders and musicians; but he had also spoken with a young female heroin addict in the back seat of a police car:

> "I've been a dope addict, a burglar, a stick-up woman—to support my habit—but I've never been a prostitute." Pride elevated her voice.
> "How old are you now?" I asked.
> "Twenty-six going on twenty-seven. And I have four children."
> "How would it be," the police officer inquired with dispassion, "if you were to find yourself up against it—maybe your children were hungry—would you get a gun and...?"
> She smiled, shaking her head. No.
> "And how about the heroin?" he persisted. "Done with that too?"
> "Yes," she answered. And summoning a remnant of a band of angels to do battle with a legion of devils, she said, "I've enrolled at college. My children are with my parents."

The *Geographic*'s New York–based Advertising Division's campaign for the Magazine centered upon the slogan: "The *National Geographic*. It's Time You Took Another Look," and it trumpeted the Harlem article in newspaper ads proclaiming, "The geography of Harlem: Poverty, dope, crime and people who wouldn't leave for a million dollars." To celebrate the article's publication, they then hired a bus to take members of the Magazine and advertising staffs and guests to Harlem for a party.

Gil Grosvenor, as Editor, had nothing to do with the Advertising Division. In theory the Division was the responsibility of the then President of the Society, Robert E. Doyle. But when Gil attended the Harlem party he was criticized by Payne for having shown a lack of judgment.

"The Harlem affair I did object to most strenuously," Payne said later. "They put a big sign on the outside of the bus, 'The Geographic Goes to Har-

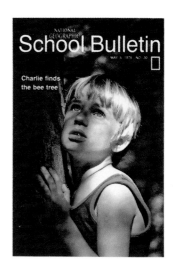

Publication of The National Geographic School Bulletin *commenced in 1919. For years this weekly magazine offering background information on people, places, natural history, and scientific developments was edited by Ralph Gray. Illustrated with maps and photographs and designed specifically for classroom use and home study,* The School Bulletin, *with its small pages, a name associated with the classroom, and a publishing schedule tied to the academic year, couldn't provide the kind of "competition" GMG felt was needed. In 1974, the idea for* World—*a new magazine—was conceived.*

"The U.S. should take the first step toward friendship," Fidel Castro (above) told Fred Ward, author of the controversial January 1977 "Inside Cuba Today" article. "We are certainly interested in improving relations. Since the U.S. practices the cold war with only a few small countries... the moral seems to be, it is a problem of size."

Publication of Ward's article distressed many conservative Trustees and members.

lem,' like a church picnic going down to the beach for the day. It just didn't fit the Geographic mode in terms of our past. It was in very poor taste."

On March 8, 1977, the National Geographic Special *The Volga* aired on television. It later won both the Television Critics Circle Award and a Council on International Non-Theatrical Events (CINE) Golden Eagle Award. It also attracted the attention of Accuracy in Media (AIM), which in its newsletter that month published "The Geographic's Flawed Picture," an article attacking both the Cuba piece and the *Volga* TV show:

> The National Geographic Society is famous for its accurate maps and lovely photographs. However, it now seems to be in danger of acquiring a new reputation—that of purveyor of inaccurate and distorted information about Communist countries such as Cuba and the U.S.S.R....

AIM took exception to *The Volga* because actor E. G. Marshall, who narrated the special, had pointed out that there had been no censorship of the film they had taken. AIM complained, "There was no need for that. Like Ward's Cuba pictures, what was shown was virtually all flattering."

The Accuracy in Media piece went on to quote excerpts from the six-page single-spaced letter AIM chairman Reed Irvine had sent Gil Grosvenor "citing numerous serious errors and omissions in the Ward article" and stated that Irvine had "asked that appropriate corrections be made by The National Geographic Magazine."

The corrections were not forthcoming, but confrontation with conservative Society Board members was.

At the Trustees meeting held on March 19, 1977, a divisive split made its first appearance before the Board—a split so serious that reports of it eventually escaped the boardroom's secrecy-cloaked deliberations and ended up on the front pages of newspapers across the land.

As the Trustees* assembled, Chairman of the Board Melvin M. Payne, upset by controversy created by the Cuba and Harlem articles, suggested that a committee be established to examine whether or not the editorial policy of the Magazine was changing direction.

Gil Grosvenor perceived Payne's proposed editorial policy committee as nothing more than an "oversight committee to throttle the Editor," a means by which the Chairman could gain editorial control. And he swiftly responded by insisting before the assembled Board of Trustees that the editorial policy had not deviated from its "89-year tradition of factual accuracy, timeliness, objectivity."

"The *Magazine* is not changing, the *world* is changing," Gil said, adding that MBG and Ted Vosburgh had both recognized that the Magazine would be doomed if it did not reflect the changing times. That was why, during their terms as Editor, sometimes controversial and critical articles had appeared on the Berlin Wall, the Berlin Airlift, the Hungarian Revolution, the Soviet Union, Vietnam, and South Africa—topics which, Gil said, under his grandfather's guiding principles would have been avoided.

"No great magazine has ever been created or maintained by a small committee," Gil insisted, adding that although he welcomed and relied upon "the Board's *collective* guidance and wisdom," he would be "terribly disappointed" if any of the Board were willing to delegate that responsibility to anyone else.

Dr. Payne reiterated his feeling that because of the concern expressed by several Trustees as to the direction in which the Magazine was going, there did need to be a discussion with the Board.

Former Editor Ted Vosburgh was so offended at the suggestion of any such editorial policy committee that to demonstrate support for Gil Grosvenor he proposed the motion "That the Board hereby expresses its confidence in the Editor and commends him."

Vosburgh's motion was voted and unanimously accepted.

For the time being, the idea of an "oversight committee" had been shunted aside.

The Magazine's April issue, the following month, contained Peter T. White's "One Canada or Two," an examination of the Quebec separatist movement. That article prompted a letter to Mel Payne from David Lank, the son of a former President of the Du Pont Company of Canada, an old friend of Trustee Crawford H. Greenewalt's. David Lank, a Canadian citizen and Society member, wrote that the piece was outrageous, that there was no such problem in Quebec, and that it was just going to stir people up over nothing.

On April 18, Trustee Greenewalt—already upset over the Castro-Cuba piece—wrote to Mel Payne:

> ...There seems to have been a rash of adverse comment on recent articles in the GEOGRAPHIC, particularly those pertaining to Cuba, the Volga, and the Quebec separatism question....I wonder...if it might not be timely to consider articles of this kind as a matter of policy....
>
> I'm inclined to believe that a rather full discussion at the Board level might be useful, and perhaps this could be based upon a written presentation by Gil Grosvenor or whoever else you think appropriate....

Greenewalt's letter gave Payne the excuse he needed at the May 12 Executive Committee meeting to raise the editorial policy committee issue again. As a result of the various complaints, Payne stated he felt it appropriate to appoint an ad hoc committee to study the controversy, review the mail the Society had received, discuss the problem with the senior officers for background, and prepare a report for the Board.

And then, when on the heels of the Cuba, Harlem, and Quebec articles, the Magazine's June issue containing "South Africa's Lonely Ordeal" angered not only the South African government, but conservative readers and some of the older Trustees as well, it must have seemed to Mel Payne to be the last straw.

"South Africa's Lonely Ordeal" contained a text by Senior Editorial Staff Writer William S. Ellis that devoted considerable attention to the impact of apartheid on blacks; it was accompanied by James P. Blair's unsentimental photographs of a black child's pathetic grave marked only by the infant's cherished white doll and a wood-and-cardboard cross, and of red-and-white gingham-clothed black maids holding a wealthy white couple's nightgowned children.

The South African Department of Information took out a large advertisement headlined "Here are some of the facts on South Africa that *National Geographic Magazine* withheld in its maiden attempt to enter the realm of advocacy reporting" and accused the *Geographic* of "anti-white racism" and serious omissions of fact.

At the June 9 Board of Trustees meeting, Mel Payne drew the Board's attention to the advertisement, reviewed the criticism generated by AIM's campaign, which had resulted in "an unusually large volume of critical mail" and "a degree of uneasiness among several Trustees," and concluded that because of these and other factors it had "seemed appropriate to take some action at the Board level to determine whether or not the charges are justified." That is why, Payne continued, at the Executive Committee meeting he had appointed an ad hoc committee "to review the situation."

Mel Payne asked for the Board's approval of the ad hoc committee, and Thomas W. McKnew moved that: The Chairman's action in appointing an ad hoc committee to study criticisms directed toward our publication of articles on Cuba and South Africa (as well as our film on the Volga River) be approved. McKnew's motion was brought to a vote and approved by a majority of the Board. Melville Grosvenor, Gil, and Ted Vosburgh clearly were not pleased.

"We *knew* we would have trouble over the South Africa article," Melville said impatiently. "We should not have to come to the Board every time there is an article like this. It is the re-spon-si-bi-lity of the *Editor*," he said, angrily tapping the top of the Board table before him. "If the Editor does not do his job, the Board can fire him. I don't believe *I* would serve as Editor with a committee like this looking over my shoulder...."

"The Magazine continues to be beautiful and interesting and informative," Mel Payne said, "*but* the question raised by all this furor is whether or not there is creeping into the Magazine some element that is *not properly a part of the National Geographic Society's mission*."

The first ad hoc committee meeting was scheduled for a little over two weeks later, on June 27 at 10 in the morning. By this time morale among the Magazine staff was low. Decisions were not being made. Work was not getting done. One Senior Editor complained that Gil was postponing everything "to keep a low profile."

But if the Magazine was in a turmoil, the rest of the Society was functioning normally.

Its 78,000-volume reference library remained open to the public. Its lectures and films continued to be well attended. Its Educational Media division, headed by Robert Breeden, was sending its filmstrips, multimedia learning kits, posters, and games to the schools. *World* magazine, Gil's updated version of the *School Bulletin* continued to be eagerly read by children.

Book Service, now under Jules B. Billard, was working on *Ancient Egypt*; a revision of *Everyday Life in Bible Times* was being published, as was *Visiting Our Past: America's Historylands*. Robert Breeden's Special Publications Division was providing books on subjects ranging from *The Mysterious Maya* to *Nature's Healing Arts*.

In the meantime, the Television Division was busy with *The Legacy of*

WORDSMITH   **By Tim Menees**

Tim Menees' June 3, 1976, "Wordsmith" comic strip lampooning a National Geographic *assignment on Harlem anticipated by eight months Frank Hercules'* "To Live in Harlem…" *article that appeared in the February 1977 issue on the heels of the Cuba piece. The above photograph from that article was captioned: "Outsiders may call it a ghetto, yet to half a million Harlemites it is home, the best-known black community in the Western world. In the heart of New York City, children play on West 138th Street, where traffic is banned for a Memorial Day block party."*

The Harlem piece also worried many conservatives.

L.S.B. Leakey. Two years before, the Division had switched from the commercial networks to Public Television; and their PBS premiere, *The Incredible Machine,* had earned the highest audience ratings in Public Television history and was the first Public Television show to score higher ratings than the commercial networks in some major markets.

The Geographic's News Service was continuing to supply feature stories relating to specific Magazine articles and discoveries of Society research grantees to well over a thousand newspapers and periodicals.

"Sign of the times: Paint sprayed by Francophones blots English from a bilingual stop sign in front of a church in Sainte-Anne-de-la-Pérade. Bilingual labels, initiated to ease language tensions, satisfy neither English- nor French-speaking Canadians. Both prefer their own language exclusively. French culture, which once rallied around the Roman Catholic Church and rural life, has survived today's urbanization and weakening of religious influence. Calls for independence continue to mount."—From Peter T. White's article "One Canada—or Two?" in the April 1977 issue

This, too, was perceived as dipping into political waters.

But despite the fact that the tempest brewing within the Magazine over the ad hoc committee had little impact on the day-to-day operations of the Society as a whole, it is still important for several reasons.

First of all, the Magazine is the Society's flagship; to its membership the Society *is* the familiar yellow-bordered Magazine. As Melvin Payne observed to his fellow Trustees, "We are known throughout the world by our Magazine, and what the Magazine says is what the National Geographic Society says."

Second, just as "Black Friday" twenty years earlier had marked a revolution within the Illustrations Division that set the photographers and picture editors free from certain constraints, the struggle for editorial integrity brought on by the Chairman of the Board's ad hoc committee served to ensure the freedom of the Magazine from outside interference.

Third, the story of the ad hoc committee provides a rare insight not only into the operations and relationship of the Society and its Trustees, but also into how the Society perceives itself.

At 10 o'clock in the morning on June 27, 1977, the committee met. The group consisted of its chairman, Lloyd H. Elliott, Melvin M. Payne, Robert E. Doyle, and Gilbert M. Grosvenor; Trustees Crawford H. Greenewalt, James E. Webb, and Louis B. Wright; and Geographic Associate Secretaries Leonard J. Grant and Edwin W. Snider.

Elliott called the meeting to order and stated, "The purpose of our committee, is 'to investigate the question of whether there has been any significant change in the editorial policy of the Society's official journal, and, if so, whether such changes meet with the Board's approval.' "

Payne reviewed why, because of the AIM letters, the threatened resignation of members, the "relentless" efforts of the South African Embassy to embarrass the Society, he had "thought it was essential to have the Board look at the criticisms."

Gil Grosvenor responded by dismissing the AIM director as a right-winger whose financial support came from an industry that had "never forgiven us for not publicly condemning our statement about DDT and its effect on birds' eggs" (in the March 1974 article "Can the Cooper's Hawk Survive?").

Gil also discounted the significance of the letters and the South African government's advertisements and said, "You cannot run a 'hearts and flowers' story from South Africa when you get daily television news about Soweto." Payne noted that the Cuba piece had produced 105 letters, whereas an average story such as the one on Mars had produced only 30. "The point is," Gil said, "we are trying to talk about the significance of the level of criticism, where it is coming from and what is motivating it."

Elliott interrupted. "The point we are considering is, 'has there been a change in editorial content, policy, or subject matter?'... The point is whether the criticism reflects any change."

"*We* don't claim that we are changing," Gil said. "The *advertising* department said in some ads that we were changing."

"In your statement prepared for the Board, you said the Magazine has to change," Elliott said.

Gil pointed out that he was talking about "evolutionary" change in layout and so forth. "It is not a change in subject matter."

Former NASA administrator James Webb asked about the genesis of the South Africa article.

"We had not published anything on it in many years," Gil said. "We felt it was time to publish a 'country story' on South Africa...it was important for our members to see an article on South Africa. We did not go in with the idea of doing a hatchet job on South Africa, but a balanced article. If we missed, we missed. If it is policy of the Board, we can cloud over political questions in a country."

"My objection," Louis Wright said, "was that [the author of the South Africa article] did not take the historical point of view but rather the contemporary American view. It is the 'missionary' instinct which is difficult for Americans to overcome.... To sum up, the desirable thing for the *Geographic* is to do what it can do best. State precisely what the current situation is, but you don't take sides; you don't even quote people on either side....The traditional *Geographic* position is historic background, geographical discussion, scientific analysis of discoveries...."

James Webb asked if it was felt that there was a need for an oversight committee.

"The Editor should be able to control the Magazine," Gil said firmly. "You cannot have a committee run editorial content. Either you like the way we have been running it or you don't; then you say change it, or if you don't, get a new Editor."

"We have been under fire from organizations that would like to see us taxed," Wright cautioned. "That is a problem if you go political."

"I don't want to go political," Gil said. "There is no input from me or anybody on this staff to go political."

"I agree with Gil that you can't have an editorial committee breathing down the neck of the Editor," Wright said. "That is not the purpose of this committee or any other committee. On the other hand, the Trustees have a responsibility to see that the policy is one that is not changed unwittingly. That is, I think, what this is all about, the fear there is unwitting change in editorial policy which the Trustees have a right to be concerned about and pass judgment on."

Suddenly, to Gil's surprise, Crawford Greenewalt said, "...it is untenable that the Board should have a committee to exercise any continuing control over the contents of a magazine. You ask the Editor to run the show, and if there is a particular situation on which the Board can give advice, it should. My feeling is...the Geographic should stay away from [those areas that are so emotionally charged] that an objective piece cannot be written....I gather Gil is not too upset with that kind of amendment to existing policy. Let him write it up and come to us to hammer it out, and then we have no problem with the Board."

Elliott agreed, "...give us that statement about a week before the next meeting so that we can digest it a bit and have suggestions and then put a piece of paper together around the table."

The second meeting of the ad hoc committee was set for Thursday, August 11, at 2:00 P.M. after the Executive Committee meeting and luncheon.

One month later reports of the internal struggle began to creep into public print. "CHANGE AT THE GEOGRAPHIC" was the front-page headline in the July 17 *Washington Post*. The subhead was "Magazine's New Look Stirs Dispute." The story carried over into the inside of the newspaper's first section where it was considered important enough to be given half a page with the head "Geographic Faces Problem of Portraying Harsh Reality." Wrote *Post* staff writer Kathy Sawyer:

The dispute goes to the basic question of how boldly the society, with its tradition of Victorian delicacy, will pick its way through a global landscape where life's harsher realities, as one staff writer put it, often "leap right out at you."

Gil Grosvenor, according to Kathy Sawyer, saw it differently:

In his book-lined office last week, Grosvenor talked with the air of a man caught in a delicate web he was loath to tear. He minimized the so-called "public reaction" noting that the "40 or 50 letters we get on a sensitive story" compared with the magazine's total readership are like walking once around the Geographic headquarters on 17th Street "and then walking to the moon."

In a nice touch, the reporter then added: "With typical Geographic attention to detail, [Grosvenor] had the society's research department check that out and later amended it to 'twice around the headquarters.' "

The following month, on August 5, the *Los Angeles Times* published a long piece headlined "Geographic: From Upbeat to Realism," which contained the following observation: "Even more startling [than the new candor of the *Geographic*'s articles] is the unseemly squabble within the genteel Geographic itself...."

The second meeting of the ad hoc committee took place not quite a week after the *Times* piece appeared.

Gil Grosvenor passed around the table the draft editorial policy statement the Committee had requested at the last meeting.

The committee members argued a bit over the wording, then Gil excused himself from the meeting to incorporate their suggestions. Together with then Senior Assistant Editor Joe Judge, Gil hammered out a revised version in about fifteen minutes, then returned to the committee to read the statement now titled "Reaffirmation of Editorial Policy at the National Geographic Society":

The mission of the Magazine is to increase and diffuse geographic knowledge. Geography is defined in a broad sense: the description of land, sea and universe; the interrelationship of man with the flora and fauna of earth; and the historical, cultural, scientific, governmental and social background of people.

The Magazine strives to present timely, accurate, factual, objective material in an unbiased presentation. Advocacy journalism is rejected.

As times and tastes change, the Magazine slowly evolves its style, format, and subject matter to reflect that change without altering the fundamental policies above.

Excellence of presentation—accuracy, technical superiority in printing and photo reproduction, and clarity of meaning—remain traditional goals against which each article is measured.

Mel Payne felt that the statement was so brief that he was not sure it adequately covered the subject.

William McChesney Martin disagreed; he thought the advantage of a short statement was that it left the Editor more flexibility.

After a brief discussion the committee agreed that the statement should be published in the Magazine.

A motion to accept tentatively the "Reaffirmation of Editorial Policy" as worked out by Gil Grosvenor was adopted without objection; it was, however, to be brought for final approval to the third and final meeting of the ad hoc committee scheduled for 11 o'clock in the morning before the September 8 meeting of the Board of Trustees. "It is my understanding," Committee Chair-

man Lloyd Elliott said, "that with our report the committee will recommend its own demise."

The meeting adjourned at 3:50 P.M.

Advance copies of the September 12, 1977, issue of *Newsweek* were already on newsstands the day before the final ad hoc committee meeting was held. Inside was an article titled "The Geographic Faces Life," with photographs of Gil Grosvenor and the controversial Harlem advertisement.

"In any other magazine," wrote *Newsweek*, "the articles on apartheid in South Africa, Cuba under Castro and life in Harlem would be considered tame—if not belated—attempts to report the issues of the day....But in a publication founded on the principle of avoiding matters of a 'partisan and controversial' nature, the stories have produced an unprecedented dispute over the Geographic's editorial direction between the magazine's earnest, socially conscious editor and its archconservative chairman of the board."

At 11 o'clock, Thursday morning, September 8,1977, the final ad hoc committee meeting was called to order.

Gil Grosvenor passed out the reworked statement along with a proposed Editor's page commenting on editorial policy. He said he hoped our members would understand it and that the press would not pick up on it.

Mel Payne stated that all the newspaper publicity had emphasized and distorted "the so-called schism." To establish the fact that the schism did not exist, to show that there was unanimity in the organization, he insisted "the

"A life apart is endured by millions of black women, such as these in a Transkei village. The only man in the house is their retired father, left; their miner husbands live far away. The men send home a meager $12 a month. When they return—once a year—to the family hearth, they often find sadness as well as warmth...."

Following articles on Cuba, Harlem, and Quebec, William S. Ellis' June 1977 article, "South Africa's Lonely Ordeal," containing the above photograph and caption, proved the last straw for those conservative Trustees already concerned over the Magazine's direction under Editor Gilbert M. Grosvenor.

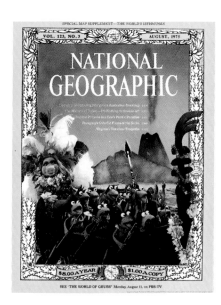

"IMAGINE BEING TOLD of a handful of lush tropical islands, peopled by handsome and hospitable natives, and ruled over by a wise and beautiful queen," begins the text for this 1981 "Miss Piggy Cover Girl Fantasy Calendar" parody of the Geographic. "Surely, you would protest, such places don't exist outside of fairy tales! And yet, as visitors to the verdant Pacific islands of Boara-Boara, Porku-Porku, and Rana-Kermi discover, this magical place does exist. And so does its storybook monarch, Queen Wahinipikki I, a lovely and sophisticated graduate of the Sorbonne and the London School of Economics, whom island legend says is descended from the sun and the moon and has the stars for cousins...."

statement should be signed by the Board, the President, and the Editor. I hope that would heal the problem."

Gil suggested reinstituting the "President's Message" column. "It could include the reaffirmation of policy, bring in some of the other things we are doing, and we could run it in the December issue."

Greenewalt thought that to hide it in the President's Message would be a mistake. "The way to avoid any suggestion of internal disagreement," he said, "is to have the Chairman of the Board sign the statement for the Board along with the President and the Editor...."

Martin agreed. "I would take Mr. Grosvenor's statement and have it in the Editor's column," he said, "and have a separate reaffirmation of editorial policy to be signed by the Chairman, the President and the Editor."

Louis Wright agreed too, and pointed out that having the separate statement "would take the wind out of the sails of the critics. We would have gone on record in the Magazine on agreement."

Crawford Greenewalt introduced the motion that Gil Grosvenor "print the reaffirmation of policy statement in a box on the Editorial page and have it signed by the Chairman, President, and Editor."

The motion was unanimously carried. Gil Grosvenor had won the first round. The second round began with Lloyd Elliott asking, "Does the Committee wish to do anything in saying to the Editor we believe certain articles have not quite lived up to the editorial policy of the magazine, or do we want to advise the Editor that this Committee finds no reason for the publicity which has been given to the magazine, or to let the 'hot potato' drop?

"I think it would be beating a dead dog if we do anything but present this statement of reaffirmation," Greenewalt said. "I don't think we need to say anything else....just say here is our policy and we'll adhere to it as well as we can...."

Since there did not seem to be any disagreement, Elliott said, "Our report then is that 'We find no evidence of any intentional change in policy and we reaffirm the policy and commit ourselves to following it.' "

The committee was in agreement and the meeting was adjourned at 12:15.

An hour and forty-five minutes later the same seven committee members were gathered together again to attend the meeting of the National Geographic Society's Board of Trustees meeting. After all the Society officers' reports and other business had been done away with Lloyd Elliott delivered his report as chairman of the ad hoc committee for editorial policy, and presented copies of the "A Reaffirmation of the Editorial Policy of the National Geographic Magazine" statement that it was the committee's recommendation that the statement be printed in an appropriate box on the Editor's page of the Magazine to be signed by the Chairman, the President, and the Editor; and that the Editor may make reference to the statement in his own column as he wishes.

There then followed a discussion of the "Reaffirmation" statement and a short time later, the "Reaffirmation of Editorial Policy" statement was voted on and unanimously adopted. Vosburgh, Melville and Gil Grosvenor exchanged satisfied glances.

The Trustees voted to commend the ad hoc committee and particularly its chairman Elliott, and then Trustee Chairman Melvin Payne declared, "The Ad Hoc Committee is now dissolved."

Vosburgh leaned forward and asked, "Mr. Chairman, does that mean the idea of *any kind* of internal editorial oversight committee has been abandoned?"

Mel Payne assured Vosburgh it did.

The Trustees moved on to further business. Gil Grosvenor reported on future issues, that the December issue would have a supplement map and articles on the history of the New World, St. Brendan's voyage, the weather of 1977, Japan's inland sea, and...bowerbirds. The National Geographic Society had returned to business as usual—or almost as usual.

What resolution of the controversy had brought about was a vindication of the *Geographic*'s Editors and the Magazine's continuing evolution.

"I think one of the problems we have at the *Geographic*—or maybe," Gil Grosvenor said, "it's a blessing—is that we're a very conservative organization. We change slowly. There are some people who don't want us to change at all. Fortunately, they don't have much of a following. Basically, some people don't like the way the world is changing, and they want *something* to stay the way it is. They want *us* to stay the same. But we can't stand still. I'm convinced that unless you constantly evolve a publication, you're in trouble. And I also think if you make revolutionary changes, you're in trouble—or you make them because you're *already* in trouble. The *Geographic* has been fortunate in its leadership because the people that have come in, have come in time to make the changes."

On the Editor's page of the January 1978 issue, the following statement appeared over Gilbert M. Grosvenor's signature:

> ...as journalists committed to objective, impartial reporting of what we can report, we accept the opportunity to reflect our times, realizing that only history can tell the full story.
>
> We often feel pressure to change that policy from those who would have journalism bear a message—their message. To some people, failure to denounce is the same as silent praise, and objective statement of fact amounts to advocacy—if fact fails to coincide with prejudice.
>
> But these voices from distant right and far left only serve to remind us to steer a course that avoids those biased and dangerous extremes of mind. Last year's record increase in new Society members, coupled with a high renewal rate, seems ample proof our course is sound.
>
> We will continue to travel the world unencumbered by ideology, to go along on a young man's walk across America, to peer into Loch Ness, and to fly to Mars and beyond. As the world goes its way, we will record it, accurately and clearly. This anniversary year is an appropriate time for us to reaffirm the principles that have served your Society so well.

The "Reaffirmation of Editorial Policy" statement approved by the ad hoc committee appeared directly below. It was signed:

Melvin M. Payne
Chairman of the Board

Robert E. Doyle
President

Gilbert M. Grosvenor
Editor

The rumblings stilled; the crisis had passed. Three years later Editor Gilbert M. Grosvenor would reluctantly give up what he called "the best job in the world" to replace Robert E. Doyle, who was stepping down as President of the Society. During Gil's decade as Editor, membership had increased from 6,402,674 to 10,771,886 despite a nearly twenty-seven percent increase in dues.

And as the decade of the 1980s commenced with a vacancy opening up in the office of Editor, for the first time in this century there wasn't another Grosvenor waiting in the wings.

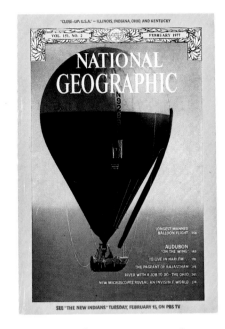

Ed Yost's "The Longest Manned Balloon Flight" [in the Silver Fox] was the cover story for the February 1977 Geographic. By the end of 1977 circulation of the Magazine had reached 9,756,312.

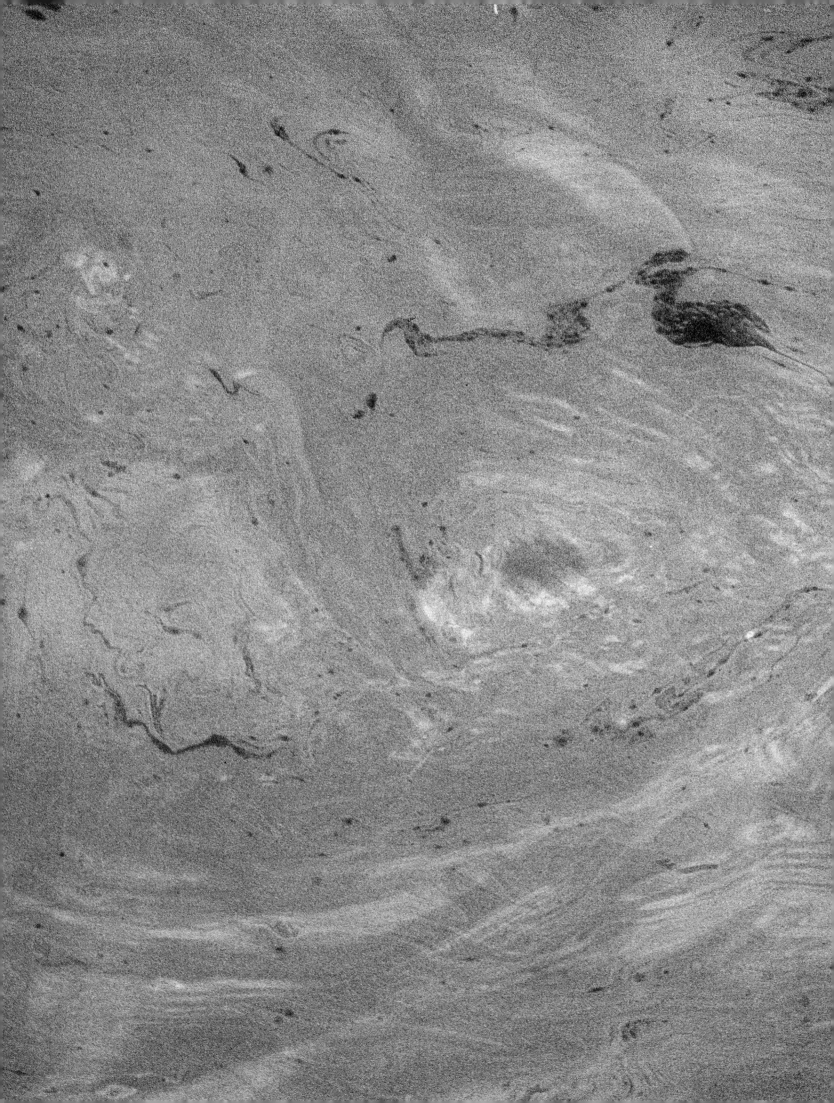

"Riders on the Earth Together"

"To see the earth as it truly is, small and blue and beautiful in that eternal silence where it floats, is to see ourselves as riders on the earth together, brothers on that bright loveliness in the eternal cold—brothers who know now they are truly brothers."

—Archibald MacLeish in *The New York Times*, December 25, 1968; partially quoted in the Editor's Note introducing Gordon Young's December 1970 *National Geographic* article, "Pollution, Threat to Man's Only Home"

The photograph that opens this chapter, a western grebe covered with a fatal coat of oil, appeared on the cover of the December 1970 *National Geographic*—the last issue edited by Frederick G. Vosburgh before his retirement. It was one he could take pride in.

Beneath the grebe on the cover was the issue's announced theme, "Our Ecological Crisis," with the subthemes "Pollution, Threat to Man's Only Home," "The World—And How We Abuse It," and "The Fragile Beauty All About Us."

The lead article contained photographer James P. Blair's grim, unromantic photographs of pollution-belching smokestacks, trash-infested New York alleyways, and Ohio's blackened Cuyahoga River "so covered with oil and debris that in July 1969 the river caught fire here in Cleveland's factory area, damaging two railroad bridges."

Gordon Young's somber text accompanying Blair's photographs began:

> We are astronauts—all of us. We ride a spaceship called Earth on its endless journey around the sun. This ship of ours is blessed with life-support systems so ingenious that they are self-renewing, so massive that they can supply the needs of billions.
>
> But for centuries we have taken them for granted, considering their capacity limitless. At last we have begun to monitor the systems, and the findings are deeply disturbing.
>
> Scientists and government officials of the United States and other countries agree that we are in trouble. Unless we stop abusing our vital life-support systems, they will fail. We must maintain them, or pay the penalty. The penalty is death.

Writer Gordon Young visited smog-bound Los Angeles (with its "witches' brew of pollutants, spewed primarily by automobiles"), Tokyo (where traffic policemen "pause regularly to breathe oxygen" and "pedestrians seek the same relief from vending machines"), Essen, Germany (where a formerly bright steel square exposed to the Ruhr's smog "for only two months, was chocolate brown and deeply corroded"). He took some comfort in London's victory over its killer fogs, the worst of which "settled over London on December 5, 1952. For four consecutive days the city's normal daily death rate of 300

Opposite: *Brig. Gen. William Mitchell (left) poses with the tiger he shot less than two miles from the palace of the Maharaja of Surguja (right), his host. "He was a beautiful creature, about 14 years old, 10 feet 4 inches between pegs," Gen. Billy Mitchell wrote in "Tiger-Hunting in India," for the November 1924* Geographic. *"His paws, one blow of which could fell the largest buffalo, were as big around as the largest soup plate....He was the biggest tiger that I have ever seen."*

Preceding spread: *The blowout of an offshore oil well near Santa Barbara, California, in early 1969 created the slick through which this doomed, oil-coated western grebe swam in this photograph for the cover of the December 1970 issue.*

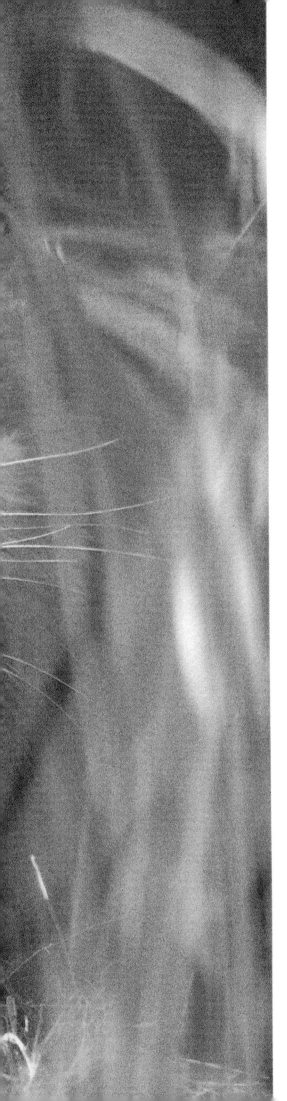

more than tripled; in all, some 4,000 extra deaths that winter were blamed on the incident." That victory had been attained after Parliament's passage of the Clean Air Act in 1956, calling for the changeover from high-sulfur-content soft coal to cleaner burning fuels. And visiting the city thirteen years after that Act had been passed, Young found "deck chairs...filled with tanning Londoners" in St. James's Park.

Still, for the most part, his article sounded a warning that our earth's air, water, and land were in trouble and something needed to be done—and done right away.

For the *National Geographic* reader of today it would seem odd for the Magazine, by 1970, *not* to have taken a stand against pollution. After all, the problem had been around long enough for the author of an 1897 *Geographic* article, "Pollution of the Potomac River," to conclude: "Until state or national legislation can be secured to regulate such matters, the Potomac...must serve as a sort of sewer into which town and manufacturing establishments empty their refuse...." But in 1970 the ghost of Gilbert Hovey Grosvenor's editorial guiding principles still lingered. Pollution was "of a controversial nature": industry polluted; industry advertised in the *Geographic*; a Du Pont director sat on the Board of Trustees—and the Magazine still shied away from controversial subjects. However, if the *Geographic* seems to have entered the battle to save man from himself somewhat tardily, it had at least long been on the side of other living things.

Interest in "The Redwood Forest of the Pacific Coast" had been expressed as early as May 1899; Gifford Pinchot's "An American Fable" (conservation of natural resources) appeared in May 1908; "Our National Parks" in June 1912. When, in 1916, the huge, thousand-year-old sequoias in the heart of Sequoia National Park were threatened with being lost to timber interests, the National Geographic Society contributed $20,000 to round out the necessary Congressional appropriation to save them—including the General Sherman, a 275-foot-tall sequoia with a circumference of 103 feet.*

Trees continue to be on the Society's mind. "Acid Rain: How Great a Menace?" was published in November 1981. In September 1982 "Our National Forests: Problems in Paradise" dealt with the fight to keep them free of mineral exploitation. A summary of the political and economic pressures threatening "Tropical Rain Forests: Nature's Dwindling Treasures" appeared in January 1983. The Society recognized that man's greed was behind these problems; it always has been.

Senior staff writer John L. Eliot noted in September 1982 that Theodore Roosevelt, whose "Forests Vital to Our Welfare" (an excerpt from a speech) was published in the Magazine's November 1905 issue, before leaving the White House in 1909, had:

Though the largest living thing in the world in total size, a sequoia is not the tallest. The tallest is a 368-foot-tall redwood discovered in California in 1963 by representatives of the Society. It and others like it are now protected within the Redwood National Park, established in part at the urging of Society members.

◄ ─────────────────────────────

"Snarling in threat, a 300-pound tigress warns the authors, in an open jeep 15 feet away, against venturing closer...."—From writer Stanley

Breeden and photographer Belinda Wright's article "Tiger! Lord of the Indian Jungle" in the December 1984 issue

Above: *With a pet mongoose on his back Dr. Iain Douglas-Hamilton is shown using photographs to identify elephant family groups in Tanzania's Lake Manyara National Park. For more than twenty years Douglas-Hamilton has fought for the elephants' survival. His Land-Rover's bent steering wheel reflects the impact of an unappreciative elephant matriarch's charge—From "Africa's Elephants: Can They Survive" in the November 1980 issue*

Right: *Using aerial survey techniques to compile a census of elephants, Oria and Iain Douglas-Hamilton photographed this gathering of elephants on a shrub-dotted savanna in Zaire, where, reportedly, ivory poachers have killed entire elephant families using poisoned fruit.*

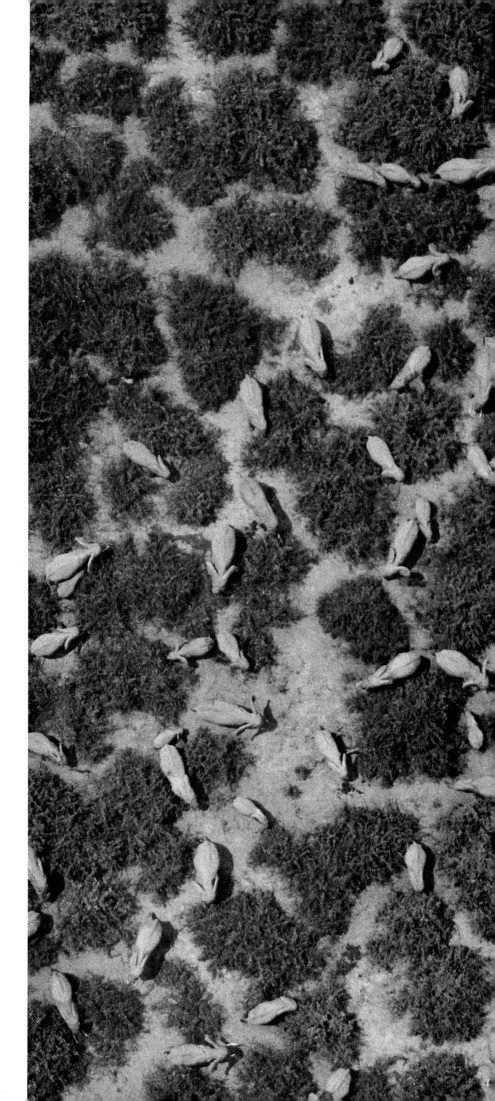

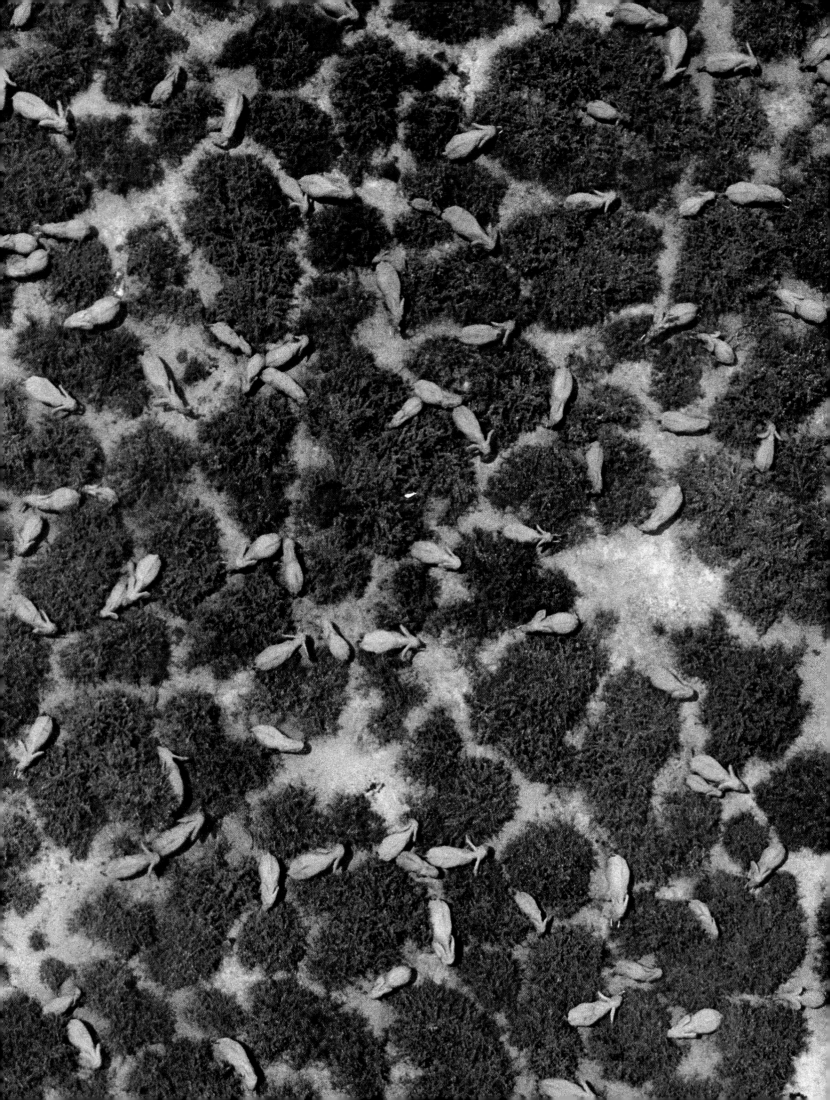

Above: *"Big Bruno, 800 pounds of muscle and claw, lies powerless from drugs as the team takes a blood sample. Red collar holds the radio that enabled the Craigheads to track No. 14 over some of the most heavily timbered and rugged country in Yellowstone. 'We followed this grizzly day and night,' Frank reported, 'and slept wherever darkness overtook us.'"*

As of August 1966, when this photograph from the Craigheads' *"Trailing Yellowstone's Grizzlies by Radio"* article appeared, the Society had contributed nine grants through its Committee for Research and Exploration to the seven-year grizzly research project then being conducted by the Craighead brothers. Since then, the brothers have received an additional nineteen grants from the Society for research on grizzly bears.

Opposite: *Like a worried mother on tiptoes, a female Alaskan grizzly rises up on her hind legs in an attempt to better locate her straying cubs.*
—From *"'Grizz'—Of Men and the Great Bear"* in the February 1986 issue

for a fledgling cause called conservation— ...enriched the public domain by approximately 230 million acres...quadrupled the existing forest reserves and proclaimed the first federal wildlife refuges, more than 50 of them. The number of national parks doubled; the first 18 national monuments came into being. The U.S. Reclamation Service was inaugurated to irrigate about three million acres in the arid West, and tens of millions of mineral-bearing acres also fell under Roosevelt's cloak.

"I hate a man who would skin the land," he bristled. He railed against corporate "timber thieves," and argued that "a live deer in the woods will attract...ten times the money that could be obtained for the deer's dead carcass." Again and again he preached, "Conservation of our natural resources is the most weighty question now before the people of the United States."

It may be hard to reconcile this conservationist Roosevelt with the T.R. who, shortly after leaving the Presidency, embarked on a hunting trip that was so thoroughly reported in the Magazine that between March 1909 and November 1910 four articles on it appeared. But his friend John Burroughs, the ardent naturalist who was instrumental in urging Roosevelt to establish wildlife refuges, was not disturbed by the President's hunting trips. "Such a hunter as Roosevelt," he wrote, "is as far removed from the game-butcher as day is from night."

Hunting did not seem at odds with loving nature. Eliza Ruhamah Scidmore wrote "The Greatest Hunt in the World" for the December 1906 issue; the Carl E. Akeley trophies described in the August 1912 article "Elephant Hunting in Equatorial Africa With Rifle and Camera" are still one of the American Museum of Natural History's major exhibits. Brig. Gen. William Mitchell went "Tiger-Hunting in India" for the *Geographic*'s November 1924 issue and photographs displayed his bag. But sixty years later, when the more conservation-minded *National Geographic* went tiger hunting again in "Tiger! Lord of the Indian Jungle" by Stanley Breeden, with photographs by Belinda Wright, there was a familiar opening but a new thrust:

> We set off at sunrise on this winter morning in central India, once again riding Pawan Mala, an old and venerable elephant. Out on the meadows the swamp deer stags are rutting. A tiger roars in the distance, a good omen.
> Mahavir, the mahout, steers the elephant toward the tiger's stirring sound. We soon find fresh tiger footprints along a sandy ravine, or nullah—the broad strong pugs of a male. They lead us into dense forest.
> We are on the right track; we can smell where the tiger has sprayed a bush to mark his territory. Belinda, my wife, sees the tiger first, an awesome vision in fiery orange and black stripes gliding through the green bamboo tracery. Ignoring us he walks on and on. We are alongside him now, about 30 feet distant. Once or twice he glances at us and snarls, pale yellow-green eyes burning.
> Suddenly the tiger stops in his tracks....

The Breedens shoot; the result—thousands of feet of film and the January 1985 National Geographic Special *Land of the Tiger*, which became the third-highest-rated evening program ever shown on public television.

"Ours has been a ten-year affair with the tiger," Stanley Breeden wrote, "a quest that lured us back again and again to the Indian jungles." The Breedens had been photographers nine years earlier for "India Struggles to Save Her

Wildlife," a September 1976 *Geographic* article written by staff writer John J. Putman, who noted:

> The years have not passed India by, but ravaged her. The great forests have shrunk under the ax and the plow, and shrunk again; the vast herds, 10,000 strong, and herd after herd, have diminished to small pockets; the predators have become the hunted. The Indian cheetah is now extinct, the tiger gravely endangered; even the birds have dwindled in both numbers and species.
>
> So great has been the slaughter of animals, so entrenched the pattern of destruction of habitat, that many came to believe that India's treasure of wildlife was doomed.

At the turn of the century 40,000 or so tigers roamed India. Big-game hunting and the destruction of their natural habitat had cut their number to fewer than 2,000 by 1972, when India began cooperating with the World Wildlife Fund in Project Tiger. By the end of 1984, the population of Indian tigers had risen to 3,000—one thousand of them in the fifteen established tiger reserves.

The African elephant, because of its ivory, has been less fortunate. A November 1980 Oria Douglas-Hamilton article pointed out that a higher proportion of them were being killed then than ever before. Poisoned arrows are used in Kenya, fires in Sudan; Pygmies use spear traps in Cameroon's forests, horsemen spears in Chad; and in Zaire pitfalls were popular—until the natives there learned it was easier to kill elephants by placing fruit poisoned with battery acid or insecticides along their paths and at their watering holes.

Ten thousand elephants were being killed every couple of years and the slaughter was permitted to continue because of greed: money gained from the sale of ivory made its way into pockets at the highest level of governments.

Between 1976 and 1979, Dr. Iain Douglas-Hamilton, who has devoted more than twenty years of his life to fighting for the elephant, was using aerial survey techniques and scientist informants in thirty-five countries to compile a census of African elephants. He concluded that about 1.3 million elephants then existed in three million square miles of continental Africa. In 1966, he noted, 8,000 elephants resided in Uganda's Kabalega Falls National Park. Ten years later ivory poachers had reduced that number to fewer than 1,700. Upon the overthrow of Idi Amin in 1979 during the Uganda-Tanzania war, his retreating troops and ivory poachers with captured automatic weapons slaughtered so many of the park's elephants that 1,360 elephants were left in the park by 1980, when Iain Douglas-Hamilton returned to Kabalega; and the number of elephants in the southern section was "a mere 160, almost all clustered in one terrified herd, moving day and night, unable to find refuge, and shedding corpses like leaves along the trail."

And in Tanzania's Selous Game Reserve, Oria Douglas-Hamilton recorded in her November 1980 article "Africa's Elephants: Can They Survive?" the poignant death of one more:

Right: *A humpback whale hurtles in a backflip breach out of Alaska's frigid Glacier Bay.* —From *"Humpbacks: The Gentle Whales"* in the January 1979 issue

Overleaf: *A bald eagle screams above the wind as it rides the air currents.* —From *"Our Bald Eagle: Freedom's Symbol Survives"* in the February 1978 issue

In the immense silence of dawn, before the sun rose and burned the skies, I watched through the opening of our tent the morning beauty of an untouched place, and dreamed how lucky we were to be here in one of the strongholds of the elephant in Africa.

Then, across the hills of [the] Reserve, a gunshot echoed and broke all dreams. A terrible stillness followed; then two, three, four shots blasted out and set my heart pounding. I could see an elephant limping away from a line of tents. People were running.

I fell into some clothes and ran down the green slope. The elephant was limping badly, his right side stained red. I joined some half-dressed game scouts. One, draped in a towel, held an empty gun. He was going back to his tent to get more cartridges.

The elephant kept walking away, and we, 15 of us, followed. He had such a sad look on his face—no anger or violence. His head was bobbing from side to side to keep watch on us, his trunk testing the ground ahead. The scout returned, fired, and missed; fired again and hit in the shoulder, and the blood ran through the crackly skin.

My husband, Iain, took the gun from the scout. He did not want to shoot the elephant, he hated shooting elephants, but there was no choice. Iain aimed and fired.

The elephant screamed and thrashed the bushes with his trunk, tottering on three legs. The bullet had not dispatched him. "There are no more cartridges," the scout said. "There is only a small gun in the lodge," and he walked away to get it. We waited in the thick wet bush as the blood-soaked elephant moved step by painful step to a little river.

He was standing looking at the water, waiting with us for death, when the gun arrived. Iain walked up to him, lifted the gun to his heart, said, "Sorry, old chap," and pulled the trigger. Instantly, his legs folded and he collapsed. No one moved, the birds were still, there was no sound now except the trickling stream. It was the saddest sight I ever saw.

The scout was standing nearby. "Why did you shoot?" I whispered.

"Because he was touching the ropes of my tent," he answered.

An August 1960 *Geographic* article opened with another shooting:

The bolt of the large-caliber rifle clicked shut. In the light of a powerful flashlight, I sighted on the neck of a grizzly cub and pulled the trigger. The cub whirled. His mother sprang to his side. But the cub neither dropped nor staggered. He merely brushed against his mother's front legs, paused under the reassuring protection of her powerful neck, then moved on. Had I overshot?

Hardly, for I heard John's assurance, "Nice shot." And now Maurice Hornocker's light beam came to rest on a dart hanging from the cub's neck. It had just injected a powerful drug. We shot to capture, not to kill.

Twin brothers Frank and John Craighead* were in Yellowstone National Park on a Society grant, beginning a long-range investigation into the habits and behavior of grizzly bears "to give conservationists the knowledge they need to save the huge bears from ultimate extinction." The Craigheads' tagging, measuring, and tattooing of the drugged bears "was like working over dynamite with a damp fuse," Frank admitted, especially when the dosage of the

*The Craighead brothers' first appearance in the Magazine came about after they showed up at the Society headquarters as teenagers in July 1935 with pictures and a manuscript about their experiences photographing hawks and training them in falconry. Their "Adventures With Birds of Prey" appeared in July 1937 and resulted in an invitation from an Indian prince to visit and see how he trained his birds. That resulted in "Life With an Indian Prince," in February 1942. Close to a dozen Craighead pieces have appeared over the years, including articles by the brothers' children.

Above: *This photograph, from Kenneth F. Weaver's April 1979 article, "The Promise and Peril of Nuclear Energy," was captioned: "The garbage problem: How to store nuclear waste that may remain radioactive for thousands of years? Until 1970 the U.S. dumped low-level solid waste in drums—such as this one off the Maryland coast—into the Atlantic and the Pacific. Now some drums are leaking radioactive materials."*

Opposite: *"A man-made sun rose over Bikini Atoll on March 1, 1954. Seen here from 50 miles away, the 15-megaton hydrogen bomb blast called Bravo ranks as the largest U.S. test, a thousand times greater than the atom bomb dropped on Hiroshima in 1945."—From "Bikini—A Way of Life Lost," in the June 1986 issue. Shifting winds scattered radioactive debris over about 50,000 square miles.*

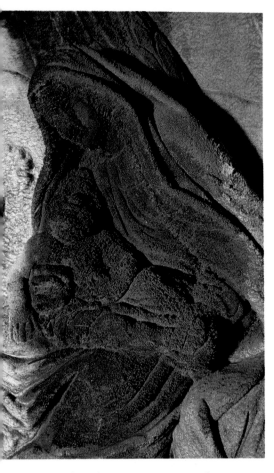

Above: *"Stone cancer spreads across a marble Madonna carved about 1650 beneath a buttress of the Cathedral of Milan. Here in Italy's industrial center, as in many world cities, the burning of high-sulphur coal and oil generates sulphur oxides. Deposited on stone, they combine with rain water to form sulphuric acid. The highly reactive fluid produces a chemical change in the stone, eventually disintegrating it under prolonged exposure. The acid has not remained long enough on the lighter areas to damage them severely. Such destruction threatens many of man's noblest monuments, from the gargoyles of Notre Dame de Paris to the Lincoln Memorial."—From "Pollution, Threat to Man's Only Home" in the December 1970 issue*

Opposite: *Man is both culprit and victim of the pollution clouding St. Louis' air.—From "Acid Rain: How Great a Menace?" in the November 1981 issue*

curare-like drug was not strong enough and a bear would sputter into consciousness while they were leaning over it.

In this article, titled "Knocking Out Grizzly Bears for Their Own Good," Frank Craighead asked:

> But what good is a grizzly? Why spend such time and research money to save an animal?...John's 12-year-old daughter [answered], "We want to save the grizzly because when he's gone, he's gone forever, and we can't make another one."
> Here was the simple answer.
> No animal species other than man has ever been known to exterminate another. But man's list is long. Do we possess the right to annihilate a fellow creature?

It was not a question that much troubled the Society during its formative years. Between September 1907 and April 1909 five bear-hunting articles appeared—in the same period a Remington Autoloading Rifle advertisement, titled "The Right of Way," depicted a surprised hunter and a snarling, bloodthirsty grizzly confronting each other on a narrow mountain ledge. The truth is that "since 1900 only 14 people have been killed by grizzlies in the lower 48 states," and those fatalities were primarily the result of hungry bears lured into campsites by garbage. Since the advent of the repeating rifle, however, the bears have stood little chance. In his February 1986 article, " 'Grizz'—Of Men and the Great Bear," Douglas H. Chadwick noted that "tens of thousands of the great bears [lived] south of Canada as late as 1850. Only 600 to 900 remain there now....Standing sheeplike, shoulder to shoulder, the current lower 48 grizzly flock wouldn't cover half an acre."

For the past two decades, the Society has focused its attention on whales. Articles have appeared on sperm whales, blue whales, humpbacks, killer whales, false killer whales, narwhals, gray whales, right whales, white whales; 10.1 million copies of the recording "Song of the Humpback Whale" were bound into the January 1979 issue; two children's books on whales have been published; and the Society's 1978 television documentary *The Great Whales* won an Emmy Award for Outstanding International Programming.

Gilbert M. Grosvenor, in a December 1976 *Geographic* editorial remarking on the international controversy surrounding efforts to halt the killing of whales and the imposition of International Whaling Commission quotas, wrote:

> On one thing, all seem agreed. The natural legions of whales have been dealt a devastating blow in the past century, under impact of the explosive harpoon and the fast catcher boat. The blue whale, like the American buffalo, nearly became extinct. No one can tell if those that remain can restore their numbers....no one...can escape the conclusion that the whale has become a symbol for a way of thinking about our planet and its creatures, and in that, at least, there is hope of a better day for both whales and men.

The "better day" has not come for the 161 members of the eleven families who lived on Bikini Atoll, no matter how patiently they have waited. As staff writer William S. Ellis reported in June 1986:

> Nearly 40 years had passed since that Sunday in 1946 when Commodore Ben Wyatt of the United States Navy met with them after church services to say that their island was needed for a project that would benefit mankind. He implied that an authority higher than any on earth would be pleased if they decided to cooperate.

416

Being both a devout and benevolent people (and not without awe over America's military power), they announced this decision, through their chief, Juda: "If the United States government and the scientists of the world want to use our island and atoll for furthering development, which with God's blessing will result in kindness and benefit to all mankind, my people will be pleased to go elsewhere."

Not quite six months after the Bikinians had been moved to Rongerik Atoll, the United States dropped the first atomic device over their lagoon. For the next twelve years, twenty-three tests were held at Bikini; and in the ten years between 1948 and 1958, forty-three more devices were exploded at Enewetak Atoll, two hundred miles to the west. In the meantime, the Bikinians had been shuttled from Rongerik to Kwajalein and finally to Kili—"a small, isolated island some 500 miles away [from Bikini], an island without a lagoon, a mere dot of land of 230 acres standing naked to the sea."

But it was at Bikini in 1954 that Bravo, "the first deliverable hydrogen bomb," was tested. It was "the most powerful bomb ever detonated by this country," wrote Ellis, a thousand times more powerful than the bomb dropped on Hiroshima. "The combined power of all the weapons fired in all the wars of history would fall short of that released by Bravo over the 242-square miles of the Bikini lagoon....A freight train carrying Bravo's equivalent in TNT would span the North American Continent."

Unfortunately the winds shifted from northward to eastward at the time of Bravo's explosion and

> showered radioactive pulverized coral and other material over...perhaps as much as 50,000 square miles. Those caught in the fallout included some 250 Marshallese from the islands of Rongelap and Utirik; 28 weather station personnel on Rongerik; and 23 crewmen of a Japanese fishing vessel, the *Daigo Fukuryu Maru (Lucky Dragon No. 5)*, one of whom died of radiation exposure.

Less than a hundred miles down-wind of the blast innocent Rongelap Atoll natives played in the curious "Bikini snow" that had fallen about them; some even tasted it; others scooped it up and were burned. As a picture caption explained, three out of four Rongelap children under age ten later developed thyroid tumors.

Bikini's radioactive soil still burns with cesium 137. It contaminated the soil, groundwater, and many food crops. It will take another eighty to ninety years before the cesium level is acceptable. Scientists and government officials are sorry. While its former inhabitants wait on inhospitable Kili, scientists are still trying to figure out how to clean up Bikini.

A clean-up problem closer to home sent staff writer Allen A. Boraiko and photographer Fred Ward to chemical dumps, lagoons, storage tank farms, and landfills in the United States to research their March 1985 article, "Storing Up Trouble...Hazardous Waste." At times wearing plastic safety suits, gloves, hard hats, and masks with backpack air tanks, the two examined firsthand the problem of how the U.S. is trying to deal with the possibly six *billion* tons of

◀

Redwoods and rhododendrons bathe in a soft coastal fog in California's Redwood National Park, a 106,000-acre preserve containing the world's tallest tree—a coast redwood 368 feet tall when it was discovered by National Geographic *naturalist Paul A. Zahl in 1963.*

Above: *Man's inhumanity to nature: "Canyons of desolation scar the landscape of Muhlenberg County, Kentucky. The gargantuan shovel, towering 250 feet high, bares seams of coal, principal fuel of the Nation's electric power plants.*

"The devastation caused by strip mining prompted a state law in 1966: Areas currently being mined must be graded and replanted. First steps of reclamation are visible above the shovel. Older, ungraded 'orphan banks' in the distance remain an eyesore."
—From *"Pollution, Threat to Man's Only Home"* in the December 1970 issue

Opposite: *Some 3,300 motorcyclists taking part in the San Gabriel Valley Hare and Hound Race at Barstow, California, race across the desert to Las Vegas, more than 150 miles away. "Cyclists extol the challenge...but environmentalists deplore the destruction of a fragile ecology."*
—From *"This Land of Ours—How Are We Using It?"* in the July 1976 issue

toxic waste it has disposed of since 1950 "in or on the land, steadily increasing our potential exposure to chemicals that can cause cancer, birth defects, miscarriages, nervous disorders, blood diseases, and damage to liver, kidneys, or genes."

Five years earlier Boraiko and Ward had reported on "The Pesticide Dilemma." As careless and indiscriminate as our dumping has been, so has been our spraying. A year in preparation, the article pointed out:

> Plagued by two thousand detrimental insects, weeds, and plant diseases, U.S. farmers and foresters devote more than 2.25 billion dollars a year to pest control. Proponents point out that pests damage one-third of our crops—nearly nine billion dollars' worth each year—and can carry death-dealing disease. But grim counterarguments emerge: pesticide-related illness and death, and warnings of dire ecological consequences.

"In poisoning the pests that each year destroy crops worth billions of dollars," the Editor asked, "are we also unwittingly poisoning ourselves?"

It was a question posed eighteen years earlier by author Rachel Carson in her book *Silent Spring*, in which she warned concerned citizens that the increasingly widespread indiscriminate use of chemical pesticides and herbicides threatened to permanently upset the earth's natural ecological balance.

Still, progress has been made.

Government and state agencies, whipped on by a host of concerned citizen groups and private organizations, are fighting daily battles to clean up the air, water, and soil. World, national, state, and local federations have united in attempting to preserve and defend endangered species and wildlife habitats. Environmental, conservation, and wildlife organizations are staffed and supported by people who care. And there have been some stunning triumphs.

One such triumph is the resurgence of the trumpeter swan, whose extinction ornithologist Edward Howe Forbush predicted in 1912 would occur within "a matter of years."

By 1932, there were but thirty-one of these lovely white swans in Yellowstone National Park, twenty-six in Montana's Red Rock Lakes, and about a dozen scattered over the rest of the nation. In 1935, the Red Rock Lakes National Wildlife Refuge was created, and the trumpeter swans had won a first-round reprieve. But in a June 1937 article, "Hunting With a Microphone the Voices of Vanishing Birds," Cornell University ornithology professor Arthur A. Allen, noting the extinction of the great auk, the Labrador duck, the passenger pigeon, and the heath hen "almost within the memory of men still living," expressed his concern that the "Carolina parakeet seems about to follow them," and warned that "It is thought that only about 75 [trumpeter swans] remain alive in all the United States, though at one time it was not an uncommon bird throughout the West."

The twenty-six trumpeters at the Red Rock Lakes refuge had become 423 thirty years later. And with similar growth in other colonies and the discovery of previously unknown breeding grounds in southern Alaska, it is now estimated that close to 12,000 trumpeter swans are alive in North America.

If we can bring back the trumpeter swan, save the tiger, fight for the elephant and the whale, it is not too much to hope we can take care of ourselves as well. If not, "If we get things wrong," as Dr. Lewis Thomas observed in his Foreword to the 1986 Book Service publication *The Incredible Machine*, "we could be leaving a very thin layer of fossils ourselves, and radioactive at that."

Above: *"Mallards at the U.S. Government's Patuxent Wildlife Research Center in Maryland were fed DDE, a breakdown product of DDT, in concentrations now common in the wild."* The result: *"On the average, three eggs in a clutch of 12 were cracked or crushed by the mother. Eight others failed to develop, and only one hatchling, rather than the usual 9 to 11 emerged alive." —From "Pollution, Threat to Man's Only Home" in the December 1970 issue*

Left: *"Strafing insects with toxic spray, a plane barrels along above a Texas cotton field. Billions of gallons of pesticide solutions are applied yearly to United States crops, ranges and forests." Gilbert M. Grosvenor's Editor's Note to Allen A Boraiko's February 1980 article "The Pesticide Dilemma," which accompanied this photograph, questioned whether we weren't also poisoning ourselves.*

Overleaf: *And now the good news: Once an imperiled species, the majestic trumpeter swan, through intelligent management and a web of protective laws, now numbers about 12,000 birds in North America.*

"Broadening the Outreach"

"There absolutely will be no radical or visible changes in the magazine. It's like being given control of apple pie and motherhood; you don't fool around with it."

—Wilbur E. Garrett, upon his appointment as the seventh Editor of the *National Geographic,* July 10, 1980

"I would hope that at our 200th anniversary, somebody will look back and say, 'Hey, you know, those people between the first and second hundred, they really broadened the outreach of the National Geographic Society!'"

—Gilbert M. Grosvenor

Preceding spread: *Shown on the display screen and a computer in the Society's new "Control Center" is a listing of the Magazine's 1986 stories by subject and area. An almost endless variety of screen display, reports, and even graphic representations can be generated from the same data. This center commenced operation in the spring of 1987, replacing the one Gilbert M. Grosvenor had "built in his basement" more than twenty years before.*

On June 27, 1977, when Dr. Melvin M. Payne, had told his fellow Trustees at the ad hoc committee meeting, "We are known throughout the world by our Magazine, and what the Magazine says is what the National Geographic Society says," his observation, even then, was only partially correct, for the Society was beginning to speak with other voices.

Perhaps twenty-five years ago the Society was, to an outsider, just the familiar yellow-bordered *National Geographic* mailed out to the Society's then 3.5 million members. To outsiders today, however, to the Society's 10.6 million members, to the Magazine's estimated 30 to 50 million readers, to the millions and millions who have watched its television specials, the National Geographic has evolved into something much, much more. And at the root of that expansion has been Gilbert M. Grosvenor's determination to make up for having to give up the Editorship of the Magazine in 1980 to become President of the Society because the ailing incumbent, Robert E. Doyle, was retiring.

Just two years before, Gil and Bob Doyle had directed the second major move of the Magazine to a new printer—and a totally different printing process. After eighteen years of printing by the letterpress method in Chicago, the Magazine was moved to a new plant in Corinth, Mississippi, where it was produced by the gravure process. As Gil Grosvenor told members in the November 1977 Editor's page, it was "the biggest technical change-over in the history of your Magazine."

When Gil Grosvenor resigned as Editor, the front-runners to fill the vacancy were Wilbur E. Garrett, Joseph Judge, and Robert L. Breeden. Both Bill Garrett and Joe Judge were Associate Editors of the Magazine. Garrett was in charge of illustrations and Judge in charge of text.

And although Bob Breeden's title was Senior Assistant Editor, his area was Related Educational Services of the Society—Special Publications and School Services. So Breeden had not, in reality, been with the Magazine since

he was asked by Melville Grosvenor to do the picture editing and layout for the White House book eighteen years before.

Because of Breeden's astounding success with the Special Publications Division, and the Society's natural reluctance to tamper with something obviously working so well, the search for Editor narrowed down to Garrett and Judge.

Garrett had joined the staff as a picture editor in 1954. He had progressed from Associate Illustrations Editor to Senior Assistant Editor and then to Associate Editor under Gil Grosvenor. Along the way Garrett had photographed, and sometimes written, almost two dozen articles, many of them reflecting his abiding interest in Southeast Asia, the preservation of antiquities—and his willingness to take risks.

In 1961, Garrett was in Laos with a U.S. Army Special Forces team and its captain, Walter Moon. Unknown to them all, they were surrounded by Pathet Laos. "It was during the monsoon season, and we couldn't get the wounded out or the supplies in because it was pouring rain," Garrett remembers:

I lived in one of those little shacks upon sticks with Walter Moon and some of his people. Finally, one rainy day, the big helicopters got in through the

The supertanker Amoco Cadiz *foundered off the coast of Brittany in March 1978, broken-backed and gushing sixty-nine million gallons of crude oil into the sea. The ship was photographed for staff writer Noel Grove's July 1978 article, "Superspill: Black Day for Brittany," by Martin Rogers, sitting in the open door of a French military helicopter crowded with other photographers. "It was impossible for me to get up," Rogers later wrote. "I just kept shooting. I nearly got pushed out. But I had plenty of time to work the situation. I used at least three cameras."*

429

clouds, and I told Moon, "I'm ready to get out of here."

"Get on this chopper," he said.

I went over and threw my bag on and started getting in, and the pilot yelled, "Get the hell out of here! I'm overloaded already." So I backed down and returned to the edge of this muddy field, and Moon said, "What are you doing back here?"

I told him the pilot said the helicopter was overloaded. Moon said, "Like hell he is!" and he walked me back to the chopper and yelled up at the pilot, "I want this guy out of here *now*! Do you understand?"

"Get on," the pilot said. And within an hour or so Moon was shot by the Pathet Laos within ten feet of where I'd said goodbye to him. He ended up in a prison in northern Laos where he eventually died of bad treatment.

By 1980, Garrett had been on the Magazine staff for twenty-six years and had received numerous awards, among them a 1963 Newhouse Citation for "significant contribution to the field of visual communication."

Joseph Judge has had his share of adventures, too. One of them took place, he recalls, when

We were flying out of Lake Minchumina in Alaska, and the radio went out, and we had no navigation cards. [Photographer] Bruce Dale and I were in a tiny Cessna 185 with a fur trapper and the pilot. Bruce, like a contortionist, had turned himself upside down in the seat—how he did it, I'll never know!—and he'd put a little flashlight in his mouth and was working up under the instrument panel trying to get the radio fixed. Meanwhile, the windows had all frosted up and the fur trapper was scraping away at the ice on the windows. Finally the trapper got a window clear, looked down, saw a cabin he recognized and said, "Hey, I know that place! *I know that place!*"

"Thank God!" the pilot said. "Where are we going?"

And the trapper said, "Right into the side of Mt. McKinley."

Joe Judge, the son and namesake of a former Washington Senators first baseman, joined the staff of the *Geographic* in 1965 after working as a *Life* magazine reporter, a producer-director for a television station in Washington D.C., and a confidential assistant to the U.S. Secretary of Labor. Judge is a "word man" with the soul of an Irish poet—which he is—and the author of more than a score of elegantly written pieces for the Magazine on subjects as varied as Alaska, John Wesley Powell's voyage down the Grand Canyon, Herculaneum, New Orleans, and the Zulus. In November 1986, he wrote the important and historic "Our Search for the True Columbus Landfall" with these wonderful opening lines:

Christopher Columbus first came to the New World at reef-girt, low, and leafy Samana Cay, a small outrider to the sea lying in haunting isolation in the far east Bahamas, at latitude 23° 05' north, longitude 73° 45' west. It has taken me five years to write that sentence.

Although both Bill Garrett and Joe Judge were eminently qualified to serve as Editor, the Trustees selected Garrett. He was, however, by no means the automatic choice—particularly since Gil Grosvenor after ten years of holding what he called "the best job in the world," had wanted, rather than resign, to be Editor and President both.

"I wanted to be both Editor and President," Gil said, "because I still feel that you are inevitably going to have conflicts when you have two people basically running an organization. In effect you now have an organization with

DIVERSIFICATION OF ANIMALS

EARLY VERTEBRATES

VERTEBRATES INVADE THE LAND

SWAMP FORESTS

INOSAURS

FLOWERING PLANTS

EMERGENCE OF MAN

MASS EXTINCTION

650 MILLION YEARS AGO

560

500

420

360

300

230

MASS EXTINCTION

180

120

65

MASS EXTINCTION

TODAY

MAN-INDUCED EXTINCTIONS?

60 MILLION YEARS FROM NOW

TRAVELING WITH NGS

"**T**ravel is an extremely important part of one's geographic education," Gil Grosvenor said in 1986. "There is no substitute for being in a place and experiencing it yourself, the sights, the sounds, the feelings, the conversations...."

For one hundred years *National Geographic* readers explored the world through the pages of the Magazine; but during the past decade, as the *National Geographic* has continued to evolve, its editorial direction has taken it further away from its traditional travel piece. Still, readers of the Magazine, glimpsing the faraway lands portrayed in its pages, frequently found their appetites whetted enough to request information on places *they* could visit themselves—without sled dogs, Sherpas, or the need of a Citroën-type half-track vehicle. The Society's response was *Traveler*, launched in 1984—its first new magazine for adults since the *National Geographic* commenced publication in 1888.

With subscribers now approaching one million, *Traveler* differs in several significant ways from its parent publication. It is news-magazine size and focuses primarily on the more accessible places American travelers like to visit. Seventy percent of the features are on destinations in the United States or Canada; the remainder are on favorite destinations in Europe, the Pacific, and the Caribbean. Articles emphasize what a particular location is really like, how to get there, and, once there, what a visitor is likely to find.

Traveler features articles on national parks, historic places, cities, specialized museums, regional crafts and cooking, vacation sites, quiet weekend retreats. "We have a huge responsibility to give our readers accurate, readable information that will broaden their experience and stimulate their minds—whether we're examining history in Williamsburg or Salzburg, a particular culture in Rajasthan or Cajun country, or the natural science of polar bear watching in Canada or bird-watching in Arizona," says editor Joan Tapper.

The magazine's photography, in the National Geographic tradition, is colorful, spectacular, and has already won awards. And with *Traveler*'s emphasis on good prose, it has attracted such freelance authors as biographer Edmund Morris, novelists Edward Abbey and Christopher Buckley, and travel writers Kate Simon and William Least Heat-Moon.

Traveler has always provided listings telling readers how to get there, what to see, and where to stay and eat. But recently, the magazine's extensive service information has included American Automobile Association (AAA) recommendations.

Subscribers have responded enthusiastically—perhaps because the Society's award-winning magazine *Traveler* recognizes that the most educational experience of traveling is in exposing minds to the unfamiliar. "Our destination is never a place, but a new way of looking at things," Henry Miller once said. *Traveler* aims to help its readers find that new way.

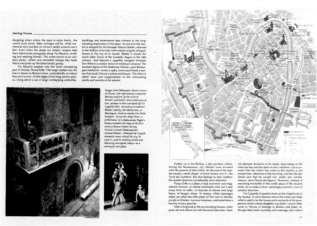

two heads: you have one person over there sending out a Magazine to 10½ million people—30 million readers—that the President has nothing to do with; and you have a President over here who's basically involved in running the Society, responsible for the books, for the educational products, but with no responsibility to the Magazine whatsoever. I feel it's very awkward, and I'm stuck with it. It's here to stay."

The appointment of the new Editor was announced on July 10, 1980. "There absolutely will be no radical or visible changes in the magazine," the *Washington Post* reported Garrett as saying. "It's like being given control of apple pie and motherhood; you don't fool around with it."

The following year, 1981, was the year that fifty-two American hostages held in the Iran embassy were freed, that Ronald Reagan took the oath as the fortieth President and escaped an assassination attempt that wounded him, his press secretary, and two others. It was the year the space shuttle *Columbia* completed its first successful test flight, Pope John Paul II was wounded by a gunman in the Vatican Square, that Charles, Prince of Wales, and Lady Diana Spencer were married, and that President Anwar Sadat of Egypt was assassinated. At the end of 1981, Gil Grosvenor took advantage of his new position as President of the Society to publish in the Magazine his "Report to the Members: It's Been a Banner Year!," a piece in which he determinedly expressed his—and the Society's—broadening interests.

The report was occasioned by the ground-breaking for the new multi-million-dollar headquarters building designed by Skidmore, Owings & Merrill architect David Childs that was to be built on the parking lot between the Society's older buildings on 16th Street N.W. and the Edward Durell Stone building on 17th Street.

"As I stood with the ceremonial shovel in hand, many memories—and hopes—crowded my mind," Gil Grosvenor wrote. "...Having relinquished my ten-year stewardship as Editor of the magazine for that of President of the Society 16 months ago, I view our mission with slightly different eyes, taking in the widening horizon of our many educational activities. What I see excites me and I wish to share with you some of the highlights of this year at National Geographic."

Gil marked the growth of the Society: "10,850,000 members in more than 180 countries...more than a third larger today than it was ten years ago"; the burgeoning staff which necessitated the new building; and the success of the television specials—"Carried by nine out of ten PBS stations, the Specials make up nearly half of the 25 top-rated PBS telecasts. Last year's series won not only four Emmys, TV's highest honor, but also a George Foster Peabody Award for 'unsurpassed excellence in documentaries.' "

"...To keep one of our most praised publications up-to-date," Grosvenor continued, "we published a completely revised and enlarged fifth edition of the *National Geographic Atlas of the World* in October. The new atlas places the earth in a galactic context. There are charts of the solar system, the visible stars, maps of the ocean floors and of the moon, and illustrations of the atmosphere, magnetosphere, plate tectonics, and climate...."

That fifth edition of the *Atlas* was not so popular as were its previous editions because all elevations and measurements were given in metric terms, prompting the sort of outcry the Geographic has come to expect ever since

First experimented with in the mid-1970s, the international publications program fully developed in the early and mid-1980s under the direction of Senior Vice President Robert L. Breeden and the coordination of William R. Gray. Under this program, publishers from other countries license the right to produce certain of the Society's books and magazines in languages other than English. Thus the Society broadens the fulfillment of its mission.

For example, the Society's 1986 large-format book The Incredible Machine *will appear in French, German, Italian, Japanese, Swedish, Finnish, and Dutch; the Society's Action ("pop-up") Books have appeared in five languages. And in June 1987, an Italian version of* Traveler *was launched by the Touring Club Italiano in Milan.*

JOURNAL OF RESEARCH

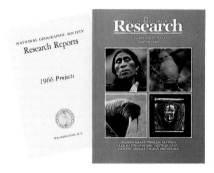

In 1890 the National Geographic Society awarded its first grant to a professional scientist in support of a field research project. The recipient was a geologist and Society founder, Israel C. Russell, of the United States Geological Survey, who was given the grant to explore the region of Mount St. Elias, Alaska. Since then, the Society has supported more than 3,000 research projects, 239 in 1986 alone.

Grant applications are subjected to peer review and subsequent consideration by the Society's prestigious Committee for Research and Exploration. Grantees are required to file preliminary and final reports on their research results. Final reports—retrospective to the first—were published, starting in 1968, in twenty-one hard-bound volumes called *National Geographic Society Research Reports*. In 1985, that series was replaced by a new quarterly scientific journal, *National Geographic Research*, edited by Harm J. de Blij, Professor of Geography at the University of Miami, member of the Society's Committee for Research and Exploration, and author of more than 25 books.

Many articles published in *National Geographic Research* are based on projects supported by research grants from the Society; however, any qualified scientist may submit a manuscript for possible publication.

Alexander Graham Bell's argument for the adoption of the metric system appeared in the Magazine seventy-five years before. The sixth edition of the *Atlas*, now in production, will retain metrics, but will also contain key measurements in inches and feet.

The Cartographic Division, responsible for all the Society's globes, atlases, maps, and supplements, is headed by John B. Garver, a former West Point professor. Garver is listed on the Magazine's masthead as a Senior Assistant Editor since, in the complex organization of the Society, the Cartographic Division falls under the direction of Wilbur E. Garrett as Editor of the Magazine.

Grosvenor's report also mentioned that "Our Special Publications and School Services Division produced 15 new books, 10 of them for children. All together, Society members received nearly four and a half million of these publications...."

The fact is the Society's books sell in such vast numbers that even an author of a book making *The New York Times Book Review*'s best-seller list would be green with envy. For example, the Book Service Division's *We Americans*, a lavishly illustrated, large-format, 456-page volume published in time for America's Bicentennial, sold more than one million copies. *Fifty States, Our World*, and *Our Universe*, the three-book set of children's atlases, have sold nearly two million copies. The Society is so conditioned by the remarkable sales records of its publications, that books which test out as selling under 150,000 copies are, as a rule, not even done.

"Our costs are between four and five million dollars to produce a book," explains Book Service Director Charles O. Hyman. "And to get that money back we've got to sell a lot of books. Break even is between 150,000 and 200,000....Our subject matter is fairly broad," Hyman said. "We don't just do 'Rivers of the United States,' we do 'Rivers of the World.' We also try to be a bit more encyclopedic, more comprehensive in our coverage, and make it more of a reference book."

Because of this, Book Service publishes books such as *Journey into China*; *Lost Empires*; *Living Tribes*; *Images of the World*; *The Adventure of Archaeology*; *Everyday Life in Bible Times*. Book Service Director Chuck Hyman reports to Bob Breeden.

Special Publications and School Services, under its Director, Donald J. Crump, publishes books that are generally more narrow-focused and tightly defined than the Book Service "encyclopedias." Established in 1965 as an outgrowth of the public service books, the Special Publications Division has grown in scope over the years to include books tailored to specific age groups and categories. For example, "Books for Young Explorers," begun in 1972, the series that carries books like *Baby Birds and How They Grow, Wonders of the Desert World, Animals That Live in the Sea*, and so forth, are designed for children four through eight years old. "Young Explorers" leads to "Books for World Explorers," started in 1977, a series with books like *Your Wonderful Body!, Amazing Mysteries of the World, Far-out Facts*, which are designed for children ages eight through twelve.

At 7 A.M. on April 12, 1981, the space shuttle Columbia *lifted off from Cape Canaveral on its maiden flight, returning Americans to space after* a six-year absence. This photograph accompanied the October 1981 article "Columbia's Astronauts' Own Story."

In "The Temples of Angkor: Will They Survive?"—his introduction to the Magazine's May 1982 coverage of Kampuchea's magnificent, eight-century-old religious complex at Angkor Wat —Geographic Editor Wilbur E. Garrett discussed his return to what had been Cambodia. "It was my first visit since 1968, when Prince Sihanouk's government still maintained a tenuous neutrality in war-ravaged Southeast Asia. That year a surplus of rice was harvested. At Angkor Wat a ballet company danced with the delicacy and grace of living apsaras, to the enjoyment of thousands of tourists. The countryside was green with life.

"What I now saw," Garrett continued, "were mass graves, barren fields, and a people trying to return from hell on earth. Only children—and only a few—were dancing."

What Garrett's photographs made evident was that not all the war's victims had been of flesh and blood, that even the temples' sculptures had been wounded.

Children who have outgrown the "World Explorer" books are then ready for the Special Publications Division's regular volumes, which the Division began publishing for the adult general membership in the mid-1960s. This series contains such popular staples as *America's Magnificent Mountains, America's Majestic Canyons, America's Seashore Wonderlands.** Special Publications Director Don Crump comes under the direction of Bob Breeden—the man he has worked with since they produced the White House Guide together in 1962.

In the same manner, children given subscriptions to *National Geographic World*, the Society's magazine for children ages eight to twelve, are expected, in time, to receive membership subscriptions to the *National Geographic*.

Gil Grosvenor's "It's Been a Banner Year" report continued: "...Countless school-age children have already been introduced to National Geographic educational materials through more than two million filmstrips so far distributed. Or perhaps they have learned about life in a pond, solar energy, or Ice Age hunters from multimedia kits prepared by our Educational Media staff, or watched a film about dinosaurs or the life of a wheat farmer...."

The report to the members further stated that the Society had given out 2.5 million dollars in grants that year to "help continue the essential discourse of research." The largest grant went to a new sky survey of the northern heavens which would probe "twice as far into space as the first survey in the 1950s." Other grants went to the archeological digs at Aphrodisias, a study of African elephant behavior and ecology, and an attempt to radio-track in the Himalayas the "elusive snow leopard, about which so little is yet known." In a conclusion that echoed Ishbel Ross' observation that membership in the Society made it possible for "the janitor, plumber, and loneliest lighthouse keeper [to] share with kings and scientists the fun of sending an expedition to Peru or an explorer to the South Pole," Gil wrote:

> These far-flung projects are the direct result of your membership in the Society and your encouragement of others to join. Each month thousands of you recommend friends by using the membership form in the front of the magazine. With the same vigor and dedication that have made the Society a unique institution for nine decades, the staff of the Geographic—and the scientists it helps support—will continue to push at the frontiers of man's knowledge of earth, sea, and sky. As a partner in these endeavors, you have good reason to feel proud.

The August 1982 issue carried the report that on April 22, 1982, Melville Bell Grosvenor had died. Bart McDowell, in writing the *Geographic*'s fond, remembrance-filled farewell, recalled:

> Once the Skipper wanted more space for a *White Mist* [MBG's yacht] story. "We need at least 55 pages," he told his friend Ted Vosburgh, MBG's successor as Editor.
> Ted objected. He knew that Mel, like most photographers, was too en-

---

**Titles with words such as "Wonderful," "Amazing," "Marvels," "Mysteries," "Splendors," "Primitive," "Adventure" and books about "America" consistently test high. So a wag at the Geographic fed various mainstays into a computer to determine what other potential best-selling titles might be. Among the hundreds of suggestions were "Marvels of the Primitive Adventure," "Isles of Sunset Paradise," "Our Mysterious Mountains," and the absolutely guaranteed, sure-fire, best-seller of all time: "America's America!"*

thusiastic about his own pictures. "That's more space than we're giving the whole solar system."

"Yes," said MBG, "but there are no *people* out there." The solar system was moved to another month.

I worked closely with the Skipper on his own stories....He loved italics and exclamation points ("Sable! Sable Island!"). I argued the virtues of understatement and finally held him to 15 exclamation points in one article. But next time around he whoopingly subverted me and used at least 40....

With his wife, Anne, at his side, he was an inveterate traveler, climbing over archaeological digs, fording forest streams, filling his pockets with notes....When President Lyndon Johnson asked him to attend the coronation of the King of Tonga, Mel Grosvenor wore striped trousers, cutaway— and a Leica camera hidden inside his silk hat; he thus made for history the only photograph of the moment of coronation....

He was, as so many friends remarked, a gentleman of the old school. And his death, on April 22 at age 80, was as gentle as the man himself. He simply went to sleep. His grave, beneath the pink bracts of a dogwood tree, was wet with a soft April rain. Sou'wester weather.

Bill Garrett said it for all of us who knew "this great, lovable man: His impact will be felt as long as there is a *Geographic*."

We wish our friend a brisk breeze for his far horizon!

Many of the *Geographic* staff consider Garrett to have inherited Melville Grosvenor's finger on the reader's pulse. "Garrett was very close to Melville," McDowell said. "He had a lot of Melville's own instinctive responses to pictures. He and Melville communicated in a terrific nonverbal way."

One long-time staffer went on to say, "Melville was perfectly willing to be the only one in the room with an opinion, and that's the one that prevailed. He didn't find this strange at all. I think Gil is more thoughtful and sensitive to other people's ideas, and sometimes hides behind a thicket of numerals to protect opinions he already has."

This same staffer feels Gil is remarkably like his father: "A kind of 'closet romantic' in a way. On the surface you don't get this from Gil. He doesn't express it the same way his father did—Melville whooped, and all his emotions were right there on the front burner; Gil inhibits his displays—but they're there. Gil really is a romantic."

Clearly the same fires that burned in Melville to move the Society into other fields burn in his son Gil. And, like his father, Gil has acted on his impulses.

On June 19, 1984, President Ronald Reagan dedicated the just-completed headquarters building on M Street, saying, "In a world that sometimes seems to have grown sated with all it knows, you still discover. You fund expeditions, you help researchers, you encourage impossible dreams—then you share the results with all the Society's members."

Earlier, as he looked around the vast new building, the President of the United States had joked, "I guess you have trouble storing your old *National Geographics*, too."

In "Our Society Opens New Doors," the Magazine's piece marking the dedication of the new M Street building, Gil Grosvenor responded, "Mr. President, rather than needing a space to store our past, we needed a space to house our future."

The six sheep that this Peruvian shepherd boy had been tending for his family had just been struck and killed by a taxi when photographer William Albert Allard happened by and took this photograph. Allard's photograph, showing the young shepherd's grief-stricken face and the terrible desolation the boy felt, appeared in the March 1982 issue accompanying Harvey Arden's article "The Two Souls of Peru."

Five months later Bill Garrett reported in his Editor's page that members had "spontaneously contributed more than $4,000 to replace the sheep [and that] through CARE in Peru, the boy, Eduardo Condor Ramos, was found, and Assistant Executive Director Ronald Burkard presented him with six new animals.... Eduardo, incredulous, broke into tears again and said, 'God will pay you.'"

Pointing out that it was "only 16 years to year 2000, and a briefer four years" to the Society's centennial, Gil's report continued:

> It is our hope that the study and science of geography will make strides in the decades ahead. We have a major commitment to this goal....
>
> The newest building, 1600 M Street N.W., will enable us to experiment with and develop educational programming utilizing the laser disk, direct broadcasting, cable networks, and technologies not yet developed. We will be able to reach directly into members' homes and classrooms of the world with printed and broadcast material....
>
> It is an exciting prospect, and especially so in the light of the many challenges, essentially geographic in nature, that confront us—how to obtain clear skies and clear streams, protect wilderness and forests where genetic pools of life are safe, while still making progress in the increase of living standards for a rapidly, in places alarmingly, growing world population.
>
> Starting with the small but elegant Hubbard Memorial Hall on a corner of our Washington complex, the Society has grown continuously over the years as it responded to the changing world. Now our headquarters is complete, our sails are trimmed, and our course is set with confidence.

And as an example of that confidence Gil Grosvenor had already launched *Traveler*, the Society's first new magazine for adults since its founding in 1888.

The steadily evolving yellow-bordered *National Geographic*, Gil felt, had been moving away from the traditional geography travel piece, and *Traveler* was designed to fill that gap. "We know our membership is deep into travel and geography," Gil said, "and since travel is a significant part of one's geographic

439

Commitment to the city of Washington —the Society's birthplace and its home for a century—has created a blockwide building complex spanning eighty-five years. Hubbard Memorial Hall, the Society's first permanent home (at far left), rose on the corner of 16th and M Streets N.W. in 1903. A connecting building along 16th Street was completed in 1913, with additions in 1932 and 1949.

The present headquarters building (at right) at 17th and M Streets N.W. was dedicated by President Lyndon B. Johnson in 1964. Designed by noted architect Edward Durell Stone, the ten-story white marble structure is now known as the Melville Bell Grosvenor Building. Its first floor contains Explorers Hall.

On June 19, 1984, President Ronald Reagan arrived to dedicate what the press delighted in referring to as the Society's new "Maya temple"-like M Street building (center), which completes this complex in the heart of the nation's capital, only blocks from the White House. Once inside, the President joked, "I guess you have trouble storing your old National Geographics, too."

EDUCATIONAL MEDIA: HELPING CHILDREN LEARN

In 1972, the Special Publications Division's Filmstrips Department's first catalogue offered thirty-four filmstrips for grades 5–12; by 1987, the Society had printed and delivered more than two million copies of nearly 450 filmstrips, which are now seen by millions of schoolchildren in grades K–12 each year.

Although filmstrips are still the cornerstone of the Society's efforts to meet the audiovisual needs of K–12 teachers, by the late 1970s and early 1980s the Society's educational programs had diversified rapidly both in media and range of subject matter. Captioned filmstrips for the hearing impaired were developed in the 1970s. So were Wonders of Learning Kits—a series of multimedia reading programs designed to teach science and social studies concepts while improving reading and listening skills. And a new teaching medium, Participatory Filmstrips

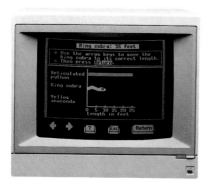

called "Look, Listen, Explore," was designed for children in grades K–2.

After extensive testing with teachers and students, new multimedia kits that included computer software—in addition to filmstrips, color-illustrated booklets, and activity sheets—were developed in 1984. And the following year full-scale production began on the Society's first computer software designed to build skills in geography and other related subjects. By then the dramatic increase in the number and diversity of materials being produced by the department led to its restructuring, also in 1985, as the Educational Media Division with George A. Peterson as its Director.

Now interactive videodiscs are under development for schools. (The Division's first research project, *Whales*, scored so highly in field tests it has already been released for general distribution.) The videodisc's principal advantage is that the disc player's laser reader permits virtually instantaneous access to any part of the videodisc. The Educational Media Division is developing an interactive videodisc and computer project to teach geography. This technology will permit students and teachers to program their own lessons and to have access to a storehouse of information literally at their fingertips.

An exciting program called National Geographic Kids Network has been developed by the Division in collaboration with the Technical Education Research Centers of Cambridge, Massachusetts. Students across the country, using classroom activity kits and computer software produced by the Society, will be able to share data and compare geographic trends from common experiments via telecommunications. The National Geographic Kids Network is par-

tially funded by a 2.2 million-dollar grant from the National Science Foundation in addition to a matching 2.5 million-dollar contribution from the National Geographic Society.

In 1985, the Society embarked on a major program to improve geography education. Pilot projects were started by the Educational Media Division at Alice Deal Junior High School in Washington, D.C., and at Audubon Junior High School in Los Angeles, California, where social studies and science teachers experimented with a variety of ways to improve geography teaching methods and materials. The Division also administers the Society's Summer Geography Institutes, a network of regional Geography Alliances, and other related programs to encourage support for local and nationwide efforts to improve geography education.

In 1888 one of the aims of the Society was stated as giving "due prominence...to the educational aspects of geographic matters." The Educational Media Division is fulfilling that aim.

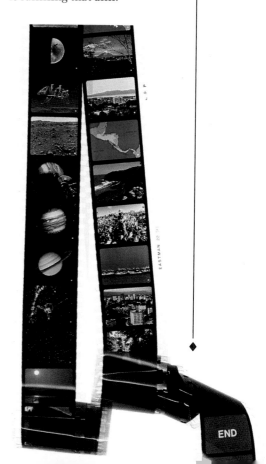

education, I feel travel is an important part of our obligation to the members. *Traveler* is still young; it's experiencing growing pains just as the parent magazine did. We're still evolving it to meet the needs of the readers."

Despite its youth, *Traveler*, under Editor Joan Tapper, has already published such fine pieces as Edward Abbey's "Big Bend: Desert Rough and Tumble," Ken Brower's "Gaspé: Quebec's Wild Peninsula," Steve Hall's "Baseball Scrapbook: Cooperstown's Hall of Fame and Museum," and Rob Schultheis' "Yellowstone: Wonderland at Zero Degrees." *Traveler* comes under the direction of Bob Breeden.

And that same year he announced he was starting yet another magazine: *National Geographic Research*, a scientific journal designed primarily as a publishing forum for the results of Society-funded research.

Ironically, seventy-six years earlier, the Geographic's Board of Trustees had passed a resolution calling for the Society to "undertake the publication of a technical journal to be separate from the *National Geographic Magazine*." But that direction faded when the Society, guided by Gilbert Hovey Grosvenor, continued to concentrate on popular geography.

Until *National Geographic Research*, summaries of grant recipients' findings appeared in the Society's annual *Research Reports*—"a ledger of stiff, dry tracts," wrote Leonard Krishtalka, the Carnegie Museum's Associate Curator in the Section of Vertebrate Fossils, "devoid of maps or photographs."

"Not so *National Geographic Research*," Krishtalka continued in his glowing review (in the September 25, 1986, issue of *Nature*) of the Society's newest magazine:

> [It is] the handsomest, most elegant professional scientific journal known to me...and it finally puts the lie to Darwin's axiom that if it's handsome, it's art; if it's science, it's dull....

National Geographic Research has already published articles on archeology, biogeography, botany, ecology, entomology, ethology, geology, historical geography, palynology, urban anthropology, vertebrate paleontology and zoology. One "unusually lucid" bit of writing from this last was Wesley W. Weathers' "Thermal Significance of Courtship Display in the Blue-black Grassquit (*Volantinia jacarina*)," which contained the following "science in the flesh" observation:

> One of the most common of a dozen or so species of small, lowland, Neotropical seed-eaters, the blue-black grassquit is remarkable...for the male's peculiar aerial display. Displaying males vault about 0.4 m into the air, emit an explosive buzzy *Dzee-we* call, and then return to the original perch. This behaviour is notable for its frequency (as many as 20 displays a minute) and because it is performed from an exposed perch, often in direct sunlight. By displaying in direct sunlight, male grassquits subject themselves to substantial heat stress. Indeed they often pant with their mouths open while displaying.

Weathers' piece begs for the kind of tongue-in-cheek response George B. Schaller's December 1981 "Pandas in the Wild" *Geographic* article inspired.

Writing of the fresh bamboo shoot diet and mating habits of pandas, Schaller had noted, "I observed [the male panda] mount her 48 times in three hours"—an observation that prompted one astonished reader to write: "Pandas—48 times in 3 hours? 180 divided by 48 = 3.75 minutes per cycle. No

Senior Associate Editor Joseph Judge's press conference on Samana Cay as the true landfall of Christopher Columbus in the New World was the most successful in the Society's history: A total of 1,016 newspapers and magazines with a combined circulation of 88,088,437 ran the story. It was on the front page of 177 newspapers and featured in all major television networks' news.

The press conference was orchestrated by one of the Society's oldest divisions: News Service—the Society's "public voice." It reaches millions of readers through the 1,500 newspapers and periodicals whose editors receive this free service from the Society, and millions of listeners through the 250 or so radio stations that regularly broadcast "Horizon" over the Associated Press network.

News Service Director is Paul Sampson.

wonder their eyes are black! P.S.: Where can I purchase fresh bamboo shoots?"

National Geographic Research falls under the direction of Bob Breeden.

Television, too, expanded during the 1980s. *The Incredible Machine*, the Society's 1975 Special which had earned the highest audience ratings in Public Television history was topped in 1982 by *The Sharks* and again by *Land of the Tiger* in 1985. In fact, by 1986 nine out of the top thirteen most-watched shows on the Public Broadcasting Service had been National Geographic Specials.

Today, under the guidance of Dennis Kane, the Geographic's Director of Television, four National Geographic Specials a year continue to be aired on Public Television; in addition these Specials are seen in syndication. And now a new, weekly award-winning *Explorer* cable television series, which draws from films produced by the National Geographic, can be viewed. Dennis Kane reports directly to Grosvenor.

Television has made such an impact on the Society that whereas a staffer, when saying he or she worked for the Geographic, might have once been told, "Oh sure, I read the piece you ran on Spiders,"—or "The Incredible Potato," or on climbing "To Torre Egger's Icy Summit"—the staffer is more likely today to hear, "Oh sure, I just saw the Special you had on pandas"—or polar bears, or wild chimpanzees, or the great whales.

This is not to suggest that the Magazine is not being read—or, at least, still being looked at. During the 1980s the *National Geographic's* evolution has been such that, once again, it *is*, in fact, being read—or should be.

"People who tell you the *Geographic* is a picture magazine are right," concedes Charles McCarry, whom Garrett hired in 1983 to be the Senior Assistant Editor in charge of Contract (free-lance) Writers. "There are lots of people here who know good pictures when they see them; historically, fewer have known good words when they see them," McCarry says. "In the past the practice was: If you need space in a story, cut the text. Nowadays nearly everyone recognizes that text is not a gray border to make the Kodachromes look brighter."

"By its history and by its nature," McCarry continues, "the *National Geographic* is meant to be a picture magazine, but there's no reason we can't have a very strong text." During the last dozen years, in fact, the text has been so strengthened that the magazine now publishes some of the finest expository writing around. Nowhere, for example, can one find such difficult scientific topics treated so comprehensively; and in no other popular magazine can one find such a depth and breadth of research on every printed page.

Under Garrett's leadership the Magazine has continued to publish such *Geographic* staples as photographs from space ("What Voyager Saw: Jupiter's Dazzling Realm," January 1980; "Voyager 1 at Saturn: Riddles of the Rings," July 1981; "The Planets: Between Fire and Ice," January 1985); wildlife (Dian Fossey's "The Imperiled Mountain Gorilla," April 1981); adventure ("*Kitty Hawk* Floats Across America," the first balloon crossing of the United States, August 1980; "First Across the Pacific: The Flight of *Double Eagle V*," April 1982); mountaineering (Reinhold Messner's "At My Limit—I Climbed Everest Alone," October 1981); bare-breasted natives ("In the Far Pacific: At the Birth of Nations," October 1986); natural disasters (volcanic eruption: "Mount St. Helens" and "Mount St. Helens Aftermath," January 1981, December 1981); war and conflict ("Along Afghanistan's War-torn Frontier," June 1985; "Nicaragua: Nation in Conflict," December 1985); and the esoteric

Opposite: *"In the heart of a Maya pyramid, excavators reach toward the roots of one of the New World's most accomplished ancient civilizations. The Maya flowered during the first millennium* A.D., *creating grand art and architecture and an advanced system of writing. Now findings at Cuello, a ceremonial center uncovered in northern Belize, push back the first stirrings of lowland Maya culture to 2400* B.C.

"The site reveals early development in building, crafts, agriculture, and trade—and unexpectedly early evidence of human sacrifices. These discoveries open a new chapter in the complicated saga of the Maya, now proven to be among the first settled peoples in the New World." —Caption from "Unearthing the Oldest Known Maya" in the July 1982 issue.

Opposite: *"Breakfast between Peshawar and Lahore is a dizzy adventure for bearers who pass trays between the dining car and first class, where locked inside doors assure security."—From Paul Theroux's article "By Rail Across the Indian Subcontinent" in the June 1984 issue*

("The Indomitable Cockroach," January 1981; "The Fascinating World of Trash," April 1983).

One of the most exciting pieces from this period was Garrett's own that appeared in the August 1984 issue. The Editor was on board a Guatemalan helicopter flying to the dramatic opening of an ancient Maya tomb that had, he wrote, "all the makings of a childhood fantasy. An ancient city lost under the green canopy of a remote jungle. Vines and roots snaking over temples and pyramids. And, beneath it all, an undiscovered tomb containing the art of a vanished civilization and the skeletal remains of a long-dead nobleman lying in undisturbed repose."

En route, the confused Guatemalan helicopter pilot—who had already landed four times in Mexico and Belize to ask directions—ran out of fuel over the jungle; the helicopter sheared off the tops of five trees on its way down. Garrett and the others in his party spent an uneasy night in the jungle, listening for guerrillas, before they were located by a second helicopter and supplied with chain saws so they could clear a landing pad and be lifted out.

Eventually Garrett reached the site of the Maya tomb—"the first to be officially reported in Guatemala in 20 years"—and recounted for the *Geographic* how the tomb had been found untouched since it had been "sealed by the Maya 1,500 years ago" in the Río Azul region of northern Guatemala:

> From the time of the fall of Rome the tomb had lain undisturbed. Even food and drink left for the deceased remained—now only powdery residue in ceramic pots and plates.
>
> For reasons still unknown, the site and the Maya culture had faded away by the tenth century. Under cover of the jungle the city—now known as Río Azul—lay for a thousand years as a vast mausoleum protecting the artifacts of the once bustling society. Palaces, homes, and pyramids of the early Classic period (A.D. 250–600) ever so slowly slumped into piles of rubble under the relentless wash of tropical rains and the insidious pushing and shoving of tree roots. As the green canopy grew over fields, canals, and plazas, only the occasional Indian chicle collector, armed with nothing more devastating than his machete, disturbed the tangle of vines and branches....

But under Garrett the Magazine has also gone into areas of advocacy journalism that would have been at odds with the sentiments of the members of the 1977 ad hoc committee. Among the more noteworthy pieces have been "Wild Cargo: the Business of Smuggling Animals" (March 1981), "Acid Rain—How Great a Menace?" (November 1981), "The Temples of Angkor" (May 1982), [this with photographs by Wilbur E. Garrett*], "Tropical Rain Forest: Nature's Dwindling Treasures" (January 1983), and "Escape From Slavery: The Underground Railroad" (July 1984), with Charles L. Blockson's moving text:

> "My father—your great-grandfather, James Blockson—was a slave over in Delaware," Grandfather said, "but as a teenager he ran away underground and escaped to Canada." Grandfather knew little more than these

---

*Garrett's "visa" for Cambodia consisted of a handwritten note torn out of a spiral notebook. He was one of the first outside journalists to reach Phnom Penh after the holocaust there and witnessed the opening of a mass grave containing more than 6,000 bodies, some with skulls still blindfolded and arms still bound by wire. The elaborate temple complex at Angkor, virtually closed to outsiders for more than a decade, had so sadly deteriorated that Garrett produced an exhibit sponsored by UNESCO which was sent to many countries in an effort to inform the world of what was happening.

PUTTING GEOGRAPHY BACK INTO AMERICA'S CLASSROOMS

"The most important thing you can do is validate teachers," Dr. Floretta McKenzie, Superintendent of the District of Columbia Public Schools and Society Trustee told Gilbert M. Grosvenor at a meeting held to discuss ways to improve geographic education in America. "Tell them they are important and that geography is important." Thus began the Geography Education Program, the Society's campaign to revitalize study of what many consider the "mother of all discipline," and what Gil's grandfather Gilbert Hovey Grosvenor realized the public regarded as "the dullest of subjects, something to inflict on schoolboys and to avoid later in life."

And yet if in recent years geography as a course of study has been inflicted at all on schoolboys and schoolgirls, it has clearly not taken hold! Consider these statistics from a recent CBS news survey: 25 percent of high school seniors in Dallas, Texas, could not identify the country that borders the United States on the south, and 63 percent of their Minneapolis–St. Paul, Minnesota, counterparts could not name the seven continents. And a statistic Gil Grosvenor finds particularly unnerving: 95 percent of the incoming freshmen at an Indiana college could not even locate Vietnam on a map!

The Geography Education Program's aim is "to elevate the status and effectiveness of geographic education," by helping to enhance geography teaching methods and materials, by developing a nationwide teacher support network, and by conducting a public awareness campaign to draw attention to geography and to recruit allies among the public, corporate, and foundation sectors.

The Society has created a network of regional Geographic Alliances—groups of teachers, administrators, and college geographers who meet to exchange and develop teaching strategies and materials, and to discuss ways to upgrade geography curricula. These alliances will number twenty-two by 1988.

Groups of five to seven topnotch teachers, recruited from each alliance, are brought to Society headquarters each July to attend a four-week Summer Geography Institute—inaugurated by the Society in 1986—before returning to their home alliances to share what they have learned and to help organize regional institutes for the benefit of their colleagues. "The ideas you are giving us are causing light bulbs to go on in our heads," one teacher wrote of her 1986 Summer Institute.

Governor Gerald Baliles of Virginia declared his stance on the need for improved geographic education: "We no longer do an effective job of teaching geography, when we teach it at all." He backed his words with a $50,000 pledge in support of alliance activities in his state. Several more governors have committed state funds for alliance-run institutes, curriculum conferences, and other professional development activities.

The Society, for the first time in its 100-year history, has also begun to seek foundation and corporate co-sponsors to provide funding and other forms of support for geographic education activities.

In re-dedicating the energies of the Society to the restoration of geography in America's schools, Gilbert M. Grosvenor is determined to fulfill the pledge of its founders "to increase and diffuse geographical knowledge."

bare details about his father's flight to freedom, for James Blockson, like tens of thousands of other black slaves who fled north along its invisible rails and hid in clandestine stations in the years before the Civil War, kept the secrets of the Underground Railroad locked in his heart until he died....

Under Garrett the Magazine also published "The Poppy" (February 1985), "Storing Up Trouble...Hazardous Waste" (March 1985), and "Vietnam Memorial," which ran in the May 1985 issue.

As part of the "Vietnam Veterans Memorial" coverage there appeared Joel L. Swerdlow's "To Heal a Nation," one of the most moving pieces ever to be printed in the Magazine. It contained this passage on the impact the Vietnam Memorial's stark, name-filled, shining black granite panels have on viewers seeing them for the first time:

> ...for an unbroken stream of months and years, millions of Americans have come and experienced that frozen moment.
>
> The names have a power, a life, all their own. Even on the coldest days, sunlight makes them warm to the touch. Young men put into the earth, rising out of the earth. You can feel their blood flowing again.
>
> Everyone, including those who knew no one who served in Vietnam, seems to touch the stone. Lips say a name over and over, and then stretch up to kiss it. Fingertips trace letters.
>
> Perhaps by touching, people renew their faith in love and in life; or perhaps they better understand sacrifice and sorrow.
>
> "We're with you," they say. "We will never forget."

Three months later Garrett's editorial page opened:

> "I wept...." "My husband cried...." "The pages were wet with my tears...." "I was crying and didn't know why...." "It was the first time I cried over this war, and it felt good."
>
> So wrote five members. They symbolize the hundreds who felt compelled to respond to the articles about the Vietnam Veterans Memorial in our May issue. The reader response may eventually be the largest in our magazine's history; there is no question that it is already the most intensely personal.

Among those who responded was the widow of Walter H. Moon, the Special Forces Army captain Garrett had last seen alive in Laos in 1961 ordering a reluctant helicopter pilot to get Garrett "out of here *now!*"

"My son...was thrilled by your Editor's column, where you said you had looked for his father's name. He was eight and my daughter four when Walt left. Though they don't remember him as well as they would like to, they loved him and are very proud of him," she wrote.

The November 1985 Magazine with its stunning holographic cover image of Africa's one- to two-million-year-old Taung child contained "The Search for Our Ancestors," a comprehensive article on fossil discoveries of prehistoric man around the world. Publication of that article led, as expected, to an outcry from creationists. A Lubbock, Texas, man wrote:

> The theory of evolution is probably the biggest hoax ever foisted on intelligent people. Even though it is widely accepted theory, it is held mostly by those who have already rejected a belief in God. I cancel my subscription.

And a man from Angleton, Texas, complained:

Early in 1981 the National Geographic Society released a special 115-page supplement to the Magazine reporting on the outlook for energy resources and technologies. Prepared during the Editorship of Gil Grosvenor under the direction of Geographic Science Editor Kenneth Weaver, the report was titled "Special Report on Energy."

In a pre-publication announcement of the energy special newly elected Editor Wilbur Garrett reported, "This edition will carry no advertising. All costs will be borne by the Society as a service to you the members."

Costs in staff effort were high enough to make "We Survived the Energy Issue" a popular T-shirt among those assigned to the project.

449

From left to right: Society Executive Vice President Owen R. Anderson, Senior Vice President Robert L. Breeden, and Chairman of the Board and President Gilbert M. Grosvenor listen from the front row as National Geographic *Editor Wilbur E. Garrett explains the state-of-the-art video projection facilities contained in the Society's new forty-seat "Control Center" to Society staff members and Trustees.*

Capable of displaying standard video signals from VCRs and videodiscs, as well as high-resolution signals from computer monitors on its 110-inch projection screen, the Control Center has the capacity to generate quickly screen displays or reports of selected data in any preferred sequence. Although the systems provide increased capabilities, they remain consistent with time-proven traditions and managing scheduling.

I'm becoming more distressed by the Society's attempt to educate everyone on its concept of the origin of the Earth. I'm tired of the attitude that evolution is a fact of life and not the controversial theory that it is. I would like to see research that promotes other theories.

To the Angleton writer, the *Geographic* responded:

Behind these *National Geographic* articles lies a vast array of studies of the fossil record, geologic strata, radiometric dating, molecular biology, embryonic development, and comparative anatomy. This observable body of facts demonstrates that earlier life-forms were ancestral to later ones. Some early forms have been so successful at adapting to various environments that they have survived into the present while others have become extinct. Scientists theorize how new species develop out of parental stock, debating the importance of random change, mutation, and isolation, but very few dispute evolution itself.

Other recent articles that have caused a stir have been the December 1985 and December 1986 pieces on finding and photographing the luxury liner S.S. *Titanic,* sunk in the North Atlantic when it struck an iceberg in 1912 with the loss of 1,522 lives; Peter Jaret's June 1986 piece, "Our Immune System: The Wars Within," which dealt, only in passing, with AIDS (and prompted a letter from a Quilcene, Washington, member who wrote: "Be advised your field of endeavor is geography not biology. Cover our planet Earth first."); and "North to the Pole," the September 1986 story by co-leader Will Steger of the first successful dogsled expedition without resupply to the North Pole since Peary in 1909.

But Steger's Polar adventure was almost lost in the rush by publications and television stations to print and comment on another piece in that issue: Boyd Gibbons' ingenuously sensuous article called "The Intimate Sense of Smell," with its lines from Baudelaire—

In bed her heavy resilient hair
—a living censer, like a sachet—
released its animal perfume,
and from discarded underclothes
still fervent with her sacred body's
form, there rose a scent of fur.

—and its shocking photograph of laboratory-jacketed middle-aged women sniffing bare-chested men's armpits: "It's the pits that produce our strongest body odors..." the photo caption began. That photograph appeared on the front page of the August 20 *New York Daily News*, prompted a *Washington Post* story called "Geographic's Call to Odor," and got Editor Garrett on the *Good Morning America* television show—along with one of the women from the laboratory.*

Clearly, the Magazine reflects Wilbur Garrett's enthusiasm for learning, his continuing concern for people and wildlife, and an unabating appreciation of adventure. But one measure of how well Garrett is doing as Editor is reflected by the *National Geographic*'s having been named recipient of the 1984 Magazine Award for General Excellence, considered the Pulitzer Prize of the magazine world.

Obviously, Garrett could not please everyone. A Monticello, Mississippi, man wrote to the Society's "Board of Governors" in April 1986:

Dear Sirs:
Because of the contents of the National Geographic for years and years and years and years, a content largely unfavorable to a better United States, and because of the imposing audacity of the title of the magazine, I recommend cessation of publication of National Geographic Magazine.

To that Mississippian's dismay, the Magazine will be around for years and years and years and years, but as the Geographic's centennial approaches, it is natural to ask where the Society will go from here.

Fittingly enough, the current emphasis is on going to the very beginning: to return to the announcement published on the opening page of that first, somewhat stiff and forbidding-looking terra-cotta October 1888 *National Geographic Magazine*, in which one found articulated the Society's aims:

The "National Geographic Society" has been organized "to increase and diffuse geographic knowledge," and the publication of a Magazine has been determined upon as one means of accomplishing these purposes...
As it is hoped to diffuse as well as to increase knowledge, due prominence will be given to the educational aspect of geographic matters, and efforts will be made to stimulate an interest in original sources of information....

Certainly the Magazine is now but *"one* of the means" by which the Society increases and diffuses knowledge, but until recently little priority has been given—other than within the Educational Media Division—"to the educational aspects of geographic matters."

All this is changing.

The opening salvo was fired by Gilbert M. Grosvenor in his June 1985 President's column, written with both a sense of pride and embarrassment at

---

Membership response to the smell survey enclosed in that issue was astonishing: more than 1.5 million returns—twenty-eight tons of mailed-back tests.

"Psst! Want to buy some National Geographic centerfolds?"

accepting an award from an honor society for professionals in education which recognized the National Geographic Society as "Educator of the Year" for its role as "a vital force in the continuous education of mankind."

Citing the appalling results of a questionnaire on geography given to 2,200 North Carolina college students (only 27 percent could name in what countries the Amazon River was mainly found and in what country the city of Manila was located, only 7 percent could name three of the thirty countries between the Sahara and South Africa, and 69 percent could not even name one!), Gil Grosvenor had written:

> How this coming generation will make any sense of a world increasingly tied together by communications, transportation, trade, and international relations, I cannot imagine....
>
> When I accepted our Society's award as "Educator of the Year," I said it would be better given for "Non-educator," considering the low state of geography in our schools. I reaffirmed my personal commitment, as well as the Society's, to help improve the education of our citizenry in geography.
>
> Mine is not an idle promise. We are increasing our efforts in developing learning materials for schools, and we are exploring joint efforts with others in the private sector. You will hear more from me on the subject of geographic education, and I would like to hear more from you. I am angry; I am embarrassed; I am determined.

Fifteen months after that President's column appeared, *Newsweek* was able to report that "Having reached a nadir, there are now signs that geography studies may be starting the long climb up." Continued the September 1, 1986, *Newsweek*:

> Much credit belongs to the National Geographic Society (NGS), which has budgeted $4 million to improve geographic education. At the NGS's recently concluded Summer Geography Institute, 50 high-school instructors from around the country gathered in Washington, D.C., to learn how to teach the subject more effectively; as part of the deal, they agreed to give at least three geographic methods workshops to colleagues back home. The society is also cosponsoring school-year pilot programs in Washington and Los Angeles. A few states, including South Dakota and California, have implemented geography requirements for graduation. Clearly, geography is finding its way back onto the map. Now it's up to teachers to make sure students can read it.

"You *can* make geography exciting," Gil Grosvenor says.

> I'm very excited about this program. I spend a helluva lot of my time on it. I feel it's important, that we can perform a service for this country, a service for the Society, and that we can materially influence the education, the geographic awareness of the average American—particularly as he or she goes through the school system.
>
> "Geography" is more than just place, location, the inner relationships between flora and fauna on this planet earth. Geography in a very broad sense is human and environmental issues, it's movement, it's regional influences. You're dealing with climate, with agriculture, with such things as desertification, soil, acid rain....Without geography you're going to have, at best, a two dimensional outlook on life. Geography permeates every aspect of our life. Pick up a newspaper and every single day you'll see how geography plays a dominant role in giving you a third dimension on life.

452

Geography *drives* history. Unless you've been to the Pass of Thermopylae you haven't the faintest idea what happened there. But the Battle of Thermopylae could only have happened at Thermopylae. Geography drove that battle. Even today geography is driving history. Look at Afghanistan: the Russians are in Afghanistan for geography. They're looking for routes to the sea, for routes down through the Himalayas, for routes to oil. They're not in Afghanistan just to expand their territories. Geographical-political implications are what's driving them! It's what drove us into Vietnam, and I submit to you that if our leaders had *really* been informed on the geography of Southeast Asia, we never would have been in a war there.

What disturbs me, is that 95 percent of the incoming freshmen at a pretty good Indiana college couldn't locate Vietnam on a map! What this tells me is that the next generation hasn't learned a damn thing about the disaster this government participated in in Southeast Asia. Not a damn thing. And that's scary. I feel it's going to be even more critical in the years ahead for a citizen to have a balanced knowledge of the planet earth. It all makes sense. What is frightening to me is why were we asleep for so long? Why did we watch geography go down the chute without really realizing it? It really is important.

What about the future?

I would hope that as time progresses the average member will perceive us as a Society above and beyond a magazine, that they will see us as a Society dedicated to the increase and diffusion of geographic knowledge across an entire age spectrum, and utilizing as many different media of communication as is possible: the printed medium, electronic communications—electronic communications, I think, are going to dramatically increase.

The day may come, 30 or 40 years down the line, when it might become prohibitively expensive to send the Magazine out in print; the day may come when the Editor will produce a thirty-page guide to what is on the "menu" for this particular month and the member will get a laser-video disc with 56,000 images on it, and maybe that guide is what will motivate the "reader" to get into this month's menu. I don't think we will reduce our Magazine, but I do think we're headed in the direction of *complementing* the Magazine, of adding to it, making it more useful and enjoyable.

"I would hope that at our 200th anniversary," said this great-grandson of Alexander Graham Bell, "somebody will look back and say, 'Hey, you know, those people between the first and second hundred, they really broadened the outreach of the National Geographic Society! It became an even more important international educational institution for the increase and diffusion of geographic knowledge."

There are currently almost eleven million members scattered all over this planet—in the Americas from Argentina to Venezuela; in Africa from Algeria to Zimbabwe; in Asia from Afghanistan to the two Yemens; in Australasia from Australia to New Zealand; in the Atlantic Ocean from Ascension Island to St. Pierre and Miquelon; in Europe from Andorra to Yugoslavia; in the Mediterranean from Corsica to Malta; in the Pacific from Fiji to Western Samoa; in the Indian Ocean from Mauritius to the Seychelles; in the British Isles from the Channel Islands to Wales—nearly eleven million members who have no doubt whatsoever that in the year 2088 the National Geographic Society will still be around.

The March 1984 issue carried the first hologram cover, an eagle. The holographic portrait of the skull of Africa's "Taung child" appeared in November 1985.

By the end of the Society's ninety-eighth year, circulation was 10,764,998 with membership in 170 of the world's 174 nations.

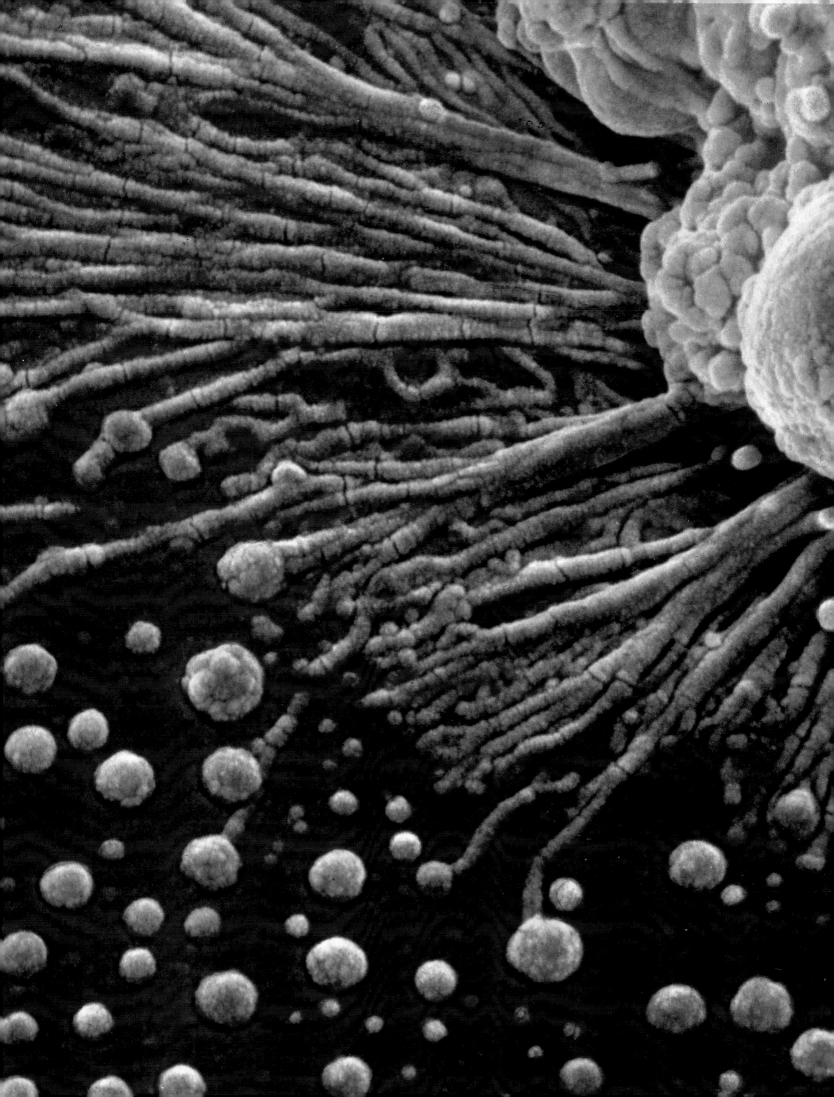

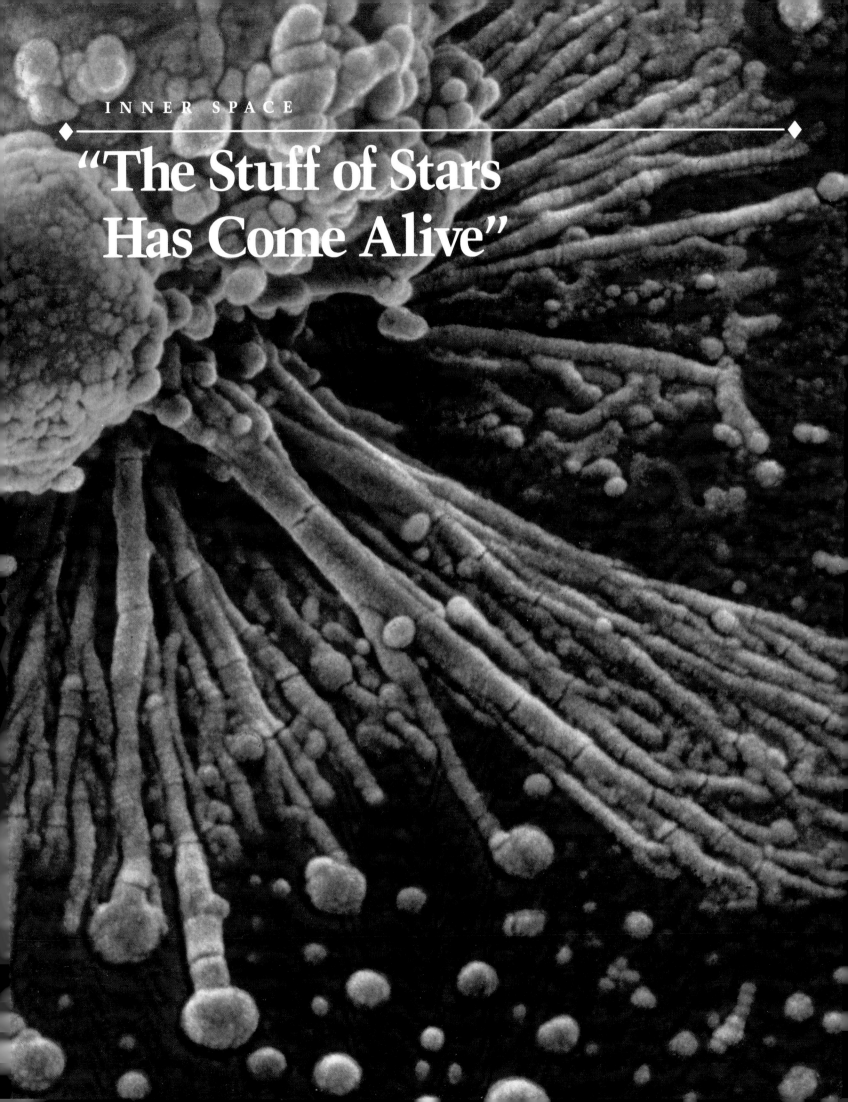

"The Stuff of Stars Has Come Alive"

Just as scientists with devices—optical and otherwise—are trying to look deeper and deeper into outer space, scientists elsewhere—also with devices, optical and otherwise—are trying to look deeper and deeper into inner space. Paradoxically, as Susan Schiefelbein writes in the National Geographic Society's striking book *The Incredible Machine*, what these scientists are finding is that:

> Within our bodies course the same elements that flame in the stars. Whether the story of life is told by a theologian who believes that creation was an act of God, or by a scientist who theorizes that it was a consequence of chemistry and physics, the result is the same: The stuff of stars has come alive. Inanimate chemicals have turned to living things that swallow, breathe, bud, blossom, think, dream.

How did that happen? What caused the "Big Bang"? Out of what did vast swirling clouds of hydrogen and helium collect together into huge clumps? Why were these riven by nuclear explosions as atom blasted into atom until, burning star bright, they burst yet again to fling even more and different combinations of atoms throughout space?

And how, out of all that nuclear debris, did the hot ball of gases we call our Sun form and burn at exactly the right temperature so that bits and chunks of matter, collecting together to become our infant planet earth, would bubble and seethe, cool and harden, wrinkle and fracture, and bathe its wounds until primordial oceans formed?

And how over those billions of years in that oceanic chemical caldron did molecules fuse into chains that mixed and mingled in such astonishing, infinite variety that suddenly somehow—because of ultraviolet light? heat? lightning?—one of those chains became unlike any other? And what made it so utterly different was that it was *alive*!

As National Book Award winner Dr. Lewis Thomas writes in his Foreword to *The Incredible Machine*, how that happened remains

> ...the greatest puzzle of all, even something of an embarrassment. Somehow or other, everything around us today—all animals, ourselves, all plants, everything alive—can trace its ancestry back to the first manifestation of life, approximately 3.5 billion years ago. That first form of life was, if we read the paleontological record right, a single bacterial cell, our Ur-ancestor, whose progeny gave rise to what we now call the natural world. The genetic code of that first cell was replicated in all the cells that occupied the Earth for the next 2 billion years, and then the code was

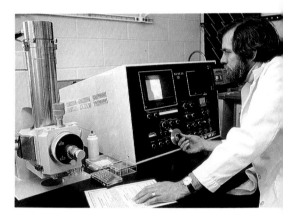

Just as the Society has reported the exploration of outer space, it has chronicled the exploration of inner space, choosing the best images technology can provide of this largely invisible world.

Above: *A technician adjusts a scanning electron microscope capable of magnifying thousands of times—it makes a human hair look like a tree limb.*

Opposite: *David Fairchild and his wife prepare "Long Tom," the twelve-foot camera with which he photographed his 1913 "monster" pictures—ordinary insects enlarged up to twenty times—that appeared in the 1914 Society publication* The Book of Monsters. *Capturing the insect's image on a 5 x 7 photographic plate requiring a one-minute exposure, though laborious, was easy compared with getting the creature to stand still.*

Preceding spread: *"Scanning electron microscope shows a macrophage, a human defense cell, seeking to engulf droplets of oil." Magnified 18,000 times.—From the June 1986 issue*

passed along to nucleated cells when they evolved, then to the earliest multicellular forms, then to the vertebrates some 600 million years ago, and then to our human forebears. The events that [have taken] life all the way from a solitary microbial cell to the convolutions of the human brain and the self-consciousness of the human mind, should be sweeping us off our feet in amazement.

We *are* amazed, of course; and it is an amazement that is in no way diminished as we develop devices and technology with which to see yet deeper, to peel back layers of the mystery, only to discover even greater mysteries still. We are at the edge of a frontier of discovery that boggles the mind; we are peering at the threshold of creation itself. How can we not approach it with wonder?

In 1913, when Department of Agriculture botanist David Fairchild began experimenting with long-lens cameras to capture close-up views of insects, his photographs were so stunning for that era that they were published in a May 1913 *Geographic* article, "The Monsters of Our Back Yards," and later as one of the Society's earliest books, *Book of Monsters*. Fairchild (Gilbert Hovey Grosvenor's brother-in-law) was a scientist first and a photographer second; but science and photography have been linked ever since Louis Daguerre's discovery in the 1830s that a real image could be fixed. Striking though they may have been, Fairchild's photographs were still only close-ups—close-ups of the *exteriors* of life forms already visible to the naked eye.

Two and a half centuries before Fairchild's experiments, Anton van Leeuwenhoek, an unschooled Dutchman, had, even with his crude instruments, been able to see and sketch protozoa and bacteria. But even now the best optical microscopes are limited, as were Fairchild's camera and van Leeuwenhoek's primitive microscope, to objects visible within the wavelengths of light. They can magnify no more than 2,000 times. That is, they can focus on objects no smaller than 1/125,000 of an inch—or 2,000 angstroms.

But with the electron microscope, then Senior Assistant Editor Kenneth Weaver wrote in "Electronic Voyage Through an Invisible World," for the February 1977 *Geographic*, "scientists have opened up a whole new realm of atoms and molecules, a world man has never seen before. Just as today's largest telescopes take us into the bizarre universe of quasars and black holes, so the electron microscope takes us into the Alice-in-Wonderland world of inner space. What we see there is sometimes hauntingly beautiful, sometimes awesome, and frequently of supreme significance."

As Weaver points out, the best electron microscope at that time could:

magnify an incredible twenty million times, with a resolution on the order of two angstroms. And even the individual atom, which has a diameter of only about one angstrom (about four-billionths of an inch), can be photographed in the same way that an invisible mote of dust can be "seen" by the light scattered when the mote floats through a bright beam of sunlight. These angstrom-wide atoms are so small that nearly a million of them lined up side by side, would fit into the thickness of this sheet of paper.

◄ ───────────────────────────────────

Sixty-four years after Fairchild's long-lensed experiment, a scanning electronic microscope in 1977 catches a quarter-inch-long velvety tree ant at its bath, capturing the insect, here magnified eighty times, in the act of cleaning its antenna with a foot.

459

The word "atom," from the ancient Greek, means, literally, "uncuttable." In his wonderfully lucid 1985 Magazine article "Worlds Within the Atom," author John Boslough both explains the derivation of the word and, like Weaver, attempts to help us comprehend its minuscule angstrom-wide size:

> Some 2,300 years ago the Greek philosophers Democritus and Leucippus proposed that if you cut an object, such as a loaf of bread, in half, and then in half again and again until you could do it no longer, you would reach the ultimate building block. They called it an atom.
>
> The atom is infinitesimal. Your every breath holds a trillion trillion atoms. And because atoms in the everyday world we inhabit are virtually indestructible, the air you suck into your lungs may include an atom or two gasped out by Democritus with his dying breath.
>
> To grasp the scale of the atom and the world within, look at a letter "i" on this page. Magnify its dot a million times with an electron microscope, and you would see an array of a million ink molecules. This is the domain of the chemist. Look closely at one ink molecule and you would see a fuzzy image of the largest atoms that compose it.
>
> Whether by eye, camera, or microscope, no one has ever seen the internal structure of an atom: Minute as atoms are, they consist of still tinier subatomic particles. Protons, carrying a positive electric charge, and electrically neutral particles called neutrons cluster within the atom's central region, or nucleus—one hundred-thousandth the diameter of the atom. Nuclear physicists work at this level of matter.

And what nuclear physicists are working at is "cutting the uncuttable." They are smashing the atom. Ironically, as Boslough points out, "exploring the smallest things in the universe requires the largest machines on earth": Machines such as Stanford University's two-mile-long linear accelerator near Palo Alto, California, which fires negatively charged electrons at atomic nuclei, and the more than four-mile-long European Laboratory for Particle Physics (CERN) near Geneva, Switzerland, which fires the heavier protons that cause more collisions. (Still, as one *Geographic* staffer noted, "To hit an atom's nucleus with a charged particle is something like playing pool in the dark on a table as big as Texas.")

There are already ten large accelerator centers in existence in the United States, Japan, Europe, and the U.S.S.R.; but they will all be dwarfed by the fifty-two-mile-long accelerator proposed to be built in the 1990s in the U.S.

In his "Worlds Within the Atom" author John Boslough interviewed the Fermi Laboratory director Dr. Leon Lederman and wrote:

> Using the CERN accelerator like an immense microscope, physicists are probing the structure of the atom, an inner cosmos of subatomic particles as remote from our daily experience as the farthest reaches of space. Yet that structure may hold an explanation of how the universe was born.
>
> During the past 50 years scientists exploring the atom's interior have solved many age-old mysteries of matter and energy. This new knowledge has brought us lasers, computers, transistors, space travel, and nuclear energy for weapons and power.

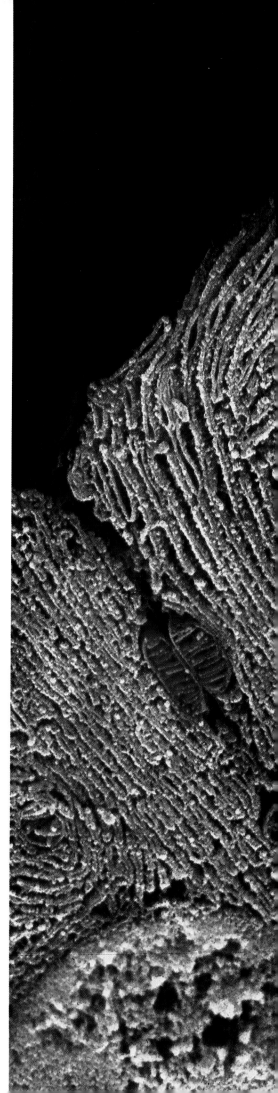

Magnified 10,000 times and opened up for an inside view, human cells display the nuclei as large yellow spheres, mitochondria colored red, ribosomes as pale dots; throughout runs the ropelike maze called the endoplasmic reticulum.—From The Incredible Machine

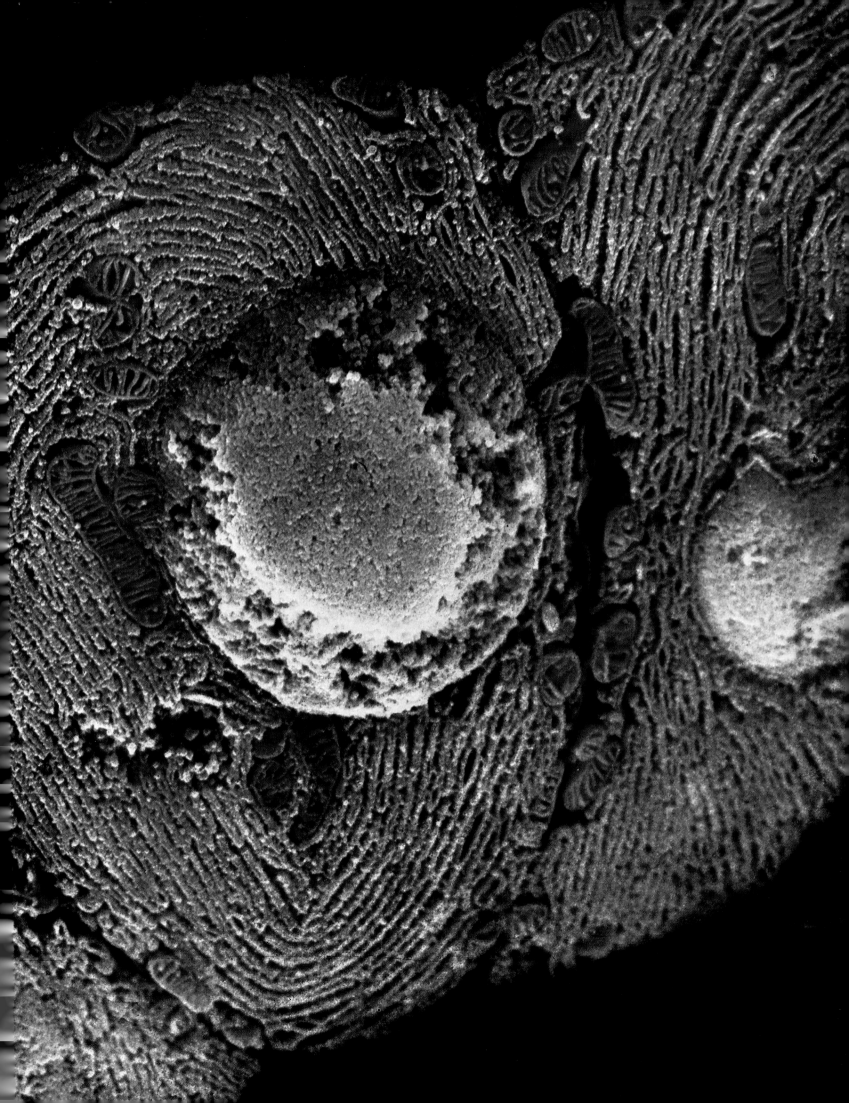

Opposite: *To take the technologically esoteric and reduce it to laymen's terms has always been the strength of Society reporting. Here, in words and images from the 1986 Book Service volume* The Incredible Machine, *the Society attacks the greatest mystery of all: the beginning of life itself.*

"Cradled in a nest of membranes, the embryo [magnified seven times] celebrates a milestone at eight weeks. Though only the size of a walnut, it has all of its basic organs and systems. Now the embryo begins to test its muscles, which are growing daily. Its heart has been beating for one month, and the embryo takes on a more familiar, babylike appearance. Its head, now almost half the embryo's total size, signifies the rapid brain development that is taking place.

"It will be known, from the ninth week until birth, as a fetus. Anchored to the life-supporting placenta by an umbilical cord, the fetus now enters a phase of dramatic growth: During the remaining seven months of prenatal life, its weight multiplies 3,600 times, and its structures will be refined."

"But what we're really after is a new concept of reality," says Leon Lederman…"We're after something akin to the revolution in thinking that followed Copernicus's announcement that the earth circles the sun."

Lederman and other physicists are searching for the ultimate building blocks from which all things—the stars, the earth, you, I, and the atom—are made. Because everything in the cosmos has been composed of these particles since the primordial big bang, scientists also hope to learn the origin of the universe, a goal that eluded even Albert Einstein. Today his successors, at CERN and elsewhere… think they are getting close.

Just as physicists hope to discover the origin of the universe, biologists are hoping to discover the origin of life.

National Geographic writer Rick Gore described himself in the September 1976 "The Awesome Worlds Within a Cell," as a "journalistic traveler in the 'new biology,' taking notes on the explosion of knowledge that in recent years has utterly transformed the life sciences." In a laboratory at the California Institute of Technology, Gore spun pure distilled deoxyribonucleic acid on the end of a glass rod. DNA, "the most celebrated chemical of our time," Gore called it, "the master choreographer of the living cell and carrier of the genetic code."

"If anything illustrates what has happened in biology," Gore wrote, "it is this profound new ability to take the very stuff of life out of the cell, to isolate it in a test tube, to dissect it, and to probe the deep mysteries borne in its fragments."

As early as 1665, the English microscopist Robert Hooke saw and named cells upon seeing their dead remains in cork shavings, but the Dutchman van Leeuwenhoek was the first to describe cells alive. "No more pleasant sight has met my eye than this of so many thousands of living creatures in one small drop of water," he wrote in 1676. A century and a half would pass before Matthias Schleiden and Theodor Schwann, two German scientists, would deduce that every living thing was made of cells. In time, scientists discovered that cells divide and multiply and manufacture chemicals, but still well into the middle years of this century little was known about the life of a cell. As Gore wrote, because of new technology this has changed:

> Now, using high-powered electron microscopes and ingenious techniques borrowed from physics and chemistry, biologists have broken through the cell's barrier of invisibility and have charted its interior. They have found a forbiddingly small yet enormously complex world; its magnitudes, like those of the cosmos, astonish and confound.
>
> Each cell is a world brimming with as many as two hundred trillion tiny groups of atoms called molecules. Even the largest molecules, like DNA, are measured in… angstroms—1/250,000,000 of an inch….
>
> The cell has turned out to be a micro-universe, science now tells us, abounding with discrete pieces of life, each performing with exquisite precision, and often in thousandths of a second, a biochemical dance its ancestors began to perfect countless generations ago in the primordial ocean.

As Max Delbruck, of the California Institute of Technology, said in Gore's article, "Any one cell, embodying as it does the record of a billion years of evolution, represents more an historical than a physical event. You cannot expect to explain such a wise old bird in a few simple words."

"The essential process by which the likeness of the parent is transmitted to the offspring," biologist William Bateson wrote in 1902, "…is as utterly mysterious to us as a flash of lightning is to a savage."

462

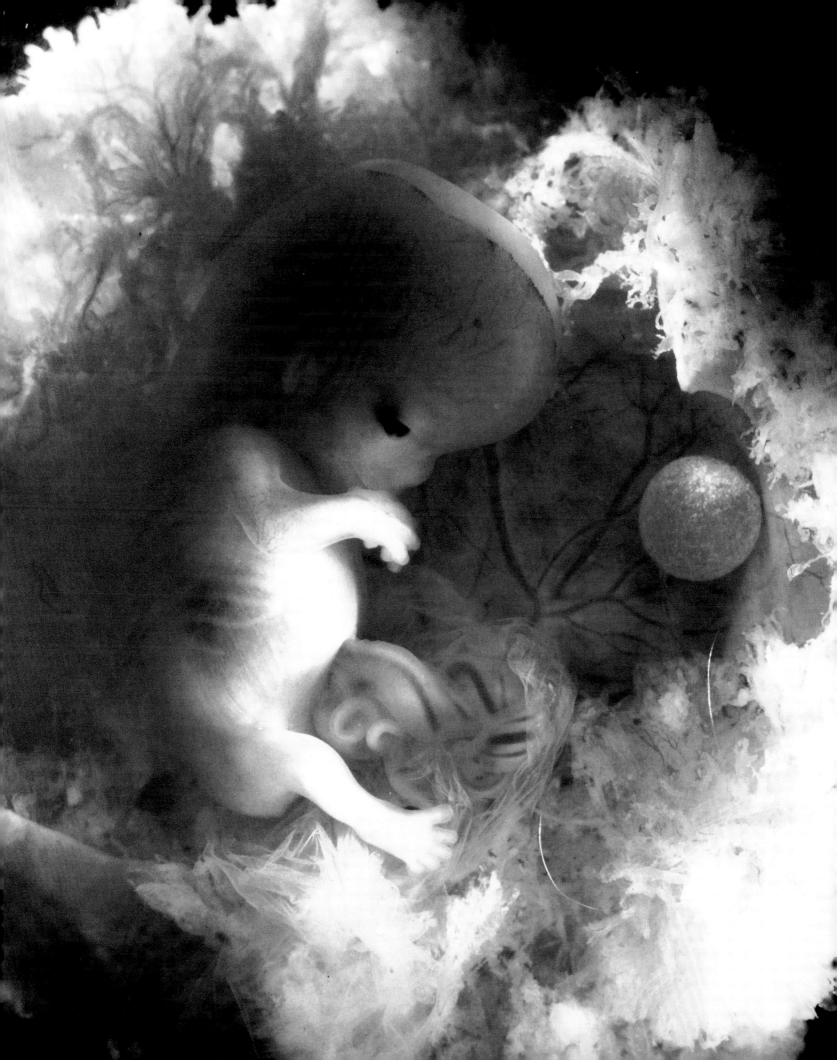

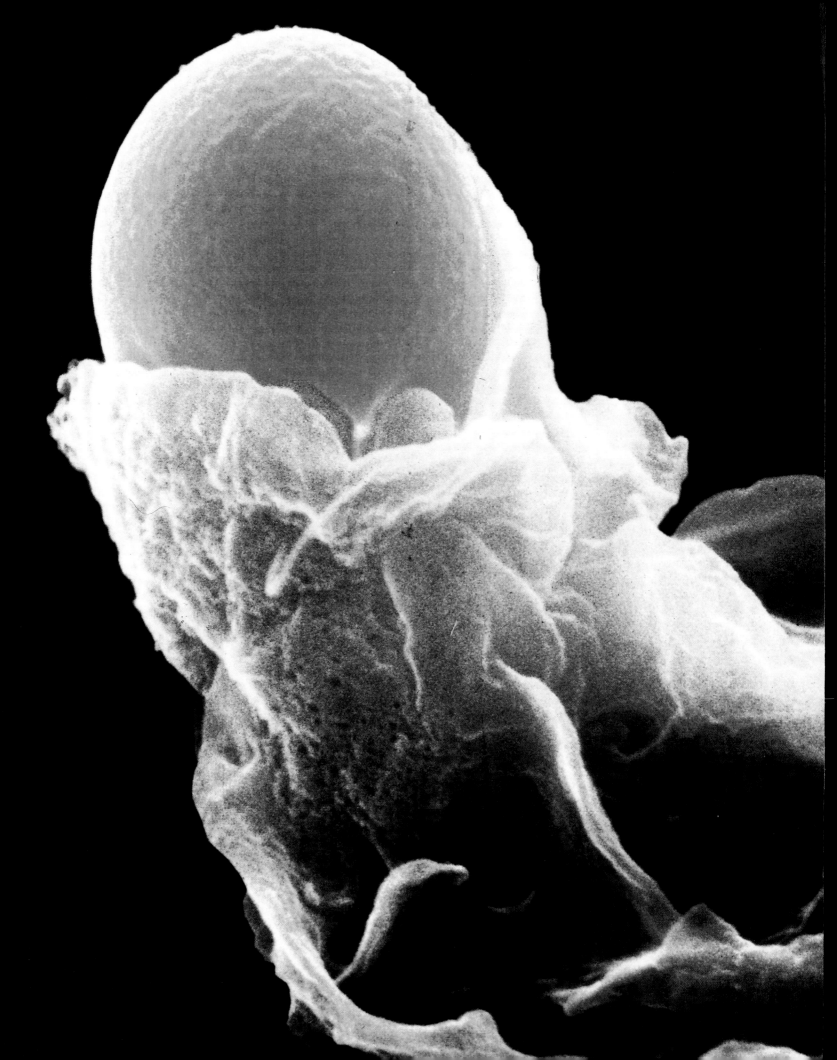

And yet in 1881 researchers studying starfish and sea urchin eggs had discovered nuclein, threadlike hereditary material that merged when egg and sperm ("DNA with a tail," Gore called it) were joined. Those researchers had unknowingly also discovered the manner in which human reproduction occurs and that life for us begins not, as had been thought for centuries, with a microscopic human being tucked into a minuscule egg, but, as Susan Schiefelbein wrote in *The Incredible Machine*, as a "fragile thread spun of chemical memory."

We have learned a great deal since 1881, as she explains:

> We know the code by which these gossamer filaments send their messages. Heredity is written on a chemical ribbon that twists like a spiral staircase, the steps built of four chemical bases attached to chains of sugars and phosphates—DNA....Thousands of these steps make up a single gene. Tens of thousands of genes, arranged along structures called chromosomes, transmit the instructions for existence, dictating eye color, hair texture, vulnerability to disease, perhaps even stuttering and altruism.

> Some six billion steps of DNA in a single cell record one life's blueprint. This DNA plan for a single human life can be stretched six feet, yet it is coiled in a repository just 1/2500 of an inch in diameter—the cell's nucleus.

And, as noted elsewhere in *The Incredible Machine:*

> The DNA molecule is a miracle of organization, structured like a twisted ladder....Elegant in structure,* DNA is also vibrant. Any still portrait of this molecule conveys only part of its nature, for motion characterizes the rest. DNA bends and twists a billion times a second while its ladder sides "breathe" in and out. This dance likely arises as DNA engages in its two key roles: to direct the creation of protein and to duplicate itself.

The creation of protein is essential to the cell and to ourselves. Although when we think of protein we tend to think of it as a specific value to be sought after in certain foods, protein is actually the name for the countless variety of chemical compounds which, when broken down, provide our cells with the nutrients needed to produce more protein. Hormones like insulin, which controls energy use, are proteins; collagen, which builds skin, is a protein; hemoglobin, which supplies oxygen, is, too; the dozens of different enzymes are proteins; as are the various structural components from which cells are built.

When cells divide, the necessary instructions for duplicating their own proteins come through the DNA in their genes. Moreover, Rick Gore wrote:

> [Scientists] have found that virtually every cell contains the entire repertoire of genes for that plant or animal. One cell in my toe, say, has all the data in its DNA for making another man physically identical to me. That many instructions, if written out, would fill a thousand 600-page books. The unique experiences of our lives, of course, make us more than a product of our genes. Yet it is our DNA that sets the basic physical limits of what we can or cannot become.

The fact remains that we are all different—even identical twins have multiple differences in palm and fingerprints and markings of their irises—but we share a common entrance into this world; and the mystery and the miracle

Opposite: *National Geographic authors have always enjoyed the "a-million-ink-molecules-are-contained-in-the-dot-at-the-top-of-this-'i'" assists at visualization; Rick Gore is no exception. In his September 1976 article "The Awesome Worlds Within A Cell," Gore wrote, "Our blood holds twenty trillion red blood cells; thirty thousand would fit in this one O."*

Beginning life deep in the bone marrow, a red blood cell divides and fills with hemoglobin. In six days the cell enters the bloodstream. At about 120 days, it begins to expire. "One theory," it was noted in The Incredible Machine, *"holds that the cell, without a nucleus to renew old parts, depletes its inner resources. Its outer membrane may also wear out, like tread on an old tire, from thousands of trips through the bloodstream. A macrophage, the white blood cell responsible for cleanup, senses that the time has come. Catching the aging cell in its embrace, the macrophage engulfs and digests it." (Magnification 10,000 times.)*

The first descriptions of the structure of DNA appeared in a one-page scientific paper published by James D. Watson and Francis Crick in 1953. In it they drew the scientific community's attention to this molecule with "novel features which are of considerable biological interest."

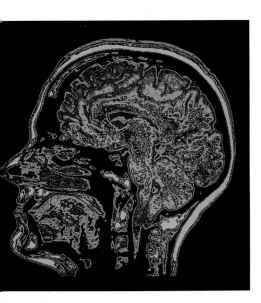

Above: *"A color-enhanced profile made by magnetic resonance imaging (MRI) shows a herniating finger of tissue from a brain slumping into the base of the skull. Frequently used to view soft tissue such as the brain's, MRI machines do not use X rays to penetrate the body but instead employ a combination of radio waves and a strong magnetic field."—From "Medicine's New Vision," January 1987 issue*

Opposite: *"Ghostly in the dark, a normal face seems otherworldly when viewed by an MRI scanner. The eerie forehead, eyebrows, cheeks, nose, and lips appear brightest because water density is higher than in other tissues. MRI reflects water because it focuses on the behavior of hydrogen atoms in water molecules....Teeth and bones, which contain little water, do not appear at all in MRI...."—From "Medicine's New Vision," January 1987 issue*

both start at our beginnings. As Schiefelbein wrote in her *Incredible Machine* chapter "Beginning the Journey":

> The newborn baby embodies innocence, yet conceals the most taunting of all riddles: the generation of human life. The story begins with sperm and egg as they combine to form a single cell. Sheltered in the mother's womb, the cell multiplies. Soon there are hundreds of different cells able to make some 50,000 different proteins to control the work of all our cells....
>
> Before long, the groups of cells are gathering into layers, then into sheets and tubes, sliding into the proper places at the proper times, forming an eye exactly where an eye should be, the pancreas where the pancreas belongs. The order of appearance is precise, with structures like veins and nerves appearing just in time to support the organs that will soon require them. In four weeks the progeny of the first cell have shaped a tiny beating heart; in only three months they are summoning reflex responses from a developing brain.
>
> Nothing more than specks of chemicals animate these nascent cells as they divide. Yet in just nine months, some twenty-five trillion living cells will emerge together from the womb; together they will jump and run and dance; sing, weep, imagine, and dream.
>
> A single cell engenders a multitude of others, but the multitude acts as an entity, as a community. The appearance of life in the womb, like the appearance of life on Earth, cannot be completely explained, even by our burgeoning scientific knowledge. Biologists have identified the sequence of reproductive events; yet they know merely *what* happens. They do not know yet exactly *how* or *why*.

And once we are born, our bodies must wage battle against a host of enemies mounting attacks without and within our bodies every sleeping and waking moment for the rest of our lives.

"The combatants are too tiny to see," Peter Jaret wrote in "Our Immune System: The Wars Within" for the June 1986 Magazine. "Some, like the infamous virus that causes AIDS, or acquired immune deficiency syndrome, are so small that 230 million would fit on the period at the end of this sentence."

The instruments that made surveillance of these wars within us possible were unavailable twenty-five years ago. The field of immunology, the study of the immune system, depended more on the science of deduction than on the science of medicine. But with the advent of the electron microscope and more sophisticated laboratory techniques, guesswork gave way to exploration and discovery of the strategies and counter-strategies of both our body's defenders and its foes. During this past decade, immunology has moved out of the backwaters of medicine and into the forefront as diseases such as AIDS have created the kinds of crises medicine mobilizes against.

"Some bacteria, such as the familiar streptococci and staphylococci, continuously swarm in legions over our skin and membranes, seeking access that can cause sore throats or boils," Jaret wrote. "Or consider the bacterium *Clostridium botulinum*, the cause of botulism. This single cell can release a toxin so potent that four hundred-thousandths of an ounce would be enough to kill a million laboratory guinea pigs."

However, of all the body's enemies, Jaret points out, both "the simplest and the most devious" is the virus:

> A virus is a protein-coated bundle of genes containing instructions for making identical copies of itself. Pure information. Because it lacks the basic machinery for reproduction, a virus is not, strictly speaking, even alive.

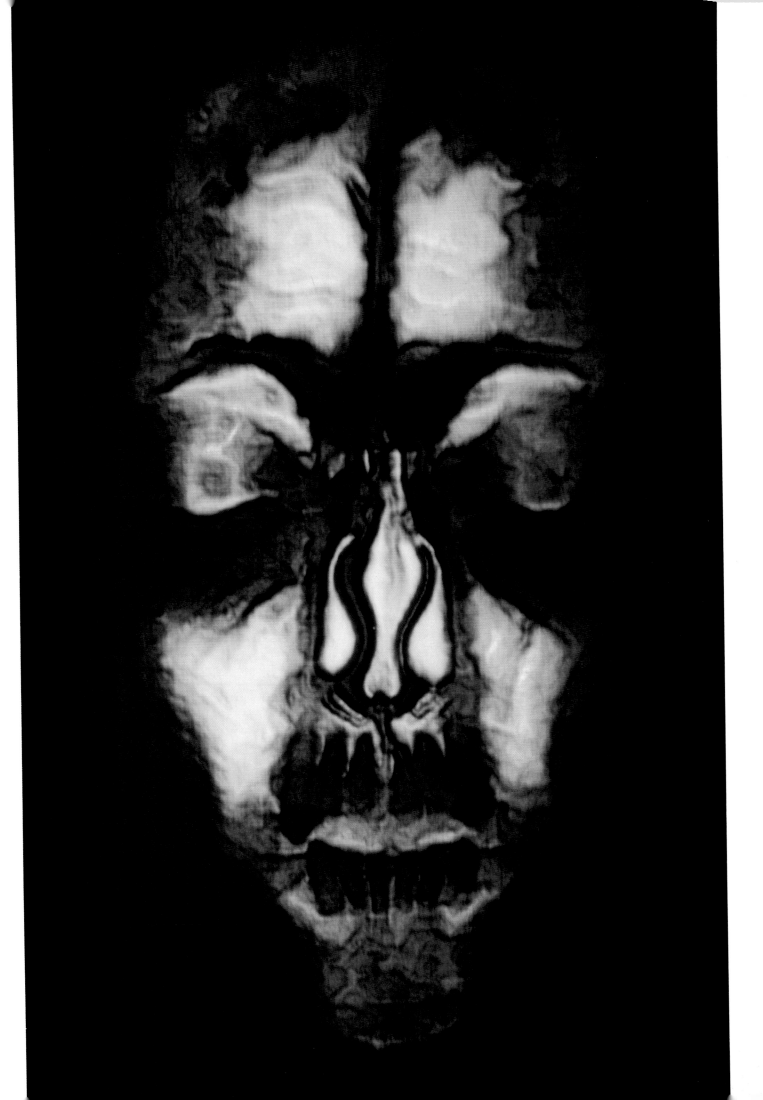

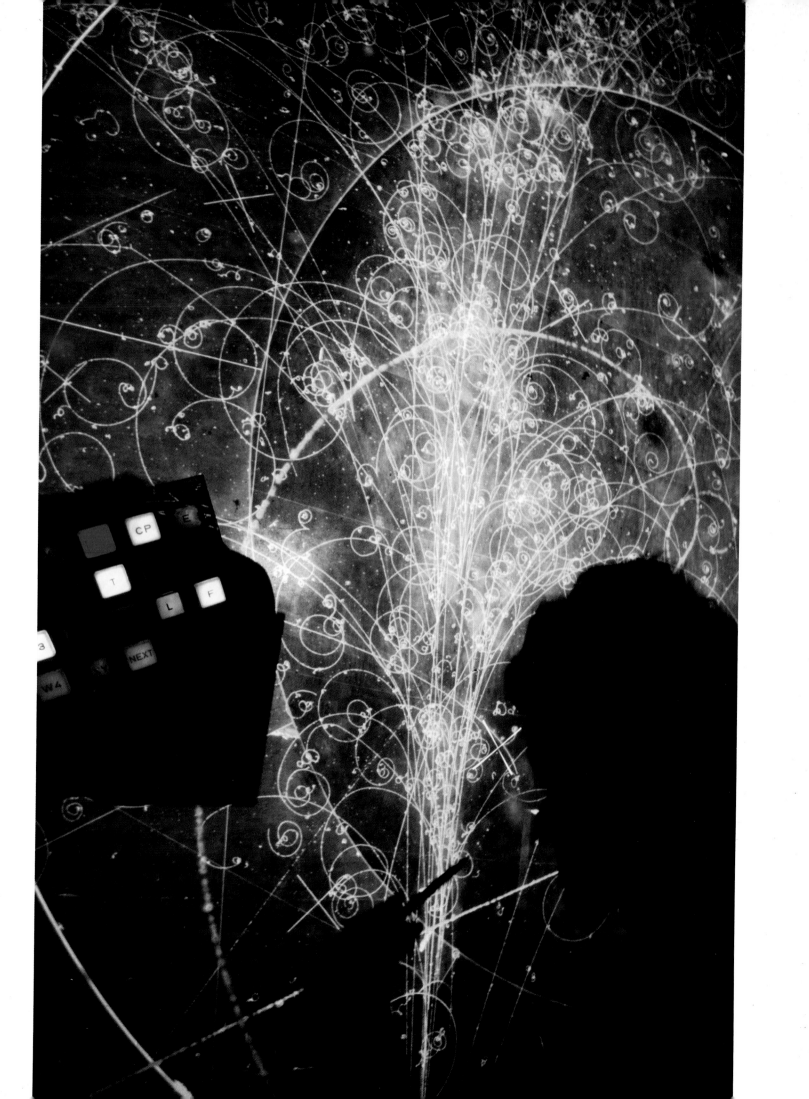

But when a virus slips inside one of our cells, that bundle of genetic information works like our cell's DNA, issuing its own instructions. The cell becomes a virus factory, producing near identical viruses. Eventually they may rupture the cell, killing it. Viral clones fan out to invade nearby cells.

"Keep in mind that a virus can create thousands of copies of itself within a single infected cell," immunologist Steven B. Mizel of Wake Forest University's Bowman Gray School of Medicine told Jaret. "Invading bacteria can double their numbers every 20 minutes. At first the odds are always on the side of the invader."

As Peter Jaret embarked on this piece he found himself at Purdue University with a sore throat, runny nose, and he was sneezing. Biologist Michael Rossmann identified the enemy battling him as "Rhinovirus 14, one of the causes of the common cold."

Jaret used his own illness as the device with which to attack the story. Taking comfort in knowing that "of the one hundred trillion cells that make up my body, one in every hundred is there to defend me," Jaret explained they were the white blood cells "that are born in the bone marrow. When they emerge, they form the three distinct regiments of warriors—the phagocytes and two kinds of lymphocytes, the T cells and B cells. Each," he pointed out, "has its own strategies of defense."

As recently as the mid-1960s it would have been impossible for Jaret to have written that piece. It wasn't until the late 1950s that immunologists began to comprehend how antibodies were produced. That there was a difference between T cells and B cells came clear only in the 1960s. Macrophages are still not completely understood. "Indeed," Jaret points out, "had AIDS struck twenty years ago, we would have been utterly baffled by it."

Among the medical advances examined in the Society's publications and films, none have been more astounding, perhaps, than those made in diagnostic medicine. Although German physicist Wilhelm Konrad Roentgen made the first X ray in 1895, and doctors were quick to see its benefit in diagnosing fractures, veteran writer for the *Geographic* Howard Sochurek points out that the "recent advances made in imaging technology, or 'machine vision,' [which] enable doctors to see inside the body without the trauma of exploratory surgery, [have resulted in] more progress [being] made in diagnostic medicine in the past 15 years than in the entire previous history of medicine."

Among the new machines and techniques Howard Sochurek discussed in "Medicine's New Vision" for the January 1987 *Geographic* was magnetic resonance imaging (MRI), which does not use X rays but rather a combination of radio waves and a strong magnetic field to produce views of soft tissue such as that of the brain.

Sochurek called MRI "an area of excitement and explosive growth. This seeing machine may prove to be as great a tool to modern medicine as the X ray....

"Magnetic resonance imaging relies on the principle that hydrogen atoms, when subjected to a magnetic field, line up like so many soldiers," Sochurek explained. "If a radio frequency is aimed at these atoms, it changes the alignment of their nuclei. When the radio waves are turned off, the nuclei realign themselves, transmitting a small electric signal. And since the body is primarily composed of hydrogen atoms, an image can be generated from the returning pulses, showing tissue and bone marrow as never seen before."

Opposite: *"Particle-track pyrotechnics from a bubble-chamber detector flash on a screen at CERN [the European Laboratory for Particle Physics near Geneva, Switzerland]. Debris from subatomic collisions leave distinctive wakes. For instance, electrons and positrons spiral tightly in opposite directions. Computers now sort through such jungles of tracks to pinpoint results significant to experimenters."—From "Worlds Within the Atom," May 1985 issue*

Digital subtraction angiography (DSA), on the other hand, uses two X-ray images. "Many of man's greatest inventions have expanded the capabilities of the human body," Sochurek wrote. "The computer has enhanced man's ability to see by making the invisible visible. This new vision lies at the heart of ...DSA, an imaging technique that produces clean, clear views of flowing blood or its blockage by narrowed vessels.

"DSA depends on the injection into the vessels of a contrast agent containing iodine that is opaque to X rays," Sochurek explained. "The shadow this opacity creates allows doctors to see the flow of blood....Before injection of the contrast substance, an X-ray image is made and stored in the computer. After injection a second image is made highlighting the flowing blood as revealed by the substance. The computer then subtracts image one from image two, leaving a sharp picture of blood vessels such as the coronary arteries, the main suppliers of blood to the heart."

Both positron emission tomography (PET) and single photon emission computed tomography (SPECT) show blood flow by imaging trace amounts of radioisotopes, but PET does it with greater accuracy and also measures metabolism, or how well the body is working.

By revolving an X-ray tube around the patient, computed tomography (CT) scanners penetrate the body with thin, fan-shaped X-ray beams to produce "slices" of the body, cross-sectional views of the tissues within.

The only body-scanning technique recommended for use with pregnant women, sonography acts on the principle of sonar by beaming high-frequency sound waves in short pulses into the body, which then bounce back and are read by a computer which translates the "echoes" into an image of the fetus.

When Sochurek asked physicians what directions they thought discoveries in diagnostic imaging would take, their answer was

> ...that in the past they often had to wait for a disease to manifest itself, as a tumor, say, or a heart defect.
>
> But diseases usually manifest themselves chemically before there is an anatomical change. In the future, MRI may be able to reveal such chemical changes by making images with elements such as phosphorous or sodium, as well as with hydrogen....Future research may perfect methods of tagging cloned antibodies with a magnetically traceable element and using them as scouts in the body to search out cancerous tumors....
>
> How often while preparing this story did I hear: "I couldn't have saved this patient ten years ago." Because of the computer and the new tools of machine vision, many more lives will be saved in the years to come, and the quality of all our lives will be improved.

Although the striking advances made in the fields of physics, chemistry, biology, and medicine reported in Society publications have already improved the quality of our lives in ways we almost take for granted, there is now an additional benefit. It is a philosophical one with both scientific and theological overtones: If these new scientific tools, techniques, and understandings cannot save us, there is some comfort to be gained from knowing it is not "ashes to ashes" but "stars to stars."

◄ ——————————————————————

Left: *"Most micro of all surgery is the technique of beaming subatomic particles at cancer cells."*—From the May 1985 issue

Overleaf: *"Streaks of interstellar dust vein the reddish glow of the Trifid Nebula."*—From The Amazing Universe

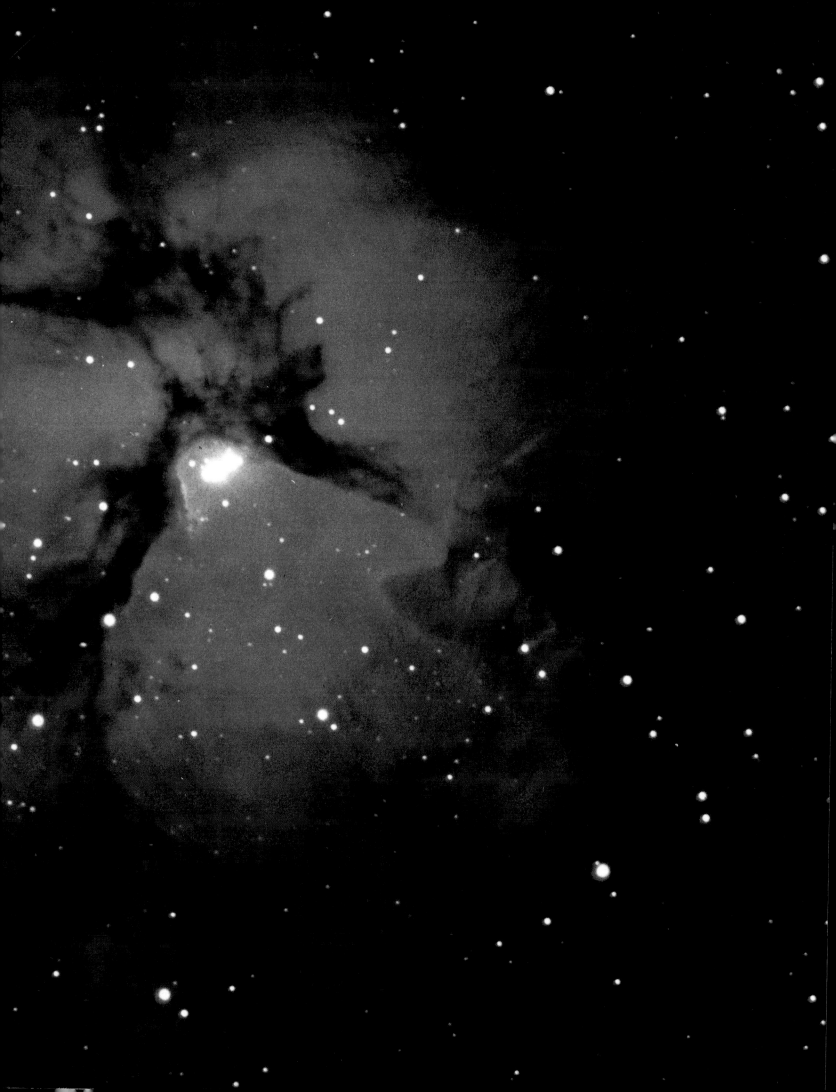

Acknowledgments

With a book such as this, it is difficult to know where to begin acknowledging one's debts. In my mind, individuals deserving precedence jostle one another like an after-theater crowd searching for taxis. Certainly, the willingness of Gilbert M. Grosvenor, President and Chairman of the Board of the National Geographic Society, to go outside the Society and choose an independent publisher of its centennial history might justify putting him at the head of the list. But since it was Paul Gottlieb, President and Publisher of Harry N. Abrams, Inc., who persuaded Gil Grosvenor that Abrams was the right organization to produce such a book and then selected me to write the text, should not his name with equal justice—and a flagon of prudence—take priority?

Robert L. Breeden, the National Geographic Society's Senior Vice President, Publications and Educational Media, has been a friend indeed. Generous with both time and advice, Bob Breeden always stepped in at the appropriate moment with the necessary energy, perspective, and experience to knit the Abrams and Society teams together when differing publishing methods threatened to unravel the project's seams. (Bob Breeden's line "The dark at the end of the tunnel is what we're finally seeing" will long be remembered.)

The enthusiastic cooperation of members of the *National Geographic* staff was amazing to this outsider. Gilbert M. Grosvenor, despite his busy schedule, sat through interview after interview, was unswervingly frank in his answers, and unhesitatingly provided supporting documents. Editor Wilbur E. Garrett, despite deadlines and staff meetings, always found time to answer questions and give advice. Senior Associate Editor Joseph Judge, writer, raconteur, and lunch companion without equal, I shall sorely miss. Associate Editor Thomas R. Smith, Senior Assistant Editors William Graves, Carolyn Bennett Patterson (ret.), Edward J. Linehan, Samuel W. Matthews, O. Louis Mazzatenta, Charles McCarry, Elizabeth A. Moize, Howard E. Paine, Mary G. Smith—thank you all for your patient, informative, and, in many cases, hilarious interviews and reminiscences.

I could not have written the sections covering the development of color photography at the Society without the counsel and advice of *Geographic* Senior Writer Priit J. Vesilind and his generous sharing of his Master's thesis. Priit deserves a special credit line here of his own.

Assistant Editors Kent Britt, Rick Gore, Bart McDowell, Merle Severy, Senior Writers Thomas J. Abercrombie, William S. Ellis, Peter T. White, and Senior Editorial Staff members Boyd Gibbons and David Jeffery—I am grateful to you as well. Photographers Dean Conger, David Alan Harvey, George F. Mobley, Steve Raymer, Robert F. Sisson, and Director of Layout Robert W. Madden, each of you is worth a thousand words—or is it ten thousand? Senior Assistant Editor John B. Garver, Jr., and his Cartography Division helped me find my way.

In daily dealings with this book, Senior Assistant Editor and Book Service director Charles O. Hyman was our friend and benefactor. Never one to let my words stand in the way of a good picture, Chuck eased us over some rough spots. Senior Assistant Editor and Special Publications Director Donald J. Crump was a major source for the early days of Special Pubs; Director of Educational Media George Peterson taught me much—and not just about the life cycle of the frog. Ralph Gray, former School Service Chief, and Executive Assistant William Gray—father and son—a special thanks to the two of you. Joan Tapper, editor of *Traveler,* and Director of Publications Art John D. Garst were giving of their time and friendship. Associate Director Tim T. Kelly was helpful about the Society's television and films, and Suzanne K. Dupré, the Society's Corporate Counsel, was open and, parenthetically, informative about the Society's efforts to protect its nonprofit-organization tax-exempt status.

Library Director Susan Fifer Canby and Reference Librarian Carolyn Locke helped me find the sources, and Ted Wojtaskik came up with exactly the right correspondence. News Service's Boris Weintraub—though in a fur-lined foxhole—showed a newspaperman's heart. And Everest conqueror and now Vice Chairman, Committee for Research and Exploration, Barry C. Bishop proved "When one can no longer see, one can at least still know."

The crusty former Senior Assistant Editor, Photography, Robert E. Gilka (in whose office had lain a pillow needlepointed, "Wipe Your Knees Before Entering") was a pleasure to talk with—if for no other reason than his line about Gil Grosvenor: "He's east, east, east, east—a Yale man. You don't get much more eastern unless you go out swimming." I am grateful to retired *Geographic* photographer Volkmar K. Wentzel and to Edwards Park, editor of the Society's 75th-anniversary book, *Great Adventures*, and, later, of *Smithsonian;* both of them were good enough to come in to talk. And former *Geographic* Editor and Society Trustee Frederick G. Vosburgh was extremely kind in submitting to lengthy telephone interviews.

Former Chairman of the Board Melvin M. Payne, Vice Chairman Owen R. Anderson, Chairman Emeritus Thomas W. McKnew, and Society Vice President Leonard J. Grant assisted me in understanding both the mood and the history of the Society's earlier decades. And it was a special honor and delight to spend as much time as I did with the Society's legendary Luis Marden.

I am grateful as well to Society Vice President Danforth P. Fales and to Dorothy (Dori) Jacobson, both of Publications and Educational Media. Their friendship and encouragement meant a great deal to me.

I suppose one always feels a special kinship with those with whom one shared the trenches. Abrams Senior Editor Edith M. Pavese's tireless good humor and supportive efforts daily saw this book through. Edith, what a pleasure it has been to work with you again!

I felt that same pleasure working with this book's Art Director—and Harry N. Abrams' Vice President, Art and Design—Samuel N. Antupit, whose work I have admired ever since we were together at *Monocle* more than twenty-five years ago. Thank you, Sam.

Thanks also to Maria Miller, who, with patient good nature, coordinated the various aspects of design and production. And to Gertrud Brehme, who organized the sometimes conflicting demands of the production and technical teams.

Illustrations Editor Anne D. Kobor's experience was invaluable. It was she and Illustrations Researcher Brooke Kane who searched for, found, and assembled all the pictures for this book—though not without the help of Leah Roberts, Linda Rinkinen, Bill Bonner, Eudora (Dori) Babyak, and Elizabeth Leader.

Edith, Sam, Maria, Anne, and Brooke, this is your book, too.

Finally, if the National Geographic Society deserves its reputation for accuracy and thoroughness—and, believe me, it does—it is due to its researchers, who tirelessly track down sources and confirm facts. I was fortunate to have had researcher Tee Loftin save me much anguish and hardship on my various literary journeys to the Poles. I thank her, and I thank, too, researcher Tori Garrett Connors, who joined this project at midpoint and spent many overtime hours seeing it through to the end.

But there is no one to whom I stand in greater debt than Sallie M. Greenwood, the researcher assigned by the Society from the beginning of the project to ensure the accuracy of my text. During the year and a half I worked on this centennial history, no one came to better embody to me the strength of the Society and what was professional and good about the men and women who work there than Sallie Greenwood. I doubt if there is a more dedicated, capable, and tactful researcher anywhere. Being the modest person she is, Sallie would demur, but I can say with utter confidence that I would have no problem finding three authoritative outside sources to confirm such an opinion. (All right, I know: "Author's observation.") Thank you, Sallie; no Sherpa could have guided me better up the mountains of material a history like this entails.

Again, to all of you whose help on this book I have appreciated so much, my thanks. I could not have done it without you!

C. D. B. Bryan
Guilford, Connecticut

Bibliography

Beebe, William. *Half Mile Down.* New York: Harcourt, Brace, and Company, 1934.

Bruce, Robert V. *Bell: Alexander Graham Bell and the Conquest of Solitude.* Boston: Little, Brown and Company, 1973.

Buckley, Tom. "With the National Geographic on Its Endless, Cloudless Voyage." *New York Times Magazine,* September 6, 1970.

Chamberlin, Anne. "Two Cheers for the National Geographic." *Esquire,* December 1963.

"Change at the *Geographic.*" *Washington Post,* July 17, 1977.

Chock, Alvin K. "J. F. Rock, 1884–1962." *Hawaiian Botanical Society Newsletter,* January 1963.

"Climbing Mount Everest Is Work for Supermen." *New York Times,* March 18, 1923.

Conaway, James. "Inside the Geographic." *Washington Post,* December 18, 19, 20, 1984.

Cornell, James. *The Great International Disaster Book.* New York: Charles Scribners' Sons, 1982.

Eber, Dorothy Harley. *Genius at Work: Images of Alexander Graham Bell.* New York: The Viking Press, 1982.

"Footnote to Geography." *The New York Times* editorial, July 26, 1942.

Frazier, Kendrick. *The Violent Face of Nature: Severe Phenomena and Natural Disasters.* New York: William Morrow and Co., Inc., 1979.

Garrett, W. E. "The Magazine Today." National Geographic Index, 1947–1983.

————. "National Geographic: The Magazine That Taught Us How to Use Pictures…and Color." *Photojournalism: Principles and Practices* (2nd ed.). Clifton C. Edom, ed. Dubuque, Iowa: Wm. C. Brown Company, 1980.

"The Geographic Faces Life." *Newsweek,* September 12, 1977.

"Geographic from Upbeat to Realism." *Los Angeles Times,* August 5, 1977.

"Geographic Moves, Like a Glacier." *San Francisco Examiner,* September 25, 1977.

"Geographic Society Raid Drill Empties Building in 3 Minutes." *Washington Star,* February 15, 1942.

"The Geographic's Flawed Picture." *Accuracy in Media,* vol.VI, no.6 (March 1977).

Great Adventures with National Geographic. Washington, D.C.: National Geographic Society, 1963.

Grosvenor, Gilbert H. "Earth, Sea, and Sky: Twenty Years of Exploration by the National Geographic Society." *The Scientific Monthly,* May 1954.

————. "First Lady of the National Geographic." *National Geographic Magazine,* July 1965.

————. *The National Geographic Society and Its Magazine: A History.* Washington, D.C.: National Geographic Society, 1957.

————. "The Romance of the Geographic." *National Geographic Magazine,* October 1963.

————. Speech at *U.S. Camera* Achievement Awards Dinner, November 6, 1951.

Grosvenor, Melville Bell. "Bringing the World to Your Fingertips." Washington, D.C.: National Geographic Society, 1970.

————. "Editor's Note." *National Geographic Magazine,* October 1963.

Grun, Bernard. *The Timetables of History: A Horizontal Linkage of People and Events.* New York: Simon and Schuster, 1975.

Hart, Scott. "Whole Earth Monthly: The Power and the Glory of Being the National Geographic." *Washington Star-News,* September 23, 1973.

Hellman, Geoffrey T. "Geography Unshackled." *The New Yorker,* September 25, October 2, October 9, 1943.

Hoffer, William. "All the World's a Page." *Regardies,* March 1986.

Huntford, Roland. *The Last Place on Earth.* New York: Atheneum, 1986.

————. *Scott and Amundsen.* London and Toronto: Hodder and Stoughton, 1979.

Images of the World: Photography at the National Geographic. Washington, D.C.: National Geographic Society, 1981.

Kirwan, L. P. *A History of Polar Exploration.* New York: Norton and Company, Inc., 1959.

LaGorce, John Oliver. *The Story of the Geographic.* Washington, D.C.: National Geographic Society, 1915.

"Maps of a Changing World." *The New York Times* editorial, January 15, 1945.

Marden, Luis. Tribute to Melville Bell Grosvenor. Address delivered at the Cosmos Club, Washington, D.C., May 1979.

McGee, W. J. "Fifty Years of American Science." *Atlantic Monthly,* September 1898.

Mott, Frank L. *A History of American Magazines, 1885–1905, vols. I–V.*

Cambridge, Massachusetts: Harvard University Press, 1930–1968.

Mowat, Farley. *The Polar Passion: The Quest for the North Pole.* Boston: Little, Brown and Company, 1967.

"The 'Mystery' Army of World War II." *Trading Post,* March–April 1956.

National Geographic Index, 1888–1946. Washington, D.C.: National Geographic Society, 1967.

National Geographic Index,1947–1983. Washington, D.C.: National Geographic Society, 1984.

"National Geographic Photographers: The Gang That Better Shoot Straight." *Potomac (The Washington Post),* July 7, 1974.

"National Geographic Still Poking Its Lens Around the World." *Chicago Tribune,* Tempo Section, September 12, 1977.

Oehser, Paul H., ed. *National Geographic Society Research Reports, 1890–1954.* Washington, D.C.: National Geographic Society, 1975.

"One of the Grosvenor Twins of Amherst to Wed the Bell Telephone Millions." *Boston Sunday Post,* October 7, 1900.

Payne, Melvin M. "The Geographic Society at 75." *Washington Star,* January 27, 1963.

————. "75 Years Expanding Earth, Sea and Sky." *National Geographic Magazine,* January 1963.

"Rose-Colored Geography." *Time,* June 15, 1959.

Ross, Ishbel. "Geography, Inc." *Scribner's,* June 1938.

Sullivan, Wilson. "National Geographic: 'The Whole World Inside!'" *Saturday Review,* March 14, 1964.

Sutton, S. B. *In China's Border Provinces: The Turbulent Career of Joseph Rock, Botanist-Explorer.* New York: Hastings House, 1974.

This Fabulous Century: 1920–1930. New York: Time-Life Books, 1968.

Vesilind, Priit Juho. "National Geographic and Color Photography." Unpublished Master's thesis in Journalism, Syracuse University, December 1977.

Vosburgh, Frederick G. "To Gilbert Grosvenor: A Monument 25 Miles High." *National Geographic Magazine,* October 1966.

"What National Geographic Can Teach You." *35mm Photography,* Spring 1970.

Index

All references are to page numbers. Text references are in roman type; pages on which illustrations appear are in *italic* type. The notations *cpn.* and *n.* following a page number refer to, respectively, information in a caption or a note on that page.

A bbe, Cleveland, *23*
Abbey, Edward, 443
Abercrombie, Thomas J., *79,* 331; photo by, *72–73*
Abruzzi, Duke of the, 309
Accuracy in Media (AIM), 390, 392, 394
acid rain, 416*cpn.*
acquired immune deficiency syndrome (AIDS), 450, 466, 469
Adams, Clifton, 169
Afghanistan, *10–11, 176–77, 182,* 183, *284, 285,* 453
Agfacolor, 206, *207,* 213
Air Force Association, 252
Akeley, Carl E., 408
Alaska, 25, *32,* 43, 298; earthquake, 99, *108–109,* 113–14; mountains, 31, *86,* 103*cpn., 133,* 173, *175, 308,* 309; *see also* McKinley; wildlife, *410–11,* 420
Albright, Horace, 133*cpn.*
Allard, William Albert, photo by, *439*
Allen, Arthur A., 261, 420; photos by, *216, 217*
Allen, Steve, *291*
Althoff, Charles, 257
Alvin, submersible, 233, 236, 239, 240
Amazon, TV documentary, 349, 385
Amin, Idi, 410
Amoco Cadiz, supertanker, *429*
Amundsen, Roald, 53, 66, 71, 76, 78, 346
Amundsen-Scott South Pole Station, *72,* 78 & *cpn.*
Andella, Dee, 305
Anders, William A., 373
Anderson, Orvil A., 261, 346, *354, 357,* 358, 360
Anderson, Owen R., 340, *450*
Andrews, Roy Chapman, *212,* 213, 216
Angkor Wat, temple, *436,* 446*n.*
Annapurna, mountain, 312, *320–21*
Antarctica, *64–65,* 66, 68, 69, 71, *72–75,* 76–78; mountains in, 298, *306–307*
Aphrodisias ruins, 154, *156,* 157, *157,* 437
Aqua-Lung, 226, *226–27,* 230
Arden, Harvey, 439*cpn.*
Armenia, famine, 109–10
Atka, icebreaker, *70–71*
Atlas of the World, 330*cpn.,* 338, *338,* 433–34
atom, explained, 460

atomic bomb, 250, 252, 289, 291–92, *293;* Bikini tests, *258,* 260, *414, 416,* 419
Atwood, Albert W., 260, 261
Augustus, Roman emperor, 157
Australia, 194*cpn.*
Autochrome process, 124, 128, 133, 167, 169–70, 173, 294; characterized, 202, 204–209; illustrations, *133, 169;* statistics, 213
avalanche, *32*
Azzi, Robert, 284

B acteria, 466, 469
Baker, Frank, *22*
Baker, Marcus, *22*
Balchen, Bernt, 76
Baliles, Gerald, Governor of Virginia, 448
Balinese women, *168,* 169
Ballard, Robert D., 21, 236, 239
balloon flights, 216–17, 261, 355, 358, 360; pictured, *215, 354–57, 399*
Bangladesh, 109, 114, *114–15,* 386
Barbour, Thomas, 121, 123 & *n.;* photos by, *126, 127*
Bard, Allan, *313*
Bartlett, John R., *23*
Bartlett, Robert, 53*cpn.,* 56, 59, 65
Barton, Otis, *222, 223,* 224
Bass, George F., underwater grid built by, *230–31*
bat, frog-eating, *8–9*
Bateson, William, 462
bears, *408, 409, 415–16*
Beck Engraving Company, 204, 213
Beebe, William, 216, 217, *222, 223–24, 226,* 346; cartoon of, *218*
Bell, Alexander Graham, 25, 34, 36, *94–95,* 330, 331*cpn.,* 382, 434, 453; articles written by, 91, 118–19; Gilbert H. Grosvenor and, 36–38, 40, 46–49, 84, 89, 90, 94, 95; as inventor, 36, *40,* 41, *44, 45, 86, 87,* 91, 355; pictured, *41–43, 85–87*
Bell, Alexander Melville, *41*
Bell, Charles J., *22*
Bell, Elsie, 37, 38, 40, 46–49, *47; see also* Grosvenor, Elsie Bell
Bell, Mabel, 36, 38, *43,* 46, 48
Bennett, Floyd, *62,* 71, 76, 164
Bennett, Ross S., 332
Berlin Airlift, 271–72, 391
Berlin Wall, *270–71,* 272, 391
Bikini Atoll, *258,* 260, *414, 416,* 419
Billard, Jules B., 392
Bingham, Hiram, 20, 133, 139, 144, 148, 151, 158, 261; pictured, *138, 139*

birds, 126, 128, 174, *216,* 217*cpn.,* 219, 259–60; conservation issues, 394, *400–401, 412–13,* 420, *423–25*
Birnie, Rogers, Jr., *23*
Bisel, Sara, 154*cpn.*
Bishop, Barry C., 20, 310, 312, 317*n.,* 337, 346, 349, 385; pictured, *317*
Blackburn, Reid, 99, 106
Blacker, L.V. S., 310
Blair, James P., 391–92, 403
Blanc, Mont, 309–10
blizzard (1888), *28–29,* 30, 109
Blockson, Charles L., 446, 449
Blockson, James, 446, 449
blood cells, *454–55, 464,* 469
blue-black grassquit, 443
Bolitho, William, 194
Bombay plague, 110, 113
Boraiko, Allen A., 106, 419, 420, 423
Borchgrevink, C. E., 86
Borman, Frank, 373, 390*n.*
Boslough, John, 460
Bostelmann, Else, paintings by, *224, 225*
Bounty, H.M.S., 339, 346
Bourdillon, Tom, 312
Brazil, Indians, 337, *343*
Breadalbane, sunken ship, *236–37*
Breeden, Robert L., 331, 333–34, *343–45,* 348, 392, 428–29, 433*cpn.,* 437, 443, 445, *450*
Breeden, Stanley, 405*cpn.,* 408
Breitenbach, Jake, 312
Brower, Ken, 443
Brown, J. Carter, 390*n.*
Bruce, Robert V., 36, 42
Buckley, Tom, 89
Buddha, world's largest sitting, *188–89*
Bumstead, Albert H., 76, 245
Burger, Warren E., 390*n.*
Burkard, Ronald, 439*cpn.*
Burrall, Jessie L., 299–300
Burroughs, John, 408
Byrd, Richard, 20, 71, 72, 76, 78, 164, 215; cartoon of, *218; Little America* base, *68–69,* 76, 78; mentioned, 261, 346; pictured, *62, 68*

C agni, Umberto, 56
California Institute of Technology, 362, 462
Calle, Paul, painting by, *88*
Calvert, James F., 346
Cambodia, 277, 331, 436*cpn.*
Canada, 391, *394;* Rocky Mts., 126
Canby, Thomas Y., 110, 113
cancer treatment, *470–71,* 471
Carmichael, Leonard, 347
Carroll, Allan, time chart designed by, *431*

Carson, Rachel, 420
Castro, Fidel, 389, *390*, 391, 397
cell physiology *454–55*, *460–61*, 462, *464*, *465–66*, 469
Cerf, Bennett, *291*
CERN accelerator, 460, 462, *468*
Cerro Torre, mountain, *326–27*
Chaco Canyon, New Mexico, *152–53*
Chadwick, Douglas H., 416
Chadwick-Onyszkiewicz, Alison, 312
Challenger, space shuttle, 364, *366–67*
Chamberlin, Anne, 304, 350
Chamberlin, Wellman, 338
Chandler, Douglas, 173*cpn.*, 219, 265
Chapelle, Dickey, 275, 279
Chapin, William W., 124; photo by, *126*
Charles, Prince of Wales, 433
Charleston, South Carolina, students, *200–201*
Chater, Melville, 109–10
Chetelat, Enzo de, 289
Chichén Itzá ruins, *146–47, 339*
Chichón, El, *102–105*, 106
children's books, 348, *348*, 416, 434, 437
Childs, David, 433
chimpanzees, 20, 337, *346–47, 347, 349,* 385; as space traveler, *358*
China, 121, *126*, 175, 194, 197, 302, 304*n.;* 310; Citroën-Haardt expedition, *176–78, 180–81, 179–94*, 216; place name spellings, 180*n.;* Qin Shi Huangdi tomb, *159*
Churchill, Winston, 246, *266*, 305
Citroën-Haardt Trans-Asiatic Expedition, *176–78, 180–81, 179–94*, 216
Clark, Eugenie, 229
Clarke, Arthur C., 355
Clatworthy, Fred Payne, 169, 173, 174
Cleveland, Ohio, *380–81*, 403
Clinch, Nicholas B., *306–307*
Cohn, Roy, *290*
Collins, Michael, 373
color photography, *202–15*, 260, 297, 341; "Red Shirt School," 288, *294–95, 296–97, 334; see also* Autochrome
Colton, F. Barrows, *291*
Columbia, space shuttle, 433, *435*
Columbus, Christopher, 194, 338, 339, 373, 430, 443*cpn.*
Conaway, James, 339
Conger, Dean, 304
Conrad, Joseph, 179
conservation, 405–21
Cook, Frederick A., 66, 123
Cooper, Gordon, 364
coral, *235*
Corey, Carol K., 266
Cornell, James, 114
Corpus Christi, hurricane, *112–13*
Courtellemont, Gervais, 169, 174, 294
Cousteau, Jacques-Yves, 20, 226, 337, 339, 346, 349, 385; Aqua-Lung and, 226, *226–27,* 230; Conshelf Two of, *220–21*
Craighead, Frank, 408*cpn.*, 415 & *n.*, 416
Craighead, John, 408*cpn.*, 415 & *n.*, 416
Crandell, Dwight, 99
Crawford, Joan, 203

Crick, Francis, 465*n.*
Crump, Donald J., 343, 345, 348, 434, 437
CT scans, 471
Cuba, 260, 389–92, 394, 397
Cuyahoga River, *380–81*, 403

Daguerre, Louis, 459
Dale, Bruce, 430
Dall, William H., *23*, 25
Darley, James M., 338
Darwin, Charles, 443
Davidson, Robyn, 194*cpn.*
Davis, Arthur P., *22*
Davis, William Morris, 90
Delbruck, Max, 462
Depression (1930s), 202, 204, 218, 295
Detroit News editorial, 265–66
digital subtraction angiography (DSA), 470
dinosaur eggs, *213*, 216
DNA, 462, 465 & *n.*, 469
dogs, *134*, 135, *204*
Donnelley & Sons, printers, 341
Doubilet, Anne Levine, 234
Doubilet, David, photos by, *228–29, 234–35*
Douglas, William O., 302
Douglas-Hamilton, Iain, *406*, 410, 415
Douglas-Hamilton, Oria, 406*cpn.*, 410, 415
Doyle, Robert C. ("T.V. Doyle"), 346, 385
Doyle, Robert E., 378, 389, 390*n.*, 394, 399, 428
Draper, William F., painting by, *268–69*
Dry Tortugas, 170, *171*
Dufay-color photography, *200–201*, 205, 206, *207*, 213
Dugan, James, 224, 230, 233
Duncan, David Douglas, 254
Dutton, C. E., *23*

Eagle, *412–13*
Earhart, Amelia, *214*, 217, 218, 355
Earle, Sylvia A., 233
earthquakes, 99, 106–109, *111*, 113–14; ensuing wave, 33–34, *34–35*
Eckener, Hugo, 215, 355
Edgerton, Harold, 230 & *n.*
Edwards, Walter Meayers (Toppy), 297, 331, 333–34
Einstein, Albert, 462
Eisenhower, Dwight D., 250, *253*, 288
elephants, *2–3, 406, 406–407,* 408, 410, 415, 437
Eliot, John L., 405
Elliott, Lloyd H., 390*n.*, 394, 395, 397, 398
Ellis, William S., 284, 391, 397*cpn.*, 416, 419
Ellis Island, 119, *120*
Ellsworth, Lincoln, 78
England, during World War Two, *255*, 266, *267, 269*
Erik, Adolf, Baron Nordenskjöld, 53
Erim, Kenan T., 157
Esquire magazine, 304, 350
European Laboratory for Particle Physics (CERN), 460, 463, *468*
Evans, Charles, 84, 312

Everest, Mt., 309, 310, 317, 321–22, 337; National Geographic expedition, 20, 312, 317, *317,* 346, 349, 385; pictured, *316–19, 324–25, 384*
Ewing, Maurice, 261
Explorer cable TV series, 445
Eyre, Lincoln, 173*cpn.*, 175

Fairchild, David, *456*, 459
Falco, Albert, 226, 230
famine, 109–10, *114–15*, 386
Findley, Rowe, 99–100, 103, 106
Finlay color process, 202, *204*, 204–206, *206*, 208, 209, 213
fish, 118–19, *171*, 224, *224, 225, 228–29*
Fisher, Franklin L., 167, 203, 205, 211, 289, 292; color photos and, 169, *173–74*, 208, 209, 213; death, 294, 331
Fitzgerald, F. Scott, 162
Fleming, Robert V., 261, 297
floods, 109, 114
Forbes-Leith, F. A. C., 164
Forbush, Edward Howe, 420
Fordney, Chester, 360
Fortescue, Granville, 133
Fossey, Dian, 346, *347, 347*
fossils, *213*, 216, 346, *347,* 449–50
Foster, John W., 43
fox hunting, *204*, 205
Francis, Arlene, *291*
Fraser, Laura Gardin, 261
Frazier, Kendrick, 106, 109
Frederika, Queen of Greece, 302
Fuchs, Vivian, Polar trek of, *74–77*, 78

Gagarin, Yuri, 358*cpn.*
Gagnan, Emile, 226
Galápagos Rift, 233, 236, *238, 239*
Galdikas, Biruté, 346, *347, 347*
Galileo space probe, *374–75*
Gannett, Henry, *23*, 25, 48, 49, 123, 124, 148
Garnett, Samuel S., *23*
Garrett, Wilbur E., 21, 90, 260, 275, 277, 331, 333–34, 428, 429–30, 436*cpn.*, 439*cpn.*, 445, 446, 449 & *cpn.*, *450*, 451
Garver, John B., 434
genetics, 462, 465
Geographic Education National Implementation Project (GENIP), 430*cpn.*
Germany, *172*, 175, 219, 265
Gibbons, Boyd, 450–51
Giddings, Al, 233
Gilbert, Grove Karl, *22*
Gilder, Richard Watson, 83
Gilka, Robert, 381
Gillespie, Vernon, 275
Gillette, Ned, 312 & *cpn.*
Glenn, John, 364, 370
globes, free-floating, *328–29*, 338
Gobi Desert, *212*, 213, 216
Golan Heights, *273*
Goodall, Jane, 20, 337, 346–47, *347,* 349, 385, 386
Goode, George Brown, *22*
Goodsell, John W., 56, 59
Gore, James Howard, *22*

Gore, Rick, 158, 462, 465 & *cpn.*
Graham, Robin Lee, 179, *195*
Grand Canyon, *18, 25, 28,* 430
Grant, Leonard J., 394
Graves, William, 113–14, 386
Gray, Ralph, 388, 389*cpn.*
Gray, William R., 433*cpn.*
Greely, A. W., *22, 29;* Arctic expedition, 24, 25, 29*cpn.*, 179
Greenewalt, Crawford H., 389, 390*n.*, 391, 394, 395, 398
Griggs, Robert F., 133, 173, 261
Grissom, Gus, 364
Grosvenor, Asa Waters, *38*
Grosvenor, Edwin A., 36–37, 48, 135
Grosvenor, Edwin Prescott, 37, *38, 39, 39,* 128
Grosvenor, Elsie Bell, 93, 94, 123, 299, 351; flag designed by, *94;* pictured, *80, 83, 85, 116; see also* Bell, Elsie
Grosvenor, Gertrude, *85*
Grosvenor, Gilbert H., 19–20, 82, 175, 215, 216, 298, 378; articles written by, 130, 133; awards received, 260, 261, 294; Bingham and, 139, 148; biographical data, 37, 38*cpn.*, 39, 46, 49, 126; caricature of, *260;* on cartoons, 218*cpn.*; color photos and, 128, 167, 169, 205, 206, 213, 294; conservation issues and, 416, 423; death, 350–51; Depression years, 202–203; 50th anniversary celebration, 244, 260–61; as first full-time editor, 37–38, 40, 42, 43, 47; guiding principles of, 90, 118, 173*cpn.*, 265, 350, 391, 405, 448; Hellman on, 257, 259; La Gorce and, 95, 121, 203, 298–302, 305; as Managing Editor, early years, 48–49, 89–95; maps and, 272; mentioned, 31, 53, 197, 382, 459; Moore and, 123–24; in 1950s, 289, 294, 297, 305; Peary and, 123; personal characteristics, 298, 299, 302, 305, 380, 386; photo illustrations and, 83–84, 86, 93–94, 114, 119, 121, 124, 174, *see also subhead* color; photos by, 44, *45, 131;* pictured, *38, 39, 43, 47, 80, 85, 116, 130, 132;* retirement, 297, 305; women staff members and, 299, 300; World War One and, 128, 130, 164; World War Two and, 244–46, 250, 252–54, 256, 257, 269
Grosvenor, Gilbert M., 31, 66, 305, 337*cpn.*, 341, 379, 428 & *cpn.*; articles by, 339, 383; as Associate Editor, 378, 379*cpn.*, 381; as Editor, 348, 350, 351, 382–84, 386–99, 428, 429, 430, 433; education, view on, 448; personal characteristics, 378, 386, 438; pictured, *379, 382, 450;* as President, 399, 428, 430 & *cpn.*, 433, 437, 439, 451–52; Tito interview, 382*cpn.*
Grosvenor, Mabel, *85*
Grosvenor, Melville Bell, 173, 297, 330, 378; articles by, 260, 269, 271;

biographical data, 330, 331*cpn.*, 437–38; on Board of Trustees, 390*n.*, 392, 398; color photos and, 204–206, 213, 288, 295; death, 437–38; as Editor, 305*cpn.*, 330–38, *335,* 340–46, 349–51, 385, 386, 391, 429; personal characteristics, 305, 334, 437–38; pictured, *85, 86, 328–29, 331;* retirement, 378, 382
Grosvenor, Melville Bell, Building, 349, 378*cpn.*, 433, *441*
Grosvenor Medal, 95, 260, *260,* 261
Grove, Noel, 429*cpn.*

Haardt, Georges-Marie, 179, 180, 183, 184, 194
Habeler, Peter, 317
Hall, G. Stanley, 90
Hall, Melvin A., 163–64
Hall, Steve, 443
Hamilton, Edith, 333
Hannigan, Ed, 294
Hanson, Arthur B., 390*n.*
Harding, Warren G., 162
Harlem, 389–90, 391, *393,* 397
Haskins, Caryl P., 390*n.*
Hayden, Edward E., *23, 25, 28,* 30
head-hunters, 126, *129,* 170
Hearn, Lafcadio, 86
Heilprin, Angelo, 86 & *n.*
Heirtzler, J. R., 226*n.*
Hellman, Geoffrey T., 257, 259
Henry, Alfred J., 48, 49
Henshaw, Henry W., 22
Henson, Matthew, *58, 59, 60,* 65
Herculaneum, 154, *154, 155,* 157, *158,* 430
Hercules, Frank, 389
Herget, H. M., paintings by, *149*
Herzat, James, painting by, *374–75*
Heurlin, Gustav, 169
Heyerdahl, Thor, 179, 197*cpn.*, *198–99*
Hildebrand, Hans, 169
Hildebrand, J. R., 289, 299
Hill, Ebenezer, Jr., 86
Hill, Robert T., 86, 89
Hillary, Edmund, 309, 312, 346
Himalaya Mountains, *12–13, 178,* 180, 183, 184, 312, 322, 437, 453; *see also* Everest
Hioki, Eki, 119
Hoffer, William, 343
Holland, Michigan, *299*
Hollis, Ralph, 236
Hollywood, 203–204, 288
hologram, *453*
Honeycutt, Brooks, 334
Hooke, Robert, 462
Hoover, Herbert, 109
Hornbein, Tom, 312, 317*cpn.*
Hornocker, Maurice, 415
Hubbard, Gardiner G., 25, 27, 34, 40, 53, 382; articles written by, 29, 33; Bell and, 25, 34, 36; pictured, *23, 26;* statue of, *91*
Hubbard, Gertrude McCurdy, 26
Hubbard Medal, 56, 75, 78 & *cpn.*, 213*cpn.*, 216, 217

Hubbard Memorial Hall, *90,* 91, *91,* 126, *167,* 330, 439, *440*
Hugo, Victor, 457
Humelsine, Carlisle H., 390*n.*
Hungarian Revolution (1956), 272, 391
Hunt, John, 312
Huntford, Roland, 71
hurricanes, *112–13,* 114
Hutchison, George W., *116*
Hyde, John, 38, 40*n.*, 42, 47–49
Hyde, Walter Woodburn, 309–10
hydrogen bomb, 288, 419
Hyman, Charles O., 332, 434

Iacovleff, Alexandre, drawings by, *184, 185*
Iceland, 96–97
immigration, 119, *120*
immune system, 466, 469
Imperial Russian Geographical Society, 93
Incredible Machine: book, 433*cpn.*, 457, 459, *460–61, 463,* 465, 466; television special, 385, 393, 445
India, 110, 113, 125; wildlife in, *402, 404–405,* 408, 410
Indians of the Americas, book, 333
industrial wastes, 419–20
insects, photographing of, *456, 458–59, 459*
Insignia and Decorations of the U.S. Armed Forces, 248–49
International Geographic Congress delegates, *80–81*
Iran, American hostages in, 433
Irvine, Andrew, 310
Irvine, Reed, 390
Israel, 277; Golan Heights, *273*
Italy, 218, 219, 271

Jackson, Carmault B., Jr., 364
Jamestown, Virginia, *298*
Janney Elementary School, Washington, D.C., *388*
Japan, *14–15,* 119, 125–26, *127;* earthquake wave, 33–34, *34–35*
Jaret, Peter, 450, 466, 469
jellyfish, *234–35*
Jenkins, Peter, 194*cpn.*
Jerstad, Lute, 312, 317*cpn.*, 346
Jerusalem, Orthodox Patriarch of, *170*
John Paul II, Pope, 433
Johnson, Lyndon B., 349, 351, 438, 440*cpn.*
Johnson, Mrs. Lyndon B., 351, 390*n.*
Johnson, W. D., *23*
Johnston, David, 99
Jonas, Lucien, drawing by, *264*
Judd, Neil M., 151*cpn.*, 175
Judge, Joseph, 277, 339, 379, 396, 428, 429, 430, *443*
Jupiter exploration, *372–73, 374* & *cpn.*

K2, mountain, *323*
Kane, Dennis, 445
Katmai, Mt., 103*cpn.*, 133, 173, 175
Keller, Helen, *41*
Kelly, Oakley G., 165

Kennan, George, *22, 25*
Kennedy, Jacqueline, 343, *344*
Kennedy, John F., 275, 343, *344,* 346n.
Kenney, Nat, 337
Kepner, William E., *357, 358,* 360
Kettering, Charles F., 261
Khomeini, Ayatollah, *276*
Kilgallen, Dorothy, *291*
Kipling, Rudyard, 179
Kirwin, L. P., 53
Kittinger, Joseph W., *342*
Klemmer, Harvey, 269
Knott, Franklin Price, 133, 169, 206;
    photo by, *133*
Kodachrome film, 205–209, *207,* 213,
    215, 294, *295,* 339
Komarkova, Vera, *321*
Korea, 121, *126*
Korean War, 254, 272, 289
Krishtalka, Leonard, 443

La Gorce, Gilbert Grosvenor, 203
La Gorce, John O., 95, 121, 123, 173*cpn.,*
    289, 351; as Editor, 297–98, 301–302,
    330, 331n., 382, 383; offices of, *300;*
    personal characteristics, 203, 298–
    302, 305, 330; photo illustrations and,
    118, 133, 167, 171, 305; pictured, *300, 301*
land conservation, *420, 421*
Langley, S. P., 86
Lank, David, 391
"Largelamb, H.A." (Bell, A.G.), 118–19
Lassen, Mt., 99
Leakey, L. S. B., *20–21,* 337, 346, *347,*
    349, 385, 393
Leakey, Richard, 346, *347*
Lebanon, 284
Lederer, Jerry, 373
Lederman, Leon, 460, 462
Leeuwenhoek, Anton van, 459, 462
LeMay, Curtis E., 390n.
Lewis, C. Day, 322
Lhasa, *92, 93,* 119
Lhotse, mountain, *318–19,* 322
Life magazine, 295
Lindbergh, Anne Morrow, 179, *193,* 217
Lindbergh, Charles, 76, 164, *165,* 179, 217,
    355; pictured, *166, 192;* reception for, *167*
Littlehales, Bates, 343
Logan, Mt., *31,* 309
London: killer fogs in, *403,* 405; during
    World War Two, *266, 267,* 269
Longley, W. H., *170, 171*
Los Angeles Times, 396
Lovell, James A., Jr., 373
Lovell, Tom, painting by, *356–57*
Lowe, George, 309
Lumière Autochrome, *see* Autochrome

MacArthur, Douglas, 250, *256*
Machu Picchu ruins, 133, 139, 144, 148,
    151, 158, 261; pictured, *136–37*
MacLeish, Archibald, 403
MacMillan, Donald, 56, 59
Macready, John A., 165
macrophage, *454–55,* 464, 469

Mad magazine parody, 350, *350*
Magee, Frank J., 164
Maggi, Giuseppe, 158
magnetic resonance imaging (MRI), *466,
    467,* 469, 471
Mallory, George Leigh, 309, 310, 322
Mallowan, M. E. L., 208*cpn.,* 215
maps, 219, 244–46, 250, *251, 252, 253,
    256,* 272, 338, 430*cpn.*
Marden, Luis, 202, 208–209, 213, 304,
    334*cpn.,* 337–38; *Bounty* and, 339, 346;
    photos by, *226–27, 339;* pictured, *339*
Mariposa Grove, California, *163*
Markwith, Carl, 258, 260
Mars explorations, *370, 371,* 374
Marshall, E. G., 390
Martin, Charles, 126, *128,* 167, 205,
    213n.; color photos and, 169, 170, 204,
    208, 209, 213; photos by, *129, 171*
Martin, William McChesney, Jr., 390n.,
    396, 398
Martinique, *see* Pelée
Marvin, Ross, 56, 65
Mason, J. Alden, 148*cpn.*
Matthews, Samuel W., 291–92, *293*
Maya Indians, *149,* 444*cpn.;* ruins of,
    146–47, 151, 338, *339,* 444, 446
McBride, Ruth Q., 218–19
McCarry, Charles, 445
McCarthy, Joseph R., 288, *290*
McCartney, Benjamin C., 271
McClure, S. S., and Company, 43, 46–49, 83
McCurdy, John A. D., *44*
McCurry, Steve, 284
McDowell, Bart, 334*cpn.,* 338, 381, 437–38
McGee, W J, 28, 33, 84
McKaye, Benton, 53
McKenzie, Floretta, 448
McKinley, Ashley, 76
McKinley, Mt., 25, 66, 310, *311,* 312, *313–
    15,* 430
McKinley, William, 82
McKnew, Thomas W., 297, 341, 378, 383,
    390n., 392
medical technology, 466–71
Meltzoff, Stanley, painting by, *22–23*
Melville, George W., *22, 25*
Membership Center Building (MCB),
    Gaithersburg, Maryland, 340, *340*
Men, Ships, and the Sea, book, 332
Menees, Tim, comic strip by, *393*
Merriam, C. Hart, *23, 25*
Messner, Reinhold, 317, 321–22, *324, 325*
Meštrović, Mrs. Maté, *382*
Mexico: earthquakes, 106, *111;* Stirling
    expedition, *142–43, 144,* 151, 154;
    volcanoes, *102–105,* 106; Yucatán
    ruins, *146–47,* 337–38, *339*
microscope, electron, *454–55, 457–59,
    459,* 462, 466
microsurgery, 470–71
Milan, Italy, *416*
Miller, Irene, *320*
Miller, Maynard M., 310
Minya Konka, mountain, 310, 312

Mion, Pierre, painting by, *240–41*
Mississippi River, 109
Mitchell, Henry, *23*
Mitchell, William (Billy), 164, *402,* 408
Mizel, Steven B., 469
Mobley, George, 284, 343
Molotov, V. M., *245*
Monfreid, Henri de, 205*cpn.,* 217
Mongolia, *169,* 216
Moon, Walter H., 429–30, 449
moon explorations, 337, 362, *362–63,*
    364, 373–74
Moore, Terris, *310*
Moore, W. Robert, 169, 205, 213; photo
    by, *209*
Moore, Willis L., 123, 124 & n.
Morley, Sylvanus Griswold, 146
Morocco, 119–21, 302
Mosier, Robert H., 272, 289
Mott, Frank Luther, 83, 294
movies, *286–87, 288–89;* educational, 386
MRI, *466, 467,* 469, 471
M Street headquarters building, 438,
    439, *440–41*
Muldrow, Robert, II, *23, 25*
Mullineaux, Donal, 99
Murdoch, Helen Messinger, 169
Murrow, Edward R., *290*
Mussolini, Benito, 218
Myers, Walter, 337

Nachtwey, Jim, photo by, *282–83*
Namibia, *2–3*
Nansen, Fridhjof, 34, 53, 56
NASA, 337, 364, 370, 373, 390n., 394
National Book Award, 457
National Geographic Magazine: advertis-
    ing in, 33, 121, 162, 301; cartoons
    lampooning, *174, 218, 304, 393;* circu-
    lation statistics, 33, 49, 95*cpn.,* 121,
    162n., 261, 297, 305, 399*cpn.;* Control
    Center, 379*cpn., 426–27,* 428*cpn., 450;*
    covers, 33, *49, 95, 125, 135, 175, 219,
    261,* 304n., *305,* 330*cpn., 337, 337,* 341,
    351, 399, 403, 449, 451, 453; first issue,
    28, 30, *30,* 451; Hellman on, 259;
    jumbled color layouts, 333, 336;
    letters to the editor of, generally, 254;
    maps in, 219, 244–46, 250, *251, 252,
    253, 256,* 272, 338; Mott on, 294;
    name changes, 337; parodies of, 350,
    350, 398; printing process, 428;
    women staff members, 299, *300,* 304
National Geographic Research magazine,
    434, 443
National Geographic Society: book
    publications, 175, 298, 330*cpn.,* 331–
    33, *332,* 345, 420, 429, 433*cpn.,* 434,
    457; Cartographic Division, 219, 244–
    46, 250, 338, 434; children's books,
    348, *348;* Committee for Research
    and Exploration, 434*cpn.;* Educational
    Films, 386; Educational Media Divi-
    sion, 442, *442,* 451; first annual
    report, *31;* flag, *94;* founding of, *22–
    23, 24–27,* 451; Geography Education
    Program, 448; globe program, *328–*

29, 338; headquarters buildings, *90, 91, 91*, 126, *167,* 330, 349, 378*cpn.,* 433, 438, *440–41;* international publications, 433*cpn.;* lecture series, 33, 43, 84, 346 & *n.,* 392; Membership Center Building, 340, *340;* membership statistics, 36, 94, 95, 125, 133, 135, 162, 175, 202, 203, 219, 301, 337, 341, 351*cpn.,* 399, 428, 453 & *cpn.;* News Service, 254, 393, 443*cpn.;* 1943 membership, categorized, 254; School Services, 254, 299, 331 & *n.,* 343, 345, 388, *389,* 392, 428, 434; seal, *376–77;* Special Publications, 343–45, *344, 345,* 348, 429, 434, 437, 442; "Special Report on Energy," *449;* staff, enumerated, 38, 95; television programs, *see under* television
Nazi party, *172,* 175, 219, 265
Nepal, *12–13,* 261
New Guinea, *6–7,* 121, 123 & *n., 125*
New Mexico, *150, 152–53,* 175
New York Daily News, 451
New York Times editorials, 244, 245, 349
New Yorker magazine, 257, 259
Newcomb, Simon, 43
Newell, Homer E., 362
Newsweek magazine, 254, 350, 397, 452–53
Nicaragua, 31, 53, 59, *282–83*
Nimitz, Chester W., 246, *256*
Norgay, Tenzing, 309, 312, 346
Normandy landing (1944), *253*
North Pole, *58, 60–61,* 65, 66, 69, 76, 123; *see also* Polar explorations
Noville, G. O., *62*
Nowak, Ronald M., 332
nuclear physics, 460, 462, *468*
nuclear wastes, *415,* 419

O akley, Thornton, paintings by, *247*
Ogden, Herbert G., *22*
oil slick, *400–401,* 429
Olmec sculptures, 142, *144, 145,* 151, 154
Orteig prize, 76
Osborne, Doug, *236–37*

P aget color process, 173, 204
Paine, Howard, 333, 334
Palomar, Observatory, 362
pandas, 443, 445
parachute leap, *342*
Park, Edwards, 21, 295, 379, 381
Passet, Stéphane, photo by, *169*
Patagonian Icecap, *326–27*
Patric, John, 219
Patterson, Carolyn, 304
Payne, Melvin M., 31, 337, 344, *347;* as President, 378, 381–83, 389–92, 394, 396–99, 428
Peabody, George Foster, Award, 433
pearl fishing, 217–18
Pearl Harbor, *242–43*
Peary, Josephine, 65*n.;* photo presented by, *56–57*

Peary, Robert E., *52,* 53–60, 65, 78, 125, 346; Cook dispute, 66, 123; Nicaragua trip, 31, 53, 59
Pelée, Mont, 84, 86, *88, 89, 98,* 114, 119
Pellerano, Luigi, 169
Perdicaris, Ion, 120–21
Pereire, General, 197
Pershing, John J., 135
Peru, *140–41,* 439*cpn.; see also* Machu Picchu
pesticide dangers, 420, *422–23*
Peterson, George A., 442
Philippine Islands, 89, *89,* 126, *129,* 170; Census Report, *93, 94,* 119
Phillips, Sam C., 373
Piccard, Auguste, 215, 216–17, 226, 355; cartoon of, *218*
Piccard, Jacques, 226, *342*
Pinchot, Gifford, 405
Pinedo, Francesco de, 162, 167
plagues, 110, 113
Poggenpohl, Andrew, 333, 334
Point Victor, 183–84
Polar explorations, 20, 21, 50–79, 179; airplanes, *62, 63, 69,* 76, 78, 215; flags, *58–60,* 72; ships, *54–57, 66, 70–71; see also* North Pole; South Pole
pollution, 403, 405, 419–20; pictured, *380–81, 400–401, 416, 417, 422–23*
Pompeii, 157, 158
Ponting, Herbert G., photo by, *67*
position emission tomography (PET), 471
Potomac River, 405
Powell, John Wesley, 23, 24–25, 28, 430
Powell, W. B., *22*
Preparedness Parade (1916), *116–17*
Pueblo Bonito ruins, *150,* 175
Putman, John J., 284, 410
Pyramids, Egypt, *16–17*

Q in Shi Huangdi tomb, *159*
Quebec, 391, *394,* 397*cpn.*

R ace prejudice, 125, 299–302
Radford, Arthur W., 272
Raymer, Steven, 109, 386; photo by, *114–15*
Reagan, Ronald, 433, 438, 440*cpn.*
redwood trees, *163,* 382, 405 & *n.*
Regardies magazine, 343
Replogle, Luther I., *328–29*
Reybold, Eugene, 250
Riggs, Arthur Stanley, 130
Ripley, Dillon, 261
Robbins, Pat, 388
Roberts, Mrs. Kenneth, 218
Rock, Joseph, 175, 179, *189,* 194, 197, 310; photos by, *186–91*
Rockefeller, Laurence S., 390*n.*
rocket planes, *352–53,* 355
Roentgen, Wilhelm K., 469
Rogers, Martin, photo by, *429*
Roosevelt, Franklin D., *245,* 245–46
Roosevelt, Theodore, 56, 121, 135, 179, 405, 408
Root, Alan, 21
Roraima, Mt., 310

Rosenborg, Pat, 343
Rosenquist, Gary, photo by, *100–101*
Ross, Ishbel, 36, 437
Ross, Kip, 21, 208, 297, 331, 334
Rossmann, Michael, 469
Rowell, Galen, photo by, *326–27*
Russell, Israel C., *22,* 31, 33*cpn.,* 86, 309, 434*cpn.*
Russell, Jane, *251*
Russia, 25, 33, 86, 130, *131*

S adat, Anwar, 433
St. Denis, Ruth, 133, *133*
St. Elias, Mt., 31, 33*cpn.,* 86, 309, 310, 434*cpn.*
St. Helens, Mt., *4–5, 99, 100–101,* 103, *103,* 106
St. Louis, pollution, *417*
Salman ibn Abdulaziz, Prince of Saudi Arabia, *274*
Sampson, Paul, 443*cpn.*
San Francisco, *160–61;* earthquake, *106– 107,* 113
San Gabriel Valley Hare and Hound Race, *421*
Santa Barbara, California, oil slick, *400–401*
Santa Rosa, quake damage, *106–107*
Saudi Arabia, *274,* 284
Sawyer, Kathy, 395–96
Schaller, George B., 443, 445
Schiefelbein, Susan, 457, 465, 466
Schledermann, Peter, 158
Schleiden, Matthias, 462
Schley, Winfield S., *23*
Schmitt, Harrison, *363*
School Bulletin, 254, 331 & *n.,* 388, *389,* 392
Schultheis, Rob, 443
Schultz, Harald, 337, 343*cpn.*
Schwann, Theodor, 462
Schwinn, Gretchen, 219
Scidmore, Eliza R., 33–34, 299, 309, 408
Scott, Robert F., 66, 71, 78, 126; ship of, *67*
Scourby, Alexander, 385
sea anemones, *234*
sea voyages, *195, 196, 198–99,* 215–16, 339
Seamans, Robert C., Jr., 390*n.*
Seidler, Ned, illustration by, *431*
Selaissie, Haile, *209*
sequoias, *132,* 382, 405 & *n.*
Settle, T. G. W., 360
Severin, Tim, 179, *196*
Severy, Merle, 331 & *n.,* 332–34, 345, 381
Shackleton, Ernest, 75, 78
Shawn, Ted, 133, *133*
Shay, Felix, 164
Shepard, Alan B., Jr., *359,* 364, 370
ships, sunken, 230*cpn., 232–33,* 233, *236–37,* 339, 346; *Titanic,* 21, 236, 239, *240–41*
Shiras, George, 3rd, 121, *122*
Shor, Franc, 302, 304 & *n.,* 343, 378
Shor, Jean, 302
Showalter, William Joseph, 163
Sigurdsson, Haraldur, 157
Sihanouk, Prince, 436*cpn.*

Silcott, Philip B., 348
Simpich, Frederick, 269
Simpich, Frederick, Jr., 269
single photon emission computed
    tomography (SPECT), 471
Siple, Paul, photo by, *78*
Sisson, Robert F., 272
Skidmore, Owings & Merrill, 433
Sky Atlas, 362
Skylab space station, *365*; sun as seen
    from, *368–69*
slavery, 29, *205*, 218
Smith, Ross, 164–65
Smith, Thomas R., 331
Smithsonian Institution, 142, 151, 261, 381
Snider, Edwin W., 394
Sochurek, Howard, 272, 275, 469, 470
Sofaer, Anna, 151*cpn.*
sonography, 471
South Africa, 284, 391–92, 394–95, 397, *397*
South Pole, 71, *72–73*, 76, *78*, 215; stars
    as seen from, *78; see also* Antarctica;
    Polar explorations
space exploration, 337, 355, 358–74
Spain, 218–19
Special Gold Medal, 215*cpn.*, 217
Spencer, Lady Diana, 433
Stanford University, 460
Steger Polar expedition, *50–51*, 179, 450
Steinwurtzel, Barbara, *388*
Stevens, Albert W., 217, 346, *354*, 355,
    357, 358, 360
Stewart, B. Anthony, 205, 213; photos by,
    200–201, *204*
Stewart, Richard H., photo by, *142–43*
Stewart, T. Dale, *347*
Stirling, Marion, *143*, 151, 154, 261
Stirling, Matthew, 142, 151, 154, 261,
    333; pictured, *142, 144, 145*
Stone, Edward Durell, 349, 378*cpn.*,
    433, 440*cpn.*
Stone, Melville E., 125
Strider, Miss (Personnel Director), 299, 304
strip mining, *420*
Sudanese slave girl, *205*, 218
Sumatra, 163–64
sun's corona, *368–69*
Suyá Indian, *343*
Swerdlow, Joel L., 265 & *cpn.*, 449

Taft, William H., 65–66, 94, 95, 124*n.*,
    135, 355
Taft, Mrs. William H., 95
Tagbanua women, 89, *89*
Tanzania, 346, 385, *406*, 410
Tapper, Joan, 432, 443
Tarbell, Ida, 299
Tate, G. H. H., 310
Taung child, 449, *453*
television, *290, 291*; National
    Geographic documentaries, 330*cpn.*,
    344, 346, 349, *384, 385, 385*, 390,
    392–93, 433, 445
Tennyson, Alfred Lord, 351
Theroux, Paul, 446*cpn.*
Thomas, Lewis, 420, 457
Thompson, A. H., 23, 27

Thompson, Gilbert, *22*
Thresher, U.S.S., submarine, 230, 233
Tibet, *92, 93, 94*, 119, *318*; Rock in, 175,
    186, 187, 189–91, 194
tigers, *402, 404–405*, 408, 410
Timchenko, Boris V., 340
Time magazine, 349–50
Titanic, S.S., 21, 236, 239, *240–41*, 450
Titicaca, Lake, *140–41*
Tito, Marshal, *382*
Tittmann, O. H., *22*, 25
Tobien, Wilhelm, 169
Toft, Jürgen, 272, 275
Tokyo, pollution, 403
Traveler magazine, 432, *432*, 433, 439, 443
Trifid Nebula, *472–73*
Truman, Harry (Mt. St. Helens
    casualty), 99, 103, 106
Truman, Harry S (President), 244, 261, 288
trumpeter swan, 420, *424–25*
Tucson, Arizona, *1*
Turkey, ruins, *see* Aphrodisias
Tutankhamun's tomb, 179

Uemura, Naomi, *60–61*
Uganda, 410
Ullman, James Ramsey, 346
Umezo, Yohijiro, *256*
Underground Railroad, 449
Underhill, Miriam O'Brien, 310
underwater exploration, 170, *171*, 216,
    220–22, 223–39, 342
U.S. Camera, 294
U.S. Capitol, book about, 343, 344, 345
U.S. Reclamation Service, 408
Unsoeld, Willi, 312, 317 & *cpn.*
Ur, *208*, 215

Valley of Ten Thousand Smokes, 133,
    173, 261
Verdun, battle, *264*
Vesilind, Priit, 169, 170, 173, 205, 213, 330
Vesuvius, Mt., 154, 157–58
Vietnam War, 254, 275, 277, *278–79*, 381,
    449; Memorial, 265*cpn.*, *280–81*, 449
Villiers, Alan J., *211*, 215, 216, 332;
    square-rigger of, *210–11*
virus, 466, 469
Voas, Robert B., 364, 370
volcanoes, 84, 86, 89, *96–98*, 99–106, 109,
    114; *see also* Katmai; Pelée; St. Helens
Volga, television documentary, 390, 392
Vosburgh, Frederick G. (Ted), 90, *303*,
    338, 341; on Board of Trustees, 390*n.*,
    391, 392, 398–99; as Editor, 351, 378–
    79, 381–83, 386, 391, 403, 437–38;
    retirement, 378, 381, 382

Wakelia, James H., Jr., 390*n.*
Walker, Joseph A., 355*cpn.*
Walker, Ronald, 215–16
Walsh, Dan, 226, *342*
Ward, Fred, 389, 390, 419, 420
Washburn, Barbara, *18*

Washburn, Bradford, *18*, 310, *310*
Washington, D.C., *46*, *440–41*
Washington Post, 395–96, 433, 451
WASP, underwater gear, *236–37*
Watson, James D., 465*n.*
Watson, Vera, 312
Weathers, Wesley W., 443
Weaver, Kenneth F., 362, 373, 449*cpn.*,
    459, 460
Webb, James H., 390*n.*, 394, 395
Weiant, C. W., 154
Weiant, Mrs. C. W., 154
Welch, Joseph, 288
Welles, Orson, 349, 385
Welling, James C., *23*
Wellman, Walter, 374
Wesly, Claude, 226
Wetmore, Alexander, 381
whales, *410–11*, 416
White, Edward H., II, 362
White, Peter T., 275, 277, 379, 391, 394*cpn.*
White House guidebook, 343, 344, *344*,
    345, 429, 437
Whittaker, James W., 317*cpn.*, 323
Wilburn, Herbert S., Jr., 297, 331, 334,
    341, 381
Williams, Garr, cartoon by, *218*
Williams, Maynard O., 38*cpn.*, 169,
    202–203, 289; on Citroën-Haardt
    expedition, 179, *179*, 180–84, 187, 193–94;
    photos by, *170, 182*
Wilson, Woodrow, 118*cpn.*, 133, 135
Wirth, Conrad L., 390*n.*
Wisherd, Edwin (Bud), 169, 213 & *n.*
Wolper, David L., 346, 385
Woodson, Leroy, 389
Woolley, C. Leonard, 208*cpn.*
Worcester, Dean C., 126, 128*cpn.*, 170
Wordsmith, comic strip, *393*
World magazine, *388, 388*, 389*cpn.*, 392, 437
World War One, 128, 130, 133, 135, 164,
    252, 266; photos, *116–17, 134, 165, 264, 265*
World War Two, 197, 244–57, 260, 269,
    271, 295; photos, *242–43, 245, 247, 251,
    253, 255, 256, 266–69*
Wright, Belinda, 408; photo by, *404–405*
Wright, Louis B., 390*n.*, 394, 395, 398
Wright, Orville, 91
Wright, Wilbur, 91, 346, 355
Wyatt, Ben, 416
Wyeth, N.C., painting by, *62–63*

X-15 rocket plane, *352–53*
X rays, 469, 471

Yale University, 133, 139*cpn.*, 148
Yellowstone National Park, *408*, 415, 420
Yorktown, U.S.S., *268–69*
Yost, Ed, 399*cpn.*
Young, Gordon, 403
Yucatán ruins, *146–47, 337–38, 339*

Zahl, Paul A., 338
zeppelin flights, 215
Ziegler Polar Expedition, 94*cpn.*
Zulu people, 34, *37*, 430

Credits

Komarkova. 323 John Roskelley. 324(l) Nena Holguin. 324, 325 Reinhold Messner. 326, 327 Galen Rowell.

328, 329 Bates Littlehales, NGP. 331 W. Robert Moore. 332 Joseph D. Lavenburg, NGP. 335 Victor R. Boswell, Jr., NGP. 336 Joseph D. Lavenburg, NGP. 338 B. Anthony Stewart. 339(tl&b) Luis Marden. 339(tr) Wilbur E. Garrett, NGS. 340(t) James L. Stanfield, NGP. 340(b) Joseph H. Bailey, NGP. 342(t) Thomas J. Abercrombie, NGS. 342(b) Volkmar Wentzel. 343 Harald Schultz. 344(tl) Winfield Parks. 344(tr&c), 345 Victor R. Boswell, Jr., NGP. 347 (tl) Gordon W. Gahan. 347 (tr) Baron Hugo Van Lawick. 347 (bl) Rod Brindamour. 347 (br) Robert M. Campbell. 348 Joseph D. Lavenburg, NGP. 350 Mad magazine © 1958 by E.C. Publications, Inc.

352, 353 Dean Conger for NASA. 354, 355 Richard H. Stewart. 356, 357 Painting by Tom Lovell for National Geographic. 358 Henry Burroughs, Wide World Photos. 359 Doug Martin. 360, 361 James A. McDivitt, NASA. 362, 363 E.C. Cernan, NASA. 364, 365 NASA. 366, 367 Scene from the IMAX®/ OMNIMAX® Film "The Dream Is Alive" © Smithsonian Institution and Lockheed Corporation 1985. 368, 369 NASA. 370, 371 U.S. Geological Survey, Flagstaff, Arizona. 372, 373 NASA. 374, 375 Painting by James Hervat for National Geographic.

376, 377 James L. Amos, NGP. 379 James L. Stanfield, NGP. 380, 381 James P. Blair, NGP. 382 Mate Mestrovic. 384(tl) Barry C. Bishop, NGS. 384(cl) Carol Hughes. 384(bl) Joseph H. Bailey, NGP. 384(tr) Daniel Tomasi, the Cousteau Society. 384(cr) Jonathan Wright. 384(br) David Doubilet. 385(l) Edgar Boyles. 385(r) Mick Coyne. 386(l) Victor R. Boswell, Jr., NGP. 386(r) Foster Wiley. 387 Bianca Lavies, NGP. 388(t) Joseph H. Bailey, NGP. 388(b) Joseph D. Lavenburg, NGP. 389 Victor R. Boswell, Jr., NGP. 390 Fred Ward, Black Star. 393(t) Leroy Woodson, Wheeler Pictures. 393(b) WORD-SMITH by Tim Menees. © 1976 Universal Press Syndicate. Reprinted by permission. 394 Winfield Parks. 397 James P. Blair, NGP. 398 © 1980 Henson Associates, Inc. Reprinted by permission.

400, 401 Bruce Dale, NGP. 402 From Brigadier General William Mitchell. 404, 405 Belinda Wright. 406, 407 Oria and Iain Douglas-Hamilton. 408 Frank and John Craighead. 409 Pat Powell. 410, 411 Al Giddings, Ocean Images. 412, 413 Glenn W. Elison. 414 U.S. Air Force. 415 Robert S. Dyer, Environmental Protection Agency. 416 James P. Blair, NGP. 417 Ted Spiegel, Black Star. 418, 419 Dewitt Jones. 420 James P. Blair, NGP. 421 Walter Meayers Edwards. 422, 423(l) Fred Ward, Black Star. 423(r) James P. Blair, NGP. 424, 425 Steven C. Wilson, Entheos.

426, 427 James L. Amos, NGP. 429 Martin Rogers. 430 Pat Lanza Field, NGS. 431 Painting by Ned M. Seidler for National Geographic. 432 Joseph D. Lavenburg, NGP. 433 Victor R. Boswell, Jr., NGP. 434 Robert S. Oakes. 435 Jon Schneeberger, Ted Johnson, Jr., and Anthony Peritore, all NGS. 436 Wilbur E. Garrett, NGS. 439 William Albert Allard. 440, 441 Claude E. Petrone, NGS. 442 Joseph D. Lavenburg, NGP. 443 Joseph H. Bailey, NGP. 444 Lowell Georgia. 447 Steve McCurry. 448 Joseph H. Bailey, NGP. 451 James L. Amos, NGP. 452 GRIN AND BEAR IT by Lichty and Wagner © 1982 Field Enterprises, Inc. by permission of North American Syndicate, Inc.

454, 455 Boehringer Ingelheim International GmbH. 456 David Fairchild. 457 Phillip Degginger, Bruce Coleman, Inc. 458, 459 David Scharf. 460–464 Lennart Nilsson. 466 Mallinckrodt Institute of Radiology, St. Louis. 467 Howard Sochurek. 468–471 Kevin Fleming. 472, 473 National Optical Astronomy Observatories, Arizona.

TEXT CREDITS

Grateful acknowledgment is made for permission to quote from the following:

William Beebe, Half Mile Down, published by Harcourt Brace, 1934. Rights reserved by Meredith Corporation, New York.

Tom Buckley, "With the National Geographic on Its Endless, Cloudless Voyage," September 6, 1970, Copyright © The New York Times.

Arthur C. Clarke, Interplanetary Flight: An Introduction to Astronautics, 1960. By permission of the author and the author's agents, Scott Meredith Literary Agency, Inc.

Michael Collins, Carrying the Fire, copyright © 1974 Michael Collins. By permission of the publishers, Farrar, Straus and Giroux, Inc.

C. Day-Lewis, "Transitional Poem" from Collected Poems 1929–1933, published by The Hogarth Press, 1938.

Kendrick Frazier, The Violent Face of Nature: Severe Phenomena and Natural Disasters, published by William Morrow & Company, Inc., 1979. By permission of William Morrow & Company, Inc., and Harold Matson Company, Inc.

Bernard Grun, The Timetables of History: A Horizontal Linkage of People and Events, copyright © 1975, 1979 Simon & Schuster, Inc. (A Touchstone Book).

Geoffrey T. Hellman, "Geography Unshackled," September 25, October 2, and October 9, 1943, published by The New Yorker. Included here by permission of Special Collections, New York University.

Tom Holzel and Audrey Salkeld, First on Everest: The Mystery of Mallory and Irvine, published by Henry Holt and Company, Inc., 1986.

Roland Huntford, Scott and Amundsen, published by Hodder & Stoughton, London and Toronto, 1979.

Roland Huntford, The Last Place on Earth, published by Atheneum, 1986.

Rudyard Kipling, The Explorer, © 1940 Elsie Kipling Bambridge, published by Doubleday & Co., Inc.

Archibald MacLeish, "A Reflection: Riders on Earth Together, Brothers in Eternal Cold," December 25, 1968, copyright © The New York Times.

Frank Luther Mott, A History of American Magazines, 1885–1905, published by The Belknap Press, Harvard University Press, 1930–1968.

New York Times editorial, "Footnote to Geography," July 26, 1942, copyright © The New York Times.

New York Times editorial, "Maps of a Changing World," January 15, 1945, copyright © The New York Times.

Time-Life Books Editors, This Fabulous Century: 1920–1930, copyright © 1968 Time-Life Books Inc.

Washington Star feature article, Sunday Section, February 5, 1942. All rights reserved.